A NEW GREEN HISTORY
OF THE WORLD

Clive Ponting was until recently Reader in Politics and
International Relations at the University of Wales,
Swansea. His most recent books are: *World History: A
New Perspective*; *Thirteen Days: The Road to the First
World War*; *The Crimean War* and *Gunpowder: From
the Alchemists of China to the Battlefields of Europe*. He
recently took early retirement and now lives on a small
Greek island where he is creating a Mediterranean
garden and cultivating olives.

ALSO BY CLIVE PONTING

The Right to Know: The Inside Story of the Belgrano Affair

Whitehall: Tragedy and Farce

Breach of Promise: Labour in Power 1964–1970

Whitehall: Changing the Old Guard

Secrecy in Britain

1940: Myth and Reality

Churchill

Armageddon: The Second World War

Progress and Barbarism: The World in the Twentieth Century

World History: A New Perspective

Thirteen Days: The Road to the Great War

The Crimean War

Gunpowder: An Explosive History

CLIVE PONTING

A New Green History of the World

The Environment and the Collapse of Great Civilisations

VINTAGE BOOKS
London

Published by Vintage 2007

2 4 6 8 10 9 7 5 3

First published in Great Britain under the title *A Green History of the World*
by Sinclair-Stevenson Limited 1991
First published in the United States of America by St Martin's Press, Inc., 1992
Reprinted by arrangement with St Martin's Press, Inc.
Published in Penguin Books 1993

Revised Vintage edition published in 2007
Vintage
Random House, 20 Vauxhall Bridge Road,
London SW1V 2SA

Y304.
28

www.vintage-books.co.uk

Addresses for companies within The Random House Group Limited
can be found at: www.randomhouse.co.uk/offices.htm

The Random House Group Limited Reg. No. 954009

A CIP catalogue record for this book is available from the British Library

ISBN 9780099516682

The Random House Group Limited supports The Forest
Stewardship Council (FSC), the leading international forest
certification organisation. All our titles that are printed on
Greenpeace approved FSC certified paper carry the FSC logo.
Our paper procurement policy can be found at:
www.rbooks.co.uk/environment

Typeset in Plantin by Palimpsest Book Production Limited,
Grangemouth, Stirlingshire

Printed in the UK by CPI Bookmarque, Croydon, CR0 4TD

Contents

List of Maps

List of Tables and Charts

PREFACE

The first edition of this book was published in 1991 and has since then been translated into thirteen languages. Over the last sixteen years much has changed in the world – the Soviet Union no longer exists and the Cold War has been replaced by the problems posed by the global dominance of the United States and the threat of terrorism. Much has also changed in the environment. Some of it has been for the better, for example the agreement to end the production of CFCs and eventually HCFCs in order to stop the destruction of the world's ozone layer. Many of the changes though have been for the worse. More than a billion extra people now live on the earth than when the first edition of this book was written. Billions of tonnes of carbon dioxide have been put into the atmosphere and the threat posed by global warming is even more starkly apparent. The destruction of the rainforests and other habitats has continued at an increasing pace.

I was therefore very pleased when Will Sulkin suggested that the time was ripe for a new edition of the book. I have taken this opportunity to completely revise the text. Every chapter, apart from the first, has been updated, revised, rewritten and expanded. One chapter has been dropped and a new one added. I have also greatly increased the number of charts and diagrams. In a world history it is inappropriate to use BC/AD and instead BCE (Before the Common Era) and CE (Common Era) have been used instead. Where there is no ambiguity CE has been dropped.

At the end of the first edition I tried to strike a balance between pessimism and optimism when I wrote: 'Past human actions have left contemporary societies with an almost insuperably difficult set of problems to solve.' In the last sixteen years the balance was clearly tipped in favour of pessimism. It has been a period of wasted opportunities on environmental problems. Machiavelli wrote in *The Prince*:

In state affairs, by foreseeing [problems] at a distance, which is only done by men of talents, the evils which might arise from them are soon cured; but when, from want of foresight, they are suffered to increase to such a height that they are perceptible to everyone, there is no longer any remedy.

How well has the world's political leadership faced up to environmental problems over the last decade and a half? World industrial output and levels of consumption have continued to rise at unprecedented rates. The consequences, in resource and energy consumption and associated pollution, have become increasingly apparent. There can be no doubt that global warming is the greatest single threat that the world now faces. Over the last few years scientists have become increasingly concerned that the world's climate is about to reach a point where dramatic changes may happen very quickly and that these changes will be irreversible and bring in their wake major social and economic disruption. The action taken to meet this threat has been minimal. The United States refuses to do anything. The Kyoto treaty requires only very small reductions in carbon dioxide output; it only involves countries responsible for a third of the world's output and they are unlikely to meet the targets set in the treaty. By the time that the effects of global warming become fully apparent, and that may be very soon, it will be too late to take action to avoid disaster. The world is now seeing the conjunction of a number of trends that emerged over the last two centuries, all of which are likely to lead to immense environmental problems in the next few years. This book tries to show just how deep-rooted these trends are in the way human societies have evolved.

I would like to thank Christopher Sinclair-Stevenson for commissioning the first edition of this book and Will Sulkin for his enthusiasm about a new edition. My greatest debt of gratitude though is to my wife, Laura, who drew all the maps, charts and diagrams for this new edition and who has given unstinting support throughout.

Clive Ponting
Greece, 2007

1
THE LESSONS OF EASTER ISLAND

Easter Island is one of the most remote inhabited places on earth. Only some 400 square kilometres in area, it lies in the Pacific Ocean, 3,200 kilometres off the west coast of South America and 2,000 kilometres from the nearest inhabitable land of Pitcairn Island. At its peak the population was only about 7,000. Yet, despite its superficial insignificance, the history of Easter Island is a grim warning to the world.

The Dutch Admiral Roggeveen, on board the *Arena*, was the first European to visit the island on Easter Sunday 1722. He found a society in a primitive state with about 3,000 people living in squalid reed huts or caves, engaged in almost perpetual warfare and resorting to cannibalism in a desperate attempt to supplement the meagre food supplies available on the island. During the next European visit in 1770 the Spanish nominally annexed the island but it was so remote, underpopulated and lacking in resources that no formal colonial occupation ever took place. There were a few more brief visits in the late eighteenth century, including one by Captain Cook in 1774. An American ship stayed long enough to carry off twenty-two inhabitants to work as slaves killing seals on Masafuera Island off the Chilean coast. The population continued to decline and conditions on the island worsened: in 1877 the Peruvians removed and enslaved all but 110 old people and children. Eventually the island was taken over by Chile and turned into a giant ranch for 40,000 sheep run by a British company, with the few remaining inhabitants confined to one small village.

What amazed and intrigued the first European visitors was the evidence, amongst all the squalor and barbarism, of a once flourishing and advanced society. Scattered across the island were over 600 massive stone statues, on average over six metres high. When anthropologists began to consider the history and culture of Easter Island early in the twentieth century they agreed on one thing. The primitive people living in such poverty-stricken and backward conditions when the Europeans first visited the island could not have been responsible for such a socially advanced and technologically complex task as carving, transporting and erecting the statues. Easter Island therefore became a 'mystery' and a wide variety of theories were advanced to explain its history. Some of the more fantastic ideas involved visits by spacemen or lost civilisations on continents that had sunk into the Pacific leaving Easter Island as a remnant. The Norwegian archaeologist Thor Heyerdahl, in his popular book *Aku-Aku* written in the 1950s, emphasises the strange aspects of the island and the mysteries that lay hidden in its history. He argued that the island was first settled from South America and that from there the people inherited a tradition of monumental sculpture and stonework (similar to the great Inca achievements). To account for the decline he introduced the idea that at a late stage other settlers arrived from the west and began a series of wars between the so-called 'long-ears' and the 'short-ears' that destroyed the complex society on the island. Whilst this theory is less extravagant than some of the others that have been put forward it has never been generally accepted by other archaeologists.

The history of Easter Island is not one of lost civilisations and esoteric knowledge. Rather it is a striking example of the dependence of human societies on their environment and of the consequences of irreversibly damaging that environment. It is the story of a people who, starting from an extremely limited resource base, constructed one of the most advanced societies in the world for the technology they had available. However, the demands placed on the environment of the island by this development were immense. When it could no longer withstand the pressure, the society that had been painfully built up over the previous thousand years fell with it.

The colonisation of Easter Island belongs to the last phase in the long-drawn-out movement of human settlement across the globe. The first people arrived sometime in the fifth century at a period when the Roman empire was collapsing in western Europe, China was still in chaos following the fall of the Han empire two hundred years earlier, India saw the end of the short-lived Gupta empire and the great city of Teotihuacan dominated most of Mesoamerica. They were Polynesians and part of a great process of exploration and settlement across the vast expanse of the Pacific Ocean. The original Polynesians came from south-east Asia and they reached the islands of Tonga and Samoa about 1000 BCE. From there they moved further east to the Marquesas Islands about 300 CE and then in two directions, south-east to Easter Island and north to Hawaii in the fifth century. The last phases of the movement were to the Society Islands about 600 and from there to New Zealand about 800. When this settlement was complete, the Polynesians were the most widely spread people on earth, encompassing a huge triangle from Hawaii in the north to New Zealand in the south-west and Easter Island in the south-east – an area twice the size of the present continental United States. Their long voyages were made in double canoes, joined together by a broad central platform to transport and shelter people, plants, animals and food. These were deliberate colonisation missions and they represented considerable feats of navigation and seamanship since the prevailing currents and winds in the Pacific are against west-to-east travel.

When the first people found Easter Island, they discovered a world with few resources. The island was volcanic in origin, but its three volcanoes had been extinct for at least 400 years before the Polynesian settlers arrived. Both temperatures and humidity were high and, although the soil was adequate, drainage was very bad and there were no permanent streams on the island; the only fresh water available was from lakes inside the extinct volcanoes. Because of its remoteness the island had only a few species of plants and animals. There were thirty indigenous species of flora, no mammals, a few insects and two types of small lizard. The waters around the island contained very few fish. The arrival of the first humans did very little to improve the situation. The Polynesians in their home islands depended on a very limited range of plants and animals for subsistence: their only domesticated animals were chickens, pigs, dogs and the Polynesian rat and the main crops were yam, taro, breadfruit, banana, coconut and sweet potato. The settlers on Easter Island brought only chickens and rats with them and they soon found that the climate was too severe for semi-tropical plants such as breadfruit and coconut and extremely marginal for the usual mainstays of their diet, taro and yam. The inhabitants were, therefore, restricted to a diet based mainly on sweet potatoes and chickens. The only advantage of this monotonous, though nutritionally adequate, diet was that the cultivation of the sweet potato was not very demanding and left plenty of time for other activities.

It is not known how many settlers arrived in the fifth century but they probably numbered no more than twenty or thirty at most. As the population slowly increased the forms of social organisation familiar in the rest of Polynesia were adopted. The basic social unit was the extended family, which jointly owned and cultivated the land. Closely related households formed lineages and clans, each of which had its own centre for religious and ceremonial activity. Each clan was headed by a chief who was able to organise and direct activities and act as a focal point for the redistribution of food and other essentials within the clan. It was this form of organisation and the competition (and probably conflict) between the clans that produced both the major achievements of Easter Island society and ultimately its collapse.

Settlements were scattered across the island in small clusters of peasant huts with crops grown in open fields. Social activities were centred around separate ceremonial centres, which were occupied for part of the year. The chief monuments were large stone platforms, similar to those found in other parts of Polynesia and known as *ahu*, which were used for burials, ancestor worship and to commemorate past clan chiefs. What made Easter Island different was that crop production took very little effort and there was therefore plenty of free time and labour which the clan chiefs were able to direct into ceremonial activities. The result was the creation of the most advanced of all the Polynesian societies and one of the most complex in the world for its limited resource base. The Easter Islanders engaged in elaborate rituals and monument construction. Some of the ceremonies involved recitation from the only known Polynesian form of writing called *rongorongo*, which was probably less a true script and more a series of mnemonic devices. One set of elaborate rituals was based on the bird cult at Orongo, where there are the remains of forty-seven special houses together with numerous platforms and a series of high-relief rock carvings. The crucial centres of ceremonial activity were the *ahu*. Over 300 of these platforms were constructed on the island, mainly near the coast. The level of intellectual achievement of at least some parts of the Easter Island society can be judged by the fact that a number of these *ahu* have sophisticated astronomical alignments, usually towards one of the solstices or the equinox. At each site they erected between one and fifteen of the huge stone statues that survive today as a unique memorial to the vanished Easter Island society. It is these statues which took up immense amounts of peasant labour. The statues were carved, using only obsidian stone tools, at the quarry at Ranu Raraku. They were fashioned to represent in a highly stylised form a male head and torso. On top of the head was placed a 'topknot' of red stone weighing about ten tonnes from another quarry. The carving was a time-consuming rather than a complex task. The most challenging problem was to transport the statues, each some six metres in length and weighing several tens of tonnes, across the island

and then erect them on top of the *ahu*.

The Easter Islanders' solution to the problem of transport provides the key to the subsequent fate of their whole society. Lacking any draught animals they had to rely on human power to drag the statues across the island using tree trunks as rollers. The population of the island grew steadily from the original small group in the fifth century to about 7,000 at its peak in 1550. Over time the number of clan groups would have increased and also the competition between them. By the sixteenth century hundreds of *ahu* had been constructed and with them over 600 of the huge stone statues. Then, when the society was at its peak, it suddenly collapsed, leaving over half the statues only partially completed around Rano Raraku quarry. The cause of the collapse and the key to understanding the 'mysteries' of Easter Island was massive environmental degradation brought on by deforestation of the whole island.

When the first Europeans visited the island in the eighteenth century it was completely treeless apart from a handful of isolated specimens at the bottom of the deepest extinct volcano crater of Rano Kao. However, recent scientific work, involving the analysis of pollen types, has shown that at the time of the initial settlement Easter Island had a dense vegetation cover including extensive woods. As the population slowly increased, trees would have been cut down to provide clearings for agriculture, fuel for heating and cooking, construction material for household goods, pole and thatch houses and canoes for fishing. The most demanding requirement of all was the need to move the large number of enormously heavy statues to ceremonial sites around the island. The only way this could have been done was by large numbers of people guiding and sliding them along a form of flexible tracking made up of tree trunks spread on the ground between the quarry and the *ahu*. Prodigious quantities of timber would have been required and in increasing amounts as the competition between the clans to erect statues grew. As a result by 1600 the island was almost completely deforested and statue erection was brought to a halt, leaving many stranded at the quarry.

The deforestation of the island was not only the death knell for the elaborate social and ceremonial life, it also had other drastic effects on the everyday life of the population. From 1500 the shortage of trees was forcing many people to abandon building houses from timber and live in caves, and when the wood eventually ran out altogether about a century later everyone had to use the only materials left. They resorted to stone shelters dug into the hillsides or flimsy reed huts cut from the vegetation that grew round the edges of the crater lakes. Canoes could no longer be built and only reed boats incapable of long voyages could be made. Fishing was also more difficult because nets had previously been made from the paper mulberry tree (which could also be made into cloth) and that was no longer available. Removal of the tree cover also badly affected the soil of the island, which would have already suffered from lack of suitable animal manure to replace the nutrients taken up by the crops. Increased exposure caused soil erosion and the leaching out of essential nutrients. As a result crop yields declined. The only source of food on the island unaffected by these problems was the chickens. As they became ever more important, they had to be protected from theft and the introduction of stone-built defensive chicken houses can be dated to this phase of the island's history. It became impossible to support 7,000 people on this diminishing resource base and numbers fell rapidly.

After 1600 Easter Island society went into decline and regressed to ever more primitive conditions. Without trees, and so without canoes, the islanders were trapped in their remote home, unable to escape the consequences of their self-inflicted environmental collapse. The social and cultural impact of deforestation was equally important. The inability to erect any more statues must have had a devastating effect on the belief systems and social organisation and called into question the foundations on which the complex society had been built. There were increasing conflicts over diminishing resources, resulting in a state of almost permanent warfare. Slavery became common and as the amount of protein available fell the population turned to cannibalism. One of the main aims of warfare was to destroy the *ahu* of opposing clans. A few survived as burial places but most were abandoned. The magnificent stone statues, too massive to destroy, were pulled down. The first Europeans found only a few still standing when they arrived in the eighteenth century and all had been toppled by the 1830s. When they were asked by the visitors how the statues had been moved from the quarry, the primitive islanders could no longer remember what their ancestors had achieved and could only say that the huge figures had 'walked' across the island. The Europeans, seeing a treeless landscape, could think of no logical explanation either and were equally mystified.

Against great odds the islanders had painstakingly constructed, over many centuries, one of the most advanced societies of its type in the world. For a thousand years they sustained a way of life in accordance with an elaborate set of social and religious customs that enabled them not only to survive but to flourish. It was in many ways a triumph of human ingenuity and an apparent victory over a difficult environment. But in the end the increasing numbers and cultural ambitions of the islanders proved too great for the limited resources available to them. When the environment was ruined by the pressure, the society very quickly collapsed with it, leading to a state of near barbarism.

The Easter Islanders, aware that they were almost completely isolated from the rest of the world, must have surely realised that their very existence depended on the limited resources of a small island. After all it was small enough for them to walk round the entire island in a day or so and see for themselves what was happening to the forests. Yet they were unable to devise a system that allowed them to find the right balance with the environment. Instead, vital resources were steadily consumed until finally none was left. Indeed, at the very time when the limitations of the island must have become starkly apparent the competition between the clans for the available timber seems to have intensified as more and more statues were carved and moved across the island in an attempt to secure prestige and status. The fact that so many were left unfinished or stranded near the quarry suggests that no account was taken of how few trees were left on the island.

The fate of Easter Island has wider implications. Like Easter Island the earth has only limited resources to support human society and all its demands. Like the islanders, the human population of the earth has no practical means of escape. How has the environment of the world shaped human history and how have people shaped and altered the world in which they live? Have other societies fallen into the same trap as the islanders? For the last two million years humans have succeeded in obtaining more food and extracting more resources on which to sustain increasing numbers of people and increasingly complex and technologically advanced societies. But have they been any more successful than the islanders in finding a way of life that does not fatally deplete the resources that are available to them and irreversibly damage their life support system?

2

THE FOUNDATIONS OF HISTORY

Human history cannot be understood in a vacuum. All human societies have been, and still are, dependent on complex, interrelated physical, chemical and biological processes. These include the energy produced by the sun, the circulation of the elements crucial for life, the geophysical processes that have caused the continental land masses to migrate across the face of the globe and the factors regulating climatic change. These constitute the essential foundations for the way in which the various types of plants and animals (including humans) form complex, interdependent communities. Research in a wide variety of disciplines is increasingly making it clear that life on earth and all human societies depend on the maintenance of a number of delicate balances within and between a whole series of complex processes. These findings also help us understand the way in which the environment has influenced the development of human societies and, just as important, the human impact on the earth.

Human history has been affected by the action of large scale geological and astronomical forces over long periods of time. Although the amount of land on the globe has remained broadly constant, its distribution has altered radically. The rocks a few miles below the earth's surface are molten and flow in convection currents caused by the heat rising from the earth's core. The flow causes movements in the earth's crust in the form of large 'plates' which move across the surface of the globe. The convection currents from the core of the earth well up in ridges under the oceans and then flow outwards, forcing the plates further apart so that the North Atlantic is currently widening at about a centimetre a year and the east Pacific by about ten centimetres a year. Material flows back down into the earth in the deep oceanic trenches. Where the plates meet there is major instability, causing earthquakes and volcanoes.

The effects of these movements can be seen in the natural disasters that have punctuated human history – volcanic eruptions such as Thera, which may have overwhelmed Minoan society on Crete, the eruption of Vesuvius that destroyed Herculaneum and Pompeii or major earthquakes such as that in the Senshi province in China in 1556 that killed over 800,000 people or those affecting Lisbon in 1755 and Tokyo in 1923 that killed tens of thousands. The impact in the long term is even greater. About 400 million years ago the earth had two supercontinents – Laurasia (North America, Europe and Asia) and Gondwanaland (South America, Africa, India, Australia and Antarctica) divided by the Tethys Sea. When this massive land mass broke up, the southern continents were situated over the south pole, causing glaciers in what is now Brazil and South Africa, and Laurasia was in the tropics. What is now North America broke away from Europe about 200 million years ago, although the major drift apart to form the Atlantic Ocean has only occurred in the last eighty million years. Gondwanaland began to break up into separate continents about 160 million years ago but the formation of most of the Indian Ocean and the break between Australia and Antarctica occurred in the last sixty million years. Africa and South America split apart 100 million years ago.

The drift of the continents across the globe has had a profound impact on human history. It has determined the distribution of resources, and it accounts for the differing flora and fauna of the continents. Material welling up from the earth's core formed part of the continental land masses and determined the location and concentration of the world's mineral resources. The position of the continents at an earlier stage of their history explains the distribution of fossil fuel reserves in the modern world. Coal, oil and natural gas come from the decomposition of the vast tropical forests present about 250–300 million years ago. Continental drift has also been a major influence in determining the current distribution of plants and animals. Some have evolved in isolation and others have been driven to extinction by competitors when suddenly brought into contact with other parts of the world. For example, marsupial mammals had a worldwide distribution about eighty million years ago. As the continents drifted apart, marsupials were replaced in Eurasia by placental mammals. They survived in South America until it joined North America about thirty million years ago and they live on in Australia, which has remained isolated. The evolution of animals in different parts of the world has also had major effects on human history. The isolation of the Americas from the Eurasian landmass meant that animals domesticated in Europe and Asia such as sheep, goats, cattle and horses were not present. This influenced both their agriculture and their transport – domesticated animals were relatively unimportant and, although societies in the Americas were aware of the principle of the wheel, they could not utilise it because no draught animals were available.

Climate has been a fundamental force in shaping human history. Year-to-year variations in the weather influence crop yields but more significantly major, long-term trends have affected the ability of humans to settle parts of the globe, influenced the way plants and animals are distributed and placed limits on the crops that can be grown. The distribution of the continents has also been one of the factors determining climate. The ice ages, which have dominated the world's climate for the last two and a half million years, depend for their effect on the current distribution of the land masses of the northern hemisphere. Large scale ice sheets could not develop, and move further south when the climate became colder, without the continents clustering towards the north pole and the formation of a land-locked Arctic Ocean about three million years ago.

The position of the continents is only one factor influencing the world's climate. Apart from the steadily increasing energy output of the sun and the levels of gases such as carbon dioxide and methane in the atmosphere, the major factor determining climate is a series of astronomical cycles affecting the earth and its orbit round the sun. In the 1920s a Yugoslavian scientist, Milhankovic, put forward his theory but was largely ignored. Only in the last thirty years, with scientific analysis of cores taken from ocean sediments and ice sheets giving information on climate stretching back over hundreds of thousands of years, have his theories been accepted. Over a period of 90–100,000 years the earth's orbit varies from being nearly circular to more elliptical. At present the orbit is becoming more circular and reducing the difference between the times when the heat from the sun falling on the earth is at its maximum and minimum. The second cycle, the timing of the earth's closest approach to the sun, is completed every 21,000 years. At the moment the earth is nearest to the sun during the northern hemisphere winter. This reduces the impact of seasonal climatic change in the north whilst increasing it in the southern hemisphere. The third cycle affects the 'tilt' of the earth, which varies over a period of about 40,000 years. At the moment the tilt is decreasing, which again reduces the difference between the seasons. Although there are other short-term cycles, such as minor variations in the sun's output over a 22–23 year period (linked to sunspot activity and reversals of the solar magnetic field), it is the combination of the three long-term cycles that largely determines the earth's changing climate.

The long-term cycles alter the distribution of the sun's energy falling on to the earth. The current closeness of the continents in the northern hemisphere to the pole is of crucial importance because only a two per cent drop in heat from the sun during the northern summer can initiate glaciation. Cool summers allow winter snow and ice to survive to the next winter and the growing snow cover causes further cooling by increasing the reflectivity of the earth's surface. This further reduces the temperature in the northern hemisphere, leading to a rapid increase in the extent of the ice sheets and glaciers. The same process cannot work in the southern hemisphere when changes in the earth's orbit produce cool summers there. Apart from Antarctica there is insufficient land near to the pole and too much water (which moderates the temperatures) to allow continental ice sheets to form. For the last two and a half million years a cycle of ice ages has affected the earth's climate; the interglacial periods have normally been short, a total of 250,000 years out of the last two million. The warmest interglacial phase was about 120,000 years ago.

The various forms of life on earth, including humans, do not exist independently, they are part of ecosystems – a term which is used to refer to a community of organisms and their environment. There are many different types of ecosystem such as a tropical forest, a grassland prairie or a coral reef but the foundation of all of them, and therefore the basis for life on earth, is photosynthesis – the process by which the energy of sunlight is used by plants and certain types of bacteria to create chemical compounds essential for life. Apart from the exotic life forms that live on the sulphur produced in deep ocean volcanic vents it is the only way that energy is introduced into the system. Very little of the sun's energy is, in fact, converted into matter and there is no way in which this efficiency can be improved since it depends on the amount of light falling on the earth, the laws of physics and the amount of carbon dioxide in the atmosphere. (Selective breeding of plants does not increase the efficiency of photosynthesis, it simply makes the plants put more of their effort into producing those parts that humans find useful at the cost of other parts.)

Within an individual ecosystem photosynthesisers (such as plants, trees and grasses) provide the basic energy input. They are the bottom of the food chain which links together all the different organisms. When photosynthesisers die they are broken down in the soil by decomposers such as fungi and their essential elements become available for other plants to use. (The pattern is similar in a marine environment.) Photosynthesisers are also eaten by animals (herbivores) which are able to extract essential nourishment from the plant. Herbivores are in turn eaten by other animals (carnivores) which are capable of extracting their food from animals. Some animals, known as top carnivores, can eat both herbivores and carnivores. When all these animals die their carcasses rot and the essential elements are recycled. Most ecosystems have complex food chains with numerous interrelationships between the various parts. However, behind this complexity there is an iron rule. The higher the animals is in the food chain, the rarer it will be. Each step up the food chain is further removed from the primary production of the photosynthesisers, is less energy efficient and consequently the numbers that can be supported get smaller. This is why only a very small number of carnivores can exist within an ecosystem compared to the number of primary producers. In the case of a deciduous wood in southern England, almost 90 per cent of the primary production by the photosynthesisers (in this case trees, plants and grasses) eventually falls to the ground and decomposes on the woodland floor and another 8 per cent is stored as dead wood, which eventually decomposes. Less than 3 per cent is available for herbivores to eat and even less for the carnivores who have to live off the herbivores.

Ecosystems are not static. Over time they develop through an orderly and predictable series of changes resulting from the modification of the environment by the plants and animals themselves to culminate in a climax system which has the maximum possible number of plants and animals for the available energy input. It takes thousands of years to move from bare rock through lichens and mosses to ferns, plants and eventually trees to create a climax forest that can survive for very long periods – provided there is no human interference. The original pioneer species that colonise the rock are adapted to coping with poor soils or with none at all. Slowly, as decayed matter is accumulated, a better soil is created, capable of sustaining annual plants and successively perennials, grasses, shrubs and trees. As the ecosystem develops and changes, so do the plants and animals that can be supported. This development of an ecosystem has occurred countless times during the earth's history – for example the retreat of an ice sheet after a glacial period exposes bare rock, which within a few thousand years is converted into a climax temperate forest. Where a climax ecosystem is destroyed (as is often the case through human intervention to clear forests) then the subsequent process of change is speeded up because good soil already exists. For example, arable land (the result of forest clearance at an earlier stage in its history) left uncultivated in England will, through a succession of arable weeds, grasses and shrubs such as hawthorn and mixed scrub, revert to an oak and ash forest within 150 years.

The different types of ecosystem depend to a large extent on the temperature and level of rainfall. They are therefore found in broad bands between the poles and the equator. Changes in the earth's climate cause these bands to change location, often by hundreds of miles over a period of several thousand years, as well as producing major local variations. At present, near the poles, low rainfall, low temperatures and permafrost produce tundra (poorly drained, acidic soil covered in low scrub). Further away from the pole in the northern hemisphere (though not in the southern because there is no land in the right place) are the great coniferous forests known as the taiga. Further away still from the poles are the temperate forests with their rich secondary flora, better soils, high leaf drop and a correspondingly large quantity of decomposers. Then there are the grasslands, which have less rain than the temperate forests and where the soils are poorer. In two bands sited about 30°N and 30°S of the equator are large areas of desert – a product of the global weather system which concentrates larges masses of very dry air in these latitudes. Finally in the tropics, around the equator, with both high rainfall and temperatures, are the vast, tropical rainforests. There are of course many local variations within this overall picture such as the tropical savannas of grassland with scattered thorny trees and the temperate rain forests of north-west America and southern New Zealand.

The productivity of the different ecosystems varies greatly. In the tundra the level of primary production is low because of the cold and lack of sunshine. This means fewer species can be supported and then not in large numbers. The food chain is therefore short and relatively simple. The open oceans are virtual deserts whereas coral reefs and estuaries teem with life, reaching levels equivalent to the most productive of all terrestrial systems – the tropical forests. These forests, which cover about 6 per cent of the world's land surface, produce about 40 per cent of all the terrestrial primary plant production and contain about half of all the plants and animals on earth. Rainforests are remarkable not just for the quantity of life found there but also for the diversity. A typical ten square kilometre patch of forest will contain the following species (not individuals) – 1,500 flowering plants, 750

trees, 125 mammals, 400 birds, 100 reptiles, 60 amphibians, 150 butterflies and probably over 50,000 insects. (In total in the tropical forests
there are probably twenty million species of insect.) Tropical forests
are, however, very different in their structure from temperate forests,
which have rich soils. Three-quarters of all the nutrients are held in
the plants and trees and only 8 per cent in the soil. Very little rain
water runs off into the soil – over half is evaporated and most of the
rest absorbed directly by the plants and trees. The soil itself is thin,
acidic and of poor quality, with very little humus. If the ecosystem is
destroyed through forest clearance most of the nutrients are destroyed
too; there is little available in the soil to support crops or grass and
the exposed ground can quickly turn into hard, baked clay.

Soil is the product of an ecosystem – it has been created by living
plants and animals and it continues to rely on them to remain fertile
and productive. Before the earliest forms of life emerged from the
sea there was no soil; all the land on earth was bare, eroded rocks
and deserts. Soils build up over thousands of years through physical,
chemical and biological processes as rock is weathered into minute
fragments and incorporated together with the remains of dead plants
and animals to form a medium that supports bigger plants and trees
as the ecosystem develops to a climax. Fertility is built up and maintained as an active process through the interaction of the plant cover,
the existing soil, the work of decomposers and other environmental
factors such as rainfall and temperature. All of these processes make
the various types of soil found in different parts of the globe one of
the most complex living systems on earth. Just half a hectare of good
soil from the temperate regions will contain one million bacteria of
just one type, 100,000 yeast cells and 50,000 fungus mycelia. Although
soils are created over time this process is, on a human timescale, so
slow that the soil is in effect a non-renewable resource. It is also a
highly fragile one. Ecosystems develop naturally in a way that protects
the soil on which they depend. On the dry grasslands it is the roots
of the grass that hold together a poor soil and in the temperate forests
it is the whole process of extensive leaf drop in the autumn, combined
with a large number of decomposers living off the dead material,
which maintains a highly fertile soil. In tropical forests, where the
soils are generally poor in nutrients and potentially vulnerable to the
high rainfall and temperatures, the ecosystem has developed in such
a way as to protect the soil. Once the trees and plants of an ecosystem
are destroyed or badly damaged then the underlying soil is very
quickly subjected to severe strain and can easily be destroyed or

eroded away by the wind and rain, leaving only a degraded remnant.

To fully understand the individual parts of an ecosystem, it is necessary to see them as part of a bigger picture. All the parts of an ecosystem are interconnected through a complex set of self-regulating cycles, feedback loops and linkages between different parts of the food chain. If one part of an ecosystem is removed or disrupted there will be knock-on effects elsewhere in the system. The extent of the reverberations will of course vary depending on the nature, scale and duration of the disruption; on the relative significance of the part or parts affected; and on the resilience of the ecosystem. For example, if one species of animal is wiped out (by disease or by being hunted to extinction) then there will be ramifications up and down the food chain. The population of those plants and animals that formed its diet will increase, while those that previously hunted it for food will decrease, and these changes will then bring further secondary disruption at other levels in the food chain. The destruction of the primary producers (as with forest fire or deliberate clearance) will constitute an attack on the base of the food chain and it will have disastrous effects on all parts of the chain.

Just as the plants and animals in an ecosystem are parts of a greater whole, so ecosystems themselves are part of a greater whole – the earth itself. For all practical purposes the earth is a closed system. Although sunlight gets in to provide the necessary energy for life all the other resources are finite. The fact that the earth is a closed system also means that nothing can get out. All waste products must go somewhere. This fact, combined with the limited resources available to all living things, means that the recycling of the materials necessary for life is an essential function of all ecosystems and other physical and chemical processes on earth. Problems can arise though when artificial wastes are 'disposed of' in ecosystems, for example by dumping at sea or discharging into the air. Many of these products cannot be recycled at all by the natural systems or not in the concentrations in which they are found as a result of human activity, and they remain as pollutants somewhere in the system. All pollution is, therefore, bound to affect natural processes and ecosystems, whether on land, in the oceans or in the atmosphere. Humans too are part of the earth's ecosystems, whether or not they are always conscious of this fact and its implications. All plants and animals tend to modify the environment as they compete and cooperate with others to survive and flourish. In their relationship to the ecosystem, two factors distinguish humans from all other animals. First, they are the only species capable of endangering and even destroying the ecosystems on which they depend for their existence. Second, humans are the only species to have spread into every terrestrial ecosystem and then, through use of technology, to have dominated them. (They have even been able to develop ways of heavily exploiting marine ecosystems too.)

The most important task in all human history has been to find a way of extracting from the different ecosystems in which people have lived enough resources for maintaining life – food, clothing, shelter, energy and other goods. Inevitably this has meant intervening in natural ecosystems. The problem for human societies has been to balance their various demands against the ability of the ecosystems to withstand the resulting pressures.

3

NINETY-NINE PER CENT OF HUMAN HISTORY

For all but the last few thousand years of their roughly two million years of existence, humans have obtained their subsistence by a combination of gathering foodstuffs and hunting animals. In nearly every case people lived in small, mobile groups. It was without doubt the most successful and flexible way of life adopted by humans and the one that caused the least damage to natural ecosystems. It enabled them to spread across the face of the globe into every terrestrial ecosystem and to survive not just in favourable areas with easily obtained food but also in the rigorous conditions of the Arctic, the tundra of ice-age Europe and the marginal dry lands of Australia and southern Africa.

The origins and earliest development of humans and their immediate ancestors have to be deduced from scanty evidence, usually the fossilised remains of partial skeletons, sometimes as little as a jaw bone or a tooth, and therefore interpretation is difficult. Not surprisingly, the topic arouses great controversy between experts, and numerous conflicting schemes have been put forward to explain the nature of particular fossils and their relationship to each other. The fact that the remains have, so far, only been found in a few areas of the world, mainly eastern and southern Africa, has naturally heavily influenced explanations of the likely geographical origins of human ancestors and their development.

Fossils found from about 2–1.5 million years ago and called *Homo erectus* are recognised as being the direct ancestors of modern humans. But much earlier fossil remains reveal evidence of certain 'human traits', notably upright posture (as early as 3.5 million years ago) and toolmaking linked to the first stone tools about two million years ago. The distinguishing characteristic of *Homo erectus* is a large brain size of about 1,100cc (about three-quarters of modern human capacity). They emerged in east Africa and they survived until about 200,000 years ago when the first anatomically modern skeletons, named in a piece of immense self-flattery *Homo sapiens*, are found in east and southern Africa. By about 30,000 years ago fully modern human types (*Homo sapiens sapiens*) were widespread throughout the world.

The earliest humans and their direct ancestors inhabited a variety of habitats within a belt of tropical and semi-tropical country stretching from Ethiopia to southern Africa. The population was small, thinly spread, living in groups which probably depended mainly on the gathering of nuts, seeds and plants, which they would have supplemented by scavenging dead animals killed by other predators and perhaps the hunting of a few small mammals. It was this basic form of subsistence – gathering and hunting – that was to last as the human way of life until the development of agriculture about 10,000 years ago.

Gathering and hunting

Gathering and hunting as a way of life is now restricted to a handful of groups in the world such as the Bushmen of south-west Africa, some pygmy groups in the equatorial forests of Africa, the Hadza of east Africa, a few groups in India and south-east Asia, some Aborigines in Australia, some Inuit in the Arctic and the native inhabitants of the tropical forests of South America. These groups now occupy essentially marginal habitats, having been gradually pushed aside by the advance of agriculture. Two-thirds of the Inuit originally lived in a much more benign climate south of the Arctic Circle, while the Aborigines used to live mainly in the productive regions of eastern Australia, not in the central and northern deserts. The commonly accepted view of gathering and hunting is that it produces a life which is, to quote Thomas Hobbes, 'nasty, brutish and short'. In the last fifty years new anthropological studies of existing gathering and hunting groups have provided a fascinating insight into how humans lived for the overwhelming majority of their history and how they were integrated into the environment. These studies have emphasised the relatively easy way in which sufficient food could be extracted from what would have been much more productive ecosystems than those now occupied by such groups. In parallel with these findings there has been a revolution in archaeological thinking and techniques for investigating early human remains. Instead of collecting large numbers of stone tools and trying to classify them into different 'cultures' on the basis of marginal differences in the way in which they were made, archaeologists have adopted a much more sophisticated approach. This emphasises trying to understand, often using contemporary groups as examples, what tools were made to do, what activities were carried out at the different sites, how human groups exploited their environment in different ways to obtain food and how their seasonal movements were integrated into this overall pattern.

What emerges from these new approaches is a much more positive view of gathering and hunting groups. In general, gatherers and hunters do not live under the constant threat of starvation. Rather, they have a nutritionally adequate diet selected from a wide range of the available food resources. This wide variety of food is normally only a small proportion of the total amount of food available in the environment. Obtaining food and other forms of work take up only a small proportion of the day, leaving large amounts of time free for leisure and ceremonial activities. Most groups survive with very few goods because their wants are few and because they would find extra goods a hindrance to their mobile way of life. Items such as hunting tools or cooking utensils have no great value because they can be easily replaced from locally available materials. The pattern of life varies during the year depending on the seasonal availability of different types of food. Most of the time they live in small groups of about 25–50 people and come together in larger groups for ceremonial purposes, marriage and other social activities at a time when food supplies allow a larger population to gather in one place. Within the group there is no concept of food ownership and food is treated as available to all. Food is not stored because that would interfere with mobility and because their experience dictates that some food will always be available even if certain items are occasionally in short supply.

The Bushmen of south-west Africa (until forcibly removed by the Botswana government) illustrated how easily gathering and hunting groups could obtain sufficient food. The mainstay of their diet was the highly nutritious mongongo nut obtained from a drought-resistant tree. It is a very reliable source which keeps for over a year. It contains five times the calories and ten times the amount of protein of an equivalent amount of cereal crops and 225 grams (about 300 nuts) has the calories of just over a kilogram of cooked rice and the protein of about 400 grams of beef. In addition eighty-four different species of food plants are available although the Bushmen normally used only twenty-three of them. There are fifty-four edible animals available although only seventeen were hunted regularly. Compared with modern recommended levels of nutrition the diet of the Bushmen was more than adequate; calorie intake was higher, protein consumption was about a third higher and there were no signs of any deficiency diseases. The amount of effort required to obtain this food was not very great – on average about two and a half days a week. The work involved was relatively constant throughout the year (unlike agriculture) and except at the height of the dry season the search for food rarely involved travelling more than ten kilometres in a day. Women and men devoted about the same amount of time overall to obtaining food but the women, who were responsible for gathering, brought in about twice as much food as the men were able to hunt. The women tended to work about one to three hours a day and spent the rest of the time in leisure activities. Hunting, which was carried out by the men, was more intermittent, involving perhaps a week of hunting followed by no activity at all for about two or three weeks. About 40 per cent of the group played no part in the work of obtaining food. About one in ten were aged over sixty and treated as honoured members and the young were not expected to provide food until they married at about the age of twenty for women and twenty-five for men. Similar patterns have been found amongst the Hadza of east Africa and the Aborigines of Australia.

All of these groups have now been pushed into marginal subsistence areas and so it is safe to assume that when similar groups were able to live in locations with more abundant resources, food gathering would have been even easier. Indeed, many contemporary groups fail to see the attractions of agriculture with its much greater workload. As one of the Bushmen told an anthropologist, 'Why should we plant when there are so many mongongo nuts in the world?' Leisure time is valued very highly and preferred to increasing food supplies (which are already more than adequate) or producing more material goods (which can be a hindrance). In the twentieth century the Siane tribe in New Guinea adopted modern steel axes instead of traditional stone tools. This reduced the amount of time needed to provide an adequate level of subsistence by about a third. The new spare time was not spent in increasing output but was devoted to ceremonies, leisure and warfare. Similarly in sixteenth-century Brazil the Portuguese found that the native tribes, if not enslaved, would only work for them until they had earned enough to buy metal tools and then they wanted to enjoy their extra leisure.

In general, gathering and hunting groups live largely by gathering. Hunting is a difficult and hazardous activity with only intermittent rewards at best. Studies of top carnivores in ecosystems (which is the role humans are trying to adopt when hunting) show that they only make a kill about once in every ten attempts. Humans, even with some help from technology, are much less adapted to this role than lions or tigers and are likely to achieve even lower success rates. In early gathering and hunting groups, operating with fairly primitive spears and bows and arrows, most of the meat in their diet is likely to have come from scavenging on animals killed by other predators. In the equatorial and tropical areas hunting rarely contributes more than a third of a group's diet. Ecosystems further away from the equator are less productive and therefore the available plant food needs to be supplemented – often through the more time-consuming task of fishing. The great grasslands pose major problems for these groups in finding food because of the lack of suitable plants for human consumption and the difficulty in hunting the large herds of grazing animals. It is only in the Arctic areas, with their almost total lack of suitable plant food, that hunting dominates subsistence. In these areas finding sufficient food is not easy and survival requires a great deal of skill and effort to make use of the limited resources available.

In order to obtain the necessary subsistence, gathering and hunting groups depend on a deep knowledge of their local areas and in particular an awareness of what types of food will be available at different places and at different times of the year. Their way of life revolves around major seasonal changes in methods of subsistence and patterns of social organisation are integrated into these changes. Contemporary gathering and hunting peoples again illustrate how historical groups would have adapted to their particular conditions. The Bushmen of south-west Africa lived in a relatively homogeneous environment, moving camp about five or six times a year but never moving on more than fifteen or twenty kilometres at a time and making longer trips only for social occasions such as marriage. The Gidjingali Aborigines of northern Australia have a clear seasonal round of varying exploitation. In the wet season, when swamps are full, they eat water lilies – the stalks are eaten raw, the seeds made into unleavened cakes and the corms cooked. In the early dry season they move to an area where large yams are found since the tubers are easy to locate at this time of the year when the tendrils are still green. Later they move to the edge of the wetlands where the men hunt geese and the women dig up spike rush corms. At the height of the dry season subsistence depends on cycad nuts which, although difficult to prepare, are plentiful and can support the large groups of people who gather together at this time for ceremonial, religious and social events. Only in a brief period before the rainy season is there any shortage of suitable food when subsistence depends on less favoured roots and plants.

An example of an extreme form of adaptation to a harsh environment influencing all forms of economic and social life is provided by the Netsilik Inuit living north and west of Hudson's Bay in Canada, who were studied in the 1920s before they had any real contact with modern technology. Their way of life depended on careful exploitation of every part of their environment. Houses and storage facilities were made from snow and ice. Clothing, kayaks, sledges and tents came from the skins of animals and their bones provided tools and weapons. Cooking utensils were made out of stone. The seasonal round of subsistence activity was highly varied. In the winter the Netsilik were totally dependent on seal hunting. A huge number of hunters had to cover the numerous holes in the ice used by the seals for breathing. This was therefore the time of year when extended social groups gathered together in great igloo communities and took part in the major religious and ceremonial activity of the year. The large winter camps broke up into small groups living in tents from June when it was possible to hunt the seals on the ice. By July the groups moved inland, fishing and occasionally hunting caribou. In August they constructed stone weirs across streams in order to catch the plentiful supply of salmon trout moving upstream to spawn. At the end of the month they again came together in large groups for the communal activity of hunting from kayaks as the caribou crossed the rivers during their annual migration. In October smaller groups of Inuit would fish for salmon before reassembling in larger groups for the winter seal hunt. In each of the phases of communal hunting there were social customs to ensure that everybody was fed and that nobody was penalised because of poor luck or lack of skill.

These modern examples of gathering and hunting groups tell us a good deal about the way in which historical groups would have operated in the various environments they inhabited around the world. All gathering and hunting groups, both contemporary and historical, seem to have tried to control their numbers so as not to overtax the resources of their ecosystem. This was achieved through a number of accepted social customs. The most widespread was infanticide involving the killing of certain categories such as twins, the handicapped and a proportion of female offspring. (Studies in the 1930s showed that Inuit groups killed about 40 per cent of their female children.) In addition, protracted weaning of infants probably provided a form of birth control and some of the old people may have been abandoned if they were ill or a burden on the group. In such ways the demand for food and, therefore, the pressure that gathering and hunting groups placed on their environment was reduced. Population densities were generally low (although the numbers would vary according to the type of environment and its natural level of productivity). The best estimate for the total population of the world about 10,000 years ago, just before the adoption of agriculture in a few areas, was not more than about four million and in earlier periods it would have been considerably less than that.

Technology and development

The gradual development of human societies and the spread of settlement across the globe into different environments can be traced to the four basic traits that distinguish humans from other primates. Fundamental to all advance was an increase in brain size. A bigger brain seems to have been important in achieving the power of abstract thought so vital to the development of technology. A second vital breakthrough (made as early as three and a half million years ago) was the ability to stand fully upright on two feet. This was important not just in increasing mobility but also in freeing the hands to undertake other tasks such as using tools. The third trait was the use of speech. Not surprisingly there is absolutely no evidence as to when speech was adopted. In a primitive form it probably existed from an early date but complex language only became possible when fully modern humans evolved about 200,000 years ago. The ability to communicate would have opened the way to increased group co-operation and more elaborate social organisation in general as well as aiding the spread of different cultural advances. The fourth trait was fundamental to the human settlement of the world – the adoption of technological means to overcome difficulties imposed by hostile environments. Although other animals use tools, humans are the only ones to make complex tools and use them on a major scale. Stone toolmaking began about two million years ago with the first crude stone choppers from pebbles, although other less durable tools that have not survived would probably have been used earlier.

Apart from stone tools, the artefacts and technologies used by the earliest humans were wooden spears (about 400,000 years ago), bolas stones for entangling animals (about 80,000 years ago), the use of wood and skins and also fire. Since fires also happen naturally, the exact date at which fire was first deliberately used is a matter of considerable controversy. There are ambiguous indications from the site of Chesowanja in east Africa dated to about one and a half million years ago but the first definite, and widely accepted, evidence comes from about 500,000 years ago. Traces of the use of fire are first found in association with animal kill sites, suggesting that it was brought to the sites in order to cook the meat, as well as at camps where it would have been used for heat and light and possibly protection. At this early stage, it is unlikely that it was used to drive animals into suitable killing sites, although this technique was certainly used at much later dates. But for at least two million years the principal technology used by humans was the stone tool. For about the first one and a half million years of toolmaking the dominant types were a chopper-like tool made from pebbles and a hand axe with a working edge round most of the perimeter. These tools were relatively easy to make and are found in huge quantities. For example, the disarticulated skeleton of a hippopotamus found at Olduvai Gorge in east Africa was surrounded by 459 blunted hand axes and choppers.

It was with this primitive tool kit that the first humans were able to move from Africa into the forest-free zones of the Middle East, India, southern China and parts of Indonesia, although the use of clothing from the skins of animals would have been necessary too. The exact chronological framework is difficult to establish because of the lack of archaeological work in many areas, but it is clear that *Homo erectus* had spread outside Africa by about one and a half million years ago, very soon after the first skeletons of this direct ancestor to be modern humans were to be found. They reached Asia about 700,000 years ago but the areas occupied were still restricted. With the skills that they possessed at this time humans could only adapt to those ecosystems found in the semi-tropical areas where there was a considerable variety of vegetable material that could be gathered easily and a wide range of small and easily hunted animals to supplement this diet. The equatorial rain forests were not penetrated and the settlement of Europe posed formidable difficulties. These problems were not solved for a very long time and so the settlement of Europe is a comparatively late phenomenon in human history despite its relatively easy access from the Near East and Africa. European ecosystems made it very difficult, even in interglacial periods, to extract enough subsistence with only a limited technological base: the plant life was less rich and the scope for gathering was more limited. The hunting of medium- and large-sized game was therefore vital, but difficult. Even if they were only scavenging and killing the sick and old members of the herds the groups would have needed to move over large areas when following the seasonal movements of the animals and they

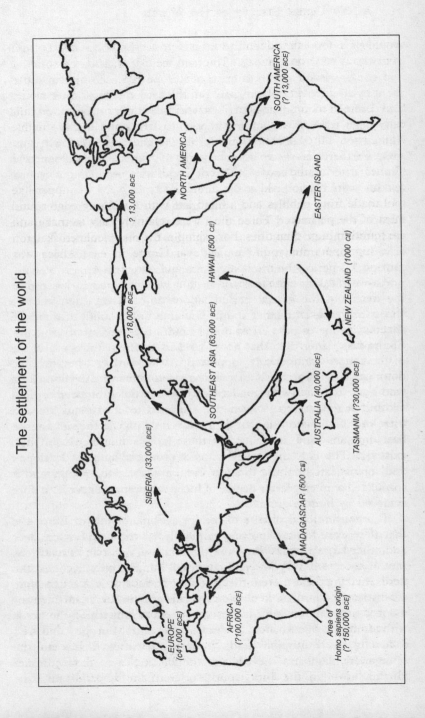

The settlement of the world

SOUTH AMERICA
(? 13,000 BCE)

NORTH AMERICA

? 13,000 BCE

? 18,000 BCE

EASTER ISLAND

HAWAII (500 CE)

NEW ZEALAND (1000 CE)

SIBERIA (33,000 BCE)

SOUTHEAST ASIA (63,000 BCE)

AUSTRALIA (40,000 BCE)

TASMANIA (?30,000 BCE)

MADAGASCAR (500 CE)

EUROPE
(c41,000 BCE)

AFRICA
(?100,000 BCE)

Area of
Homo sapiens origin
(? 150,000 BCE)

would have found it difficult to keep in touch for social and cultural activities. The first evidence of human occupation is dated to about 730,000 years ago and most areas of Europe had seen human settlement by about 350,000 years ago. But this settlement was intermittent and confined to interglacial periods when the climate of Europe would have been equable enough to support gathering and hunting with a limited tool kit. The conditions during the glacial periods, when the great northern ice sheets advanced and the climate of even southern France deteriorated into semi-arctic conditions producing a type of tundra, would have been too severe.

It was not until the last, long, glacial period that began about 80,000 years ago and lasted until about 12,000 years ago that the first permanent occupation of Europe took place. It marked a major advance in the ability of humans to adapt to a harsh ecosystem. During this period the whole of Scandinavia, north Germany, Poland, north-west Russia and most of Britain were covered in ice and at the height of the glacial period, about 20,000 years ago, the ice sheets moved even further south. The area to the south of these ice sheets was one of permafrost and a tundra-type vegetation. But this tundra was richer than that now found in northern Europe because of the longer summers. It supported a wide variety of animal life dominated by large herds of reindeer, woolly mammoth, bison and wild horse together with smaller numbers of woolly rhinoceros, giant elk and saiga antelope. Given the restricted level of plant life and therefore the limited role that gathering could play in the subsistence base, humans were dependent on these large animal herds for their existence. This challenging environment produced a highly developed and sophisticated culture to control the available food supply and it stimulated a much greater degree of social integration then ever before achieved by human groups.

The conventional picture of the ice-age inhabitants of Europe is that they were hunters preying randomly on the herds of reindeer and other large animals. But pure hunting is a high risk strategy: the rate of success is very low and continual hunting only frightens the herds, making them more difficult to follow and attack. A much more sophisticated approach to obtaining subsistence in a difficult environment was in fact employed. This was based on managing the herds with minimum disturbance. In eastern and central Europe it involved following the migrating herds from their winter ranges on the Hungarian plain and the edge of the Black Sea to their summer grazing areas in the Jura, southern German highlands and the

Carpathians. Human habitation sites are found along the natural migration routes and on the edges of the natural grazing grounds of the reindeer. The herds were not hunted at random but culled on a selective basis to remove the sick and the old. Enough animals to provide meat for the season were herded away from the main group, driven to areas such as natural basins to be killed as required. The numbers of humans that could be supported in this way though were very small. A herd of about 1,500 reindeer would perhaps be sufficient for only about three families or fifteen or so individuals. These groups would also have been forced to be highly mobile, coming together in larger numbers, essential for ceremonial and social activities, for only short periods of the year.

A very different way of life developed in south-west France and northern Spain at the height of the last glaciation about 25–20,000 years ago. When the climate was at its most severe, northern Europe seems to have been almost abandoned and a relatively dense population developed further south. Subsistence was based on the large herds of reindeer and red deer that passed through the Dordogne area and northern Spain. Within this region it was possible for a moderately dense human population to sustain a reasonable level of food supplies without undertaking long migrations following the herds. They were able to utilize different parts of the area at different times of the year and food from the herds was supplemented by a plentiful supply of salmon and other fish in the rivers. In these semi-settled conditions a highly integrated society emerged that produced the great cave paintings found at sites such as Lascaux in south-west France and Altamaria in northern Spain (roughly contemporaneous paintings and rock art have also been found at the Apollo cave in South Africa and in Australia). The exact function and meaning of the European cave paintings is still unclear but their religious and ceremonial nature is not in dispute and there was almost certainly some magical element involved in seeking to control the herds on which the way of life of the community depended.

Europe was one of the areas to see a significant advance in human technology involving the development of new techniques in the production of tools and also the use of new materials. They constitute one of the most significant changes made by humans and the greatest burst of innovation before the invention of pottery and the use of metal. They began about 40–30,000 years ago and appear to be linked to the spread of fully modern humans, *Homo sapiens sapiens*. The number of different types of stone tools manufactured rose from six to eighty and the nature of the tools changed significantly. Before about 40,000 years ago tools tended to be large – mainly hand axes or flakes from a prepared core – with only a minimal investment of time and effort in their preparation. After this date the emphasis changed to the production of very thin, parallel-sided blades from the core and later still, after about 20,000 years ago, small light blades used as points for projectiles. These new tools required different and more complicated manufacturing techniques involving heat treatment and pressure flaking from the core. Mastery of these techniques not only needed higher motor abilities and co-ordination but also greater mental skills to be able to cope with the number of separate stages needed to manufacture these artefacts.

For the first time, previously available materials such as bone, antler and ivory were worked into tools, some of which were extremely complex to make, such as barbed harpoons. Spears were improved through the use of bone or ivory points to replace stone and by the use of the spear thrower to increase range. Hunting was also made easier and less manpower-intensive by the invention, about 23,000 years ago, of the bow and arrow and the likely use about this time of snares, traps and nets, which would have widened the resource base available for human exploitation. Although clothes had been made from skins for hundreds of thousands of years, life in Europe, at the height of the last glaciation, required major improvements in survival techniques. Hoods, gloves and foot mittens were produced and by 20,000 years ago eyed needles and fine thread (a product of fur trapping) were being used. Good insulation from the cold provided by warm clothing meant that the level of calorie intake necessary for survival in the harsh conditions was low enough for it to be extracted from the difficult environment. The development of new techniques was probably accompanied by a greater degree of specialisation within the gathering, hunting and herding groups. The use of increasingly high quality materials that could only be found in a small number of locations led to the creation of regional networks for their exchange.

The permanent settlement of Europe at a time of extremely severe climatic conditions was a major human achievement and a sign of increasing human control over the environment. It was made possible by a combination of the adoption of new technologies and more sophisticated animal management. The settlement of Australia did not require such elaborate adaptations because of the relatively benign climate in the eastern part of the continent and the ease with which gathering and hunting groups could find food. It could only have come about, however, after one major invention – the boat – because

Australia, although joined to New Guinea at the height of the last glaciation, was never linked to the Asian mainland. Australia was settled about 40,000 years ago when a voyage of about 100 kilometres would have been needed. Tasmania was linked to Australia until about 27,000 years ago and it was settled shortly before this date. The initial settlement of Australia was probably by a small group, perhaps as few as twenty-five people, but the population grew rapidly in what had been an undisturbed environment to reach about 300,000 – the same level as when Australia was first reached by Europeans.

The settlement of America was almost the last stage in the movement of humans across the globe. This was because it depended on the ability of human groups first to survive the harsh climate of Siberia and then advance eastwards to the Bering Strait. The crossing into Alaska was made at the height of the last glaciation when the reduced sea levels turned the Bering Strait into a land bridge. The climate in the area is likely to have been less severe than now with reasonable opportunities for hunting and herding the large animals of the area. But movement south out of Alaska could only have taken place at a slightly warmer stage when the two major ice sheets of North America, centred on the Rocky Mountains and the Laurentian shield, would have retreated and separated enough to open up the passes to the south-east. This could have occurred either 30–23,000 years ago or about 13,000 years ago. Although this is a matter of great debate in early American archaeology, the latter date is more likely. Once the first human settlers were able to move through the passes to the south they found an enormously rich environment that provided plenty of opportunities for relatively easy subsistence. The human population multiplied rapidly and within a few thousand years had spread to the tip of South America.

A whole series of adaptations was required in order to extract food from the wide variety of ecosystems found in the Americas. On the plains of North America, given the lack of a large variety of plants for gathering, subsistence depended on exploiting large herds of bison and other animals. These were often killed in a crude and highly wasteful way by driving them into narrow canyons or over cliffs. At Caspar in Wyoming about 10,000 years ago a single kill involved at least seventy-four animals and in a roughly contemporaneous kill in south-east Colorado hunters appear to have started a stampede into a canyon causing about 200 corpses, most of which could not be used because they were squashed at the bottom of a large pile of bodies. In eastern North America the spread of the forest after the ice sheets retreated changed the ecosystem and removed most of the large animals suitable for hunting. Societies adapted to these new conditions in much the same way as those in post-glacial Europe – by exploiting smaller animals such as deer, by fishing, and by placing a greater emphasis on gathering. Further north in the Arctic areas settlers were attracted by the abundant meat from caribou, arctic foxes and hares and only later moved on to exploit marine resources, especially the seals. The deserts of the south-west required a different adaptation with the emphasis on mobility so as to exploit a wide range of plants and animals in a difficult environment. In the tropical areas of Central and South America a way of life based on the widely available plant resources supplemented by a small amount of hunting was possible.

Perhaps the most extraordinary development took place on the north-west Pacific coast with its abundant marine resources of seals, sea lions, sea otters and in particular the salmon that came to spawn in the rivers. This relatively plentiful supply of food probably involved more effort in storage than in hunting. The various animals were dried during the summer or smoked in the autumn to provide sufficient food for the winter. Although there were fluctuations the supply of food was sufficiently reliable to avoid the need for mobility and this area produced one of the few examples of a settled society not based on agriculture. Villages developed, each with a population of about 1,000 living in long, communal houses. There were village chiefs, considerable social stratification and specialisation of labour together with complex mechanisms for barter and gifts of food as a way of obtaining prestige and ensuring adequate subsistence was available for all. The complex society even produced a hereditary caste of slaves. Extensive food storage meant that the winter was a time when the effort required for subsistence was minimal and elaborate ceremonial activities took up most of the spare time. It was a highly stable way of life that survived until Europeans arrived in the area.

By about 10,000 years ago, with the movement of the human frontier through the Americas, nearly all parts of the globe had been settled. The final phase in the human settlement of the world took place relatively late in the Pacific and Indian Oceans. This settlement was carried out not by pure gatherers and hunters but by groups that had obtained their subsistence through a primitive form of agriculture although they still relied on stone tools and supplemented their diet through occasional hunting. In the Pacific, the Micronesian people settled islands such as the Marshalls and Carolines but it was the Polynesians who carried out the most extensive voyages. From New

Guinea they reached Tonga and Samoa about 1000 BCE and moved further east to the Marquesas about 300 CE. From there they sailed to Easter Island and Hawaii about a century or two later. The last two major islands of the world were settled by humans about 800 CE in the Pacific and the Indian Oceans. This was at the time when Charlemagne's empire was at its peak in western Europe and the Vikings were beginning their epic voyages, when Islam dominated the Mediterranean and the Near East, and China was ruled by the T'ang dynasty. The Polynesians reached New Zealand and people moving west from Indonesia settled the small island groups in the Indian Ocean together with Madagascar.

Gathering, hunting and the environment

Every major area of the world (except Antarctica) had now been settled by humans. Gathering and hunting groups had, over hundreds of thousands of years, adapted to every possible environment in the world from the semi-tropical areas of Africa to ice-age Europe, from the Arctic to the deserts of south-west Africa. The subsistence techniques used in these differing environments varied widely from dependence on gathering and hunting small animals to the herding of reindeer, hunting of bison and the highly complex mixtures of strategies required in the Arctic. It is often assumed that these groups lived in close harmony with the environment and did minimal damage to natural ecosystems. The gathering of food did require very detailed knowledge and considerable understanding of where resources could be found at different times of the year so that the annual round of subsistence activity could be organised. The herding and hunting of animals similarly required close study of their habits and movements. There is also evidence that some of these groups did try to conserve resources in the interest of maintaining subsistence over a long period. Totemic restrictions on hunting particular species at certain times of the year or a pattern of only hunting an area every few years would have helped to maintain population levels in the hunted animals. Apart from specific cultural restrictions one of the main reasons why gathering and hunting groups usually avoided over-exploiting the available natural resources was that their numbers were small and therefore the pressure they placed on the environment was very limited.

However, gatherers and hunters are by no means passive in their acceptance of ecosystems and many of their activities do alter the environment. The modern Hazda of east Africa are known to destroy wild beehives in order to obtain a small amount of honey and other groups often destroy many of the wild plants on which they depend by carelessly uprooting large numbers. Moreover, gathering and hunting groups do alter the conditions in which wild 'crops' grow, intervening in order to benefit some favoured plants at the expense of others that they do not require. One of the most effective ways of doing this is by burning and this practice was widespread among gathering and hunting groups. Fire alters the habitat significantly, by favouring annual plants that grow well in new ground and by increasing nutrient recycling. The Aborigines used fire regularly to encourage an edible bracken on Tasmania and the Maoris on New Zealand used the same technique to increase the spread of another bracken, the rhizome of which formed a substantial part of their diet. In New Guinea from about 30,000 years ago, not long after it was first settled, there is widespread evidence of forest clearance by felling, ring barking and the use of fire. This opening up of the forest cover was to provide room for the sago tree and to encourage food plants such as yams, bananas and taro to grow. In post-glacial Britain patches of woodland were cleared by burning in order to encourage the growth of forage for red deer. Most groups also tended wild plants by transplanting and sowing in their natural habitats and by removing competing plants. Some even used techniques such as irrigation on a small scale to improve the habitat of favoured plants. Although these interventions in natural ecosystems are very different from agriculture, which involves creating an artificial ecosystem, they reveal humans modifying the environment, if only on a small scale and in limited locations.

The most dramatic impact that gathering and hunting groups had on their environment though was through hunting animals. It is much easier to damage this part of an ecosystem because the number of animals at the top of the food chain is small and they usually take a long time to recover from any over-hunting. Although there is some evidence of attempts by groups not to over-hunt, there is far more of uncontrolled hunting and even the extinction of species. The great bison hunts on the plains of North America could kill hundreds of animals in one drive even though only a few were required. The bison population was huge (about 50–60 million) so that even a large number of kills a year on this scale would not significantly reduce numbers. Smaller populations, especially those on islands, could be badly affected. On the Aleutian Islands in the North Pacific the population concentrated on killing the sea otter for over a thousand years after the settlement of the islands in about

500 BCE until it was almost extinct and the subsistence base of the community was destroyed.

The impact humans could have on animal numbers is well illustrated by the examples of Madagascar, Hawaii and New Zealand, islands with unique fauna which were suddenly subjected to severe stress. Because no large mammals had been able to reach such isolated islands large flightless birds had evolved, in the absence of major predators, to become the dominant animals. They were defenceless against human predation. Within a few hundred years of the settlement of Madagascar many of the larger animals, including a big flightless bird and a pygmy hippopotamus, were extinct. On Hawaii within a thousand years of human settlement 70 per cent of all bird species were extinct. In New Zealand the Maoris were faced with a temperate environment where many of their traditional crops such as banana, breadfruit and coconut from the sub-tropical Polynesian islands could not be grown and even yams and taro could only be grown on the North Island. This forced a radical shift in their normal subsistence pattern towards wild plants such as bracken fern and the leaf heads of the cabbage tree together with marine resources. Hunting also became more important. The large number of flightless birds such as kiwi, weka and the many species of moa (most of them about 180 cm high though one type was over 350 cm tall) were hunted ruthlessly and their eggs were eaten too. Within 600 years of the first settlement twenty-four species of moa were extinct together with twenty other types of bird.

Gathering and hunting groups could even have had an impact on animal populations on a continental scale. A number of species became extinct around the end of the last glaciation at a time when climatic change and the consequential shift in vegetation types were adversely affecting the large mammals that had inhabited the tundra of northern and central Europe. In Eurasia five large animals – the woolly mammoth, woolly rhinoceros, giant Irish elk, musk ox and steppe bison – together with a number of carnivores became extinct within a period of a few thousand years as the ice sheets retreated and the tundra was replaced by forest. The changing environment put the greatest strain on these large animals but the hunting by humans would have had a significant impact on a population already in decline and may have tipped the balance between extinction and survival.

The extinction of species in Eurasia was on a relatively small scale. Elsewhere in the world it was massive. In Australia, over the last 100,000 years, over 80 per cent of the large animals have become extinct in an area where the climatic impact and therefore the effect on animal habits of the ice ages was minimal. The most likely explanation is hunting by Aboriginal groups in the last 40,000 years. Even if the largest animals were not themselves hunted on a large scale, disruption of the ecosystem as a result of human intervention – by destroying habitats or killing the smaller herbivores on which the carnivores depended – could easily lead to extinction. Equally remarkable is the 80 per cent loss of large animals in South America and the loss of three-quarters of the large animals in the north of the continent. Unlike Eurasia, where only the animals on the steppe tundra were affected, the extinctions in the Americas involved every type of ecosystem. Although some occurred at the end of the last glaciation, climatic changes had not produced such massive extinctions in the past and there is little doubt that human intervention played a major part. As the first settlers in America moved south out of Alaska and the Rocky Mountains they would have found a rich, undisturbed environment and their numbers would have risen quickly. These first American settlers left a trail of destruction across the continent. Two-thirds of the large mammals present when humans first arrived were driven to extinction. Some of these were archaic types, such as the plains camel (found only in North America because of its isolation); others were giant species particularly sensitive to both climatic change and over-hunting. Overall the extinctions included three genera of elephants, six of giant edentates (armadillos, anteaters and sloths), fifteen of ungulates and a large number of giant rodents and carnivores.

By about 10,000 years ago humans had spread from their original area of southern and eastern Africa to every continent. The slow expansion of human settlement, which had taken about two million years, depended upon a number of linked developments. Growth in brain size gave increased capacity for abstract thought and speech and an ability to provide increasingly sophisticated cultural and technological solutions to the challenges posed by a wide range of difficult and even hostile environments. These changes occurred on several different fronts, starting with the production of increasingly sophisticated stone tools and the introduction of new weapons such as the bow and arrow but also including the use of fire, furs and skins for clothing, the construction of shelters from a wide variety of materials and the adoption of more complex food processing techniques – cooking in pit hearths rather than open fires and grinding nuts and seeds. The rate of development was of course very slow and also patchy. Not until about 40,000 years ago did the pace of technological change increase rapidly, at least in comparison with earlier periods.

These developments were of fundamental importance for the rest of human history and the future of the earth. Humans had become the only animals to dominate and exploit every terrestrial ecosystem. Yet at this stage the overall impact of the gathering and hunting groups on the environment was small because of the low, thinly spread population and their limited technology. Even so they were already making their presence felt as a number of animals were hunted to extinction and the environment was modified in subtle ways. The gathering and hunting way of life was highly stable and very long lasting. For hundreds of thousands of years it was the only way in which humans were able to extract the necessary subsistence from the environment. The number of people that could survive in any one area was constrained by their position at the top of the food chain. Only in exceptional cases such as the Pacific coast of North America were resources so abundant that settled populations could develop in sizeable villages. Then, about 12,000 years ago, the methods humans used to obtain their food began to change in a number of locations across the globe. The pace of change was still slow but far faster than in the past. Its consequences were far more radical than anything that had gone before. It brought about the most fundamental alteration in human history – and one which made possible all the subsequent developments in human society.

4

THE FIRST GREAT TRANSITION

The radically different way in which humans obtained their food was based on major alterations to natural ecosystems in order to produce fields to grow crops and pasture to feed animals. This more intensive system of food production was adopted not once but on a number of occasions across the world using different crops and animals. The key centres were south-west Asia, China, Mesoamerica, the Andes and the tropical areas of Africa and south-east Asia. These changes marked the most important transition in human history. The extra food that was produced made possible the evolution of settled, complex, hierarchical societies and everything we grandly call 'civilisation'. It also allowed a much faster growth in human population once the constraints imposed by a hunting and gathering way of life were removed. About 12,000 years ago, during the early stages of the evolution of agriculture, the population of the world was about four million (about half of the population of modern London). It probably increased slightly in the period up to 5000 BCE as agricultural settlements developed. Then it began doubling every millenium to reach about 50 million in 1000 BCE (still smaller than the current population of Britain). Then, during the period of the first great empires across the world, it rose to about 250 million by 200 CE. The upward trend has continued ever since, though not at a steady rate and often interrupted by the impact of famine and disease, so that agriculture now supports a world population of just over six billion.

Agriculture: How and Why

The combined phenomena of the transition to agriculture, the growth of settled societies, the eventual emergence of cities, the development of craft specialisation and the rise of powerful religious and political elites, are often referred to as the 'Neolithic Revolution'. However,

World population, 10,000 BCE–BCE/CE

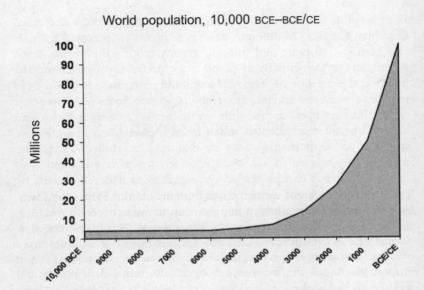

although the consequences of all these changes were clearly revolutionary – both in their impact on the way of life and on the environment – it is highly misleading to describe the process itself as a revolution. These changes took several thousand years to evolve and the contribution of any one generation would have been small. Moreover the idea of a revolution implies action undertaken with the aim of bringing about change. This was not the case with agriculture and its consequences. Human societies did not set out to invent agriculture and produce permanent settlements. Instead a series of marginal changes were made gradually in existing ways of obtaining food as a result of particular local circumstances. The cumulative effect of the various alterations was important because they acted like a ratchet. Changes in subsistence methods often allowed a larger population to be supported but this made it difficult and eventually impossible to return to a gathering and hunting way of life because the extra people could not then be fed. Many different ways of obtaining food would have been tried; some would have failed and others would have been only partially successful. Only slowly and unconsciously did a radically new solution to the human problem of finding enough food emerge.

The processes involved in the emergence of agriculture can best be understood by abandoning any idea of a clear distinction between gathering and hunting on the one hand and agriculture on the other.

Both ways of obtaining food should be seen as parts of a spectrum of human activities of different degrees of intensity to exploit animals and plants. Gathering and hunting groups modify the environment to encourage the growth of plants they prefer through controlled burning, the creation of 'irrigated' areas and replanting. It is also clear that groups across Eurasia from about 30,000 years ago were not simply hunters preying randomly on the herds of animals they came across. Instead they adopted much more sophisticated and effective strategies of herd management so that they carefully selected the animals they wished to kill. Such relatively intensive exploitation of animals does not require settled communities as is demonstrated by present-day groups of nomadic herders such as the Sami (reindeer), Masai (cattle) and various groups in central Asia (horse).

There is a whole range of problems involved in studying the slow transition to agriculture. Although archaeological techniques have become increasingly sophisticated over the last few decades it is still impossible to tell the difference between the remains of plants and grain gathered in the wild and those from the same varieties that have merely been planted and tended in specially created fields. It is usually possible to distinguish the characteristics of plants during the domestication process as they gradually alter from their wild progenitors to the fully cultivated variety but such changes take place over a very long period. These problems are compounded in tropical and subtropical areas where plant remains are rarely well preserved because of the warm, wet climate. Plants such as yams and potatoes as well as trees such as coconut and sago show almost no changes when domesticated and therefore it is almost impossible to date changes in subsistence methods. There are also major problems in studying the domestication of animals. It is almost impossible to tell directly from the archaeological record whether wild animals were being herded. The best indirect clue is the presence of a high percentage of bones from young animals which suggests that highly selective predation techniques were being used. Exactly how animals change when they are domesticated remains a controversial subject although it is generally agreed that they become smaller and retain more of their juvenile characteristics. However, these changes can only take place over long periods of time, making it difficult to identify exact domestication dates.

Other aspects of the so-called 'Neolithic Revolution' are unreliable indicators of changes in subsistence methods. Small villages can be found among some gathering and hunting groups such as the coastal communities of north-west America, where food supplies were abundant throughout the year. Pottery can also be made by gathering and hunting groups. Indeed the first known pottery in the world, dating to 10,000 BCE, is associated with the Jomon gatherers and hunters in Japan. (They continued to make it for ten thousand years without any agriculture.) Similarly the technologies used to process nuts and seeds gathered from the wild (grinding stones, pestles and mortars and containers) were all made about twenty thousand years before agriculture developed. Conversely, many tools associated with gathering and hunting groups (such as burins and scrapers) are also found in the remains of the earliest agricultural communities.

All of these factors reinforce the view that a fundamental distinction between agriculture, gathering, herding and hunting should not be drawn. No radically new techniques or relations between humans and plants and animals emerged about 12,000 years ago. Human subsistence techniques had been evolving for hundreds of thousands of years, albeit slowly. The techniques available to groups living in late glacial Europe or at a similar period in south-west Asia were far in advance of anything used by earliest hominids in east Africa. What was new in the earliest forms of agriculture was an intensification and combining of existing ways of obtaining food. This process did not stop with the first adoption of agriculture. New plants and animals continued to be domesticated, others were discarded. It was several thousand years after domestication that the 'secondary products revolution' took place – the use of cattle, sheep and goats to provide milk and other dairy products. This process still continues with the development of new varieties of domesticated plants (for example the 'Green Revolution' of the second half of the twentieth century) and may be taken even further with the use of genetically modified crops. In addition, domesticated varieties of crops and animals were later transported around the world and grown or tended on continents far removed from the original sites of their domestication.

There is, however, an even more fundamental problem in studying the emergence of agriculture and that is why it was ever adopted in the first place. Gathering and hunting groups had developed a wide range of methods for obtaining food and in most cases (apart from the most marginal areas) this did not involve using up large amounts of time and effort. By exploiting a wide range of resources they were able to reduce the risks involved by over-dependence on a single animal or crop. Selective herding of animals could provide plenty of meat, and wild grains were as good as the earliest cultivated varieties. Recent

experiments using stone sickles to harvest the large stands of the wild
ancestors of current crops that still grow in south-west Asia have
shown that yields of 800 kilos per hectare could be obtained. (This
rate is as high as that of wheat in medieval England.) Not only were
yields adequate but the wild varieties are usually more nutritious than
the cultivated types. Similarly, in Mexico, teosinte, a wild form of
maize, has been found to be highly productive. Three and a half hours'
gathering provided enough food for two people for five days.

The major drawback with agriculture is the amount of work involved
in clearing land, weeding, sowing, tending and harvesting crops and
in looking after domesticated animals. It does not necessarily provide
more nutritious food, nor does it offer greater security because it
depends on a far smaller range of plants and animals. In a poor season
food shortages and even famine are much more likely. Crops also
have to be stored for much of the year leading to wastage. The one
advantage of agriculture is that in return for this greater effort it can,
in most circumstances, provide more food from a smaller area of land.

Given these disadvantages why did agriculture evolve? Many of the
earliest explanations assumed that agriculture offered such obvious
advantages that it was adopted as soon as human knowledge and cultural
achievements had reached a sufficiently advanced level. Work in the last
half century on the sophisticated way in which gathering and hunting
groups operate and their ability to obtain food relatively easily has meant
that this explanation is no longer tenable. Another theory has linked
agriculture to the climatic changes at the end of the last major glaciation.
These do seem to have had an important impact in south-west Asia
and possibly China but their impact would have been minimal in the
tropics, and agriculture in Mesoamerica and Peru emerged thousands
of years after the end of the last ice age. It has also been argued that
mounting population pressure might be the main cause behind the shift
to more intensive forms of obtaining food. Although gathering and
hunting groups take a number of measures to limit their population
these are not, over a long period, completely successful. Normally when
the population becomes too big for the existing territory that the group
exploits during the year a number of families split away, form a new
group, and move to exploit a new area. It is possible that over a period
of tens of thousands of years the human population in some areas
gradually grew to a level that was the maximum that gathering and
hunting and the existing technology could support. The impact of climatic
change could have intensified this process by reducing the area available
for easy exploitation. If population growth continued beyond this point

then some groups might be forced into less favourable habitats where more intensive ways of gaining food were essential. Over time these methods would have become even more intensive, and population would have slowly risen, until eventually what we now recognise as agriculture would have evolved. The changes from one generation to another would, in most cases, have been slight but eventually there would have been no going back – knowledge of gathering and hunting techniques would gradually have been lost and the existing population of what had become a small village could only be supported through agriculture. It is likely that the impact of climatic change and population pressure varied from area to area and it is probably impossible to provide a single, simple explanation of why agriculture was adopted in the different areas of the world.

Although the animals and plants domesticated across the world varied and depended on what was available in nature there are a number of common patterns. Probably the first animal to be domesticated was the wolf. This was first done by the Aborigines of Australia and New Guinea, with this process spreading across the northern hemisphere in the late glacial period. (South-west Asia was comparatively late in this development.) Its agricultural significance was slight – the dogs that evolved from the wolf seem to have been eaten only rarely. They provided companionship, probably helped with hunting and provided some protection and were fed by scavenging among the debris left by humans. Humans generally domesticated placid, slow-moving animals that ate a wide range of food and already existed in highly social groups with a submissive herd structure – sheep and goats are the obvious examples. Domestication isolated the animals from the wild and gradually led to the selection of characteristics that the early farmers thought desirable.

The wild ancestors of the plants domesticated for their seeds were weedy types well adapted to growing in disturbed and open sites. Before they were cultivated they survived in poor thin soils in areas with marked wet and dry seasons and with only limited competition. They produced relatively large seeds that germinated easily and they could grow quickly and survive in long, dry summers. All of these characteristics would have been useful for the earliest human gatherers and cultivators. Continual harvesting of these plants would unconsciously select those seeds with less effective dispersal mechanisms. This would quickly (within a few years) have further isolated the cultivated varieties through the loss of natural dispersal ability. Eventually this would lead to types where the seeds only became detached after

threshing, making the plant entirely dependent on humans to reproduce. Most of the grains that were domesticated (especially wheat
and barley) are predominantly self-pollinating (unlike the majority of
plant species which are cross-pollinated). The advantage of this characteristic for the earliest farmers was that the cultivated types would
easily have become independent and not subject to swamping by the
more numerous wild varieties. Occasional cross-pollination would have
provided just enough genetic variation for new forms to adapt to new
conditions and produce slightly different characteristics.

Despite the similarities in the overall structure in the evolution of
agriculture across the world the varieties of plants and animals domesticated and the very different timescales involved mean that it is best
to study the most fundamental change in human history on a regional
basis. Each region of the world – south-west Asia, China, Mesoamerica,
the Andes and the tropics – had its own unique characteristics. Most
important of all, in each of these regions agriculture developed in
isolation from the rest of the world.

South-west Asia

It was in this area (running from the Levant through south-eastern
Turkey to the Zagros mountains in Iran) that agriculture first developed. It is also the area that has been studied most and where the
clearest picture of the transition to farming is apparent. Genetic studies
of the early plants used by the farmers strongly suggest that each
plant was only domesticated once. Exactly where this happened and
whether they all happened in the same place is unknown. Animals
appear to have been domesticated separately, mainly in the Zagros
mountains, and then spread to existing farming groups further west.

What plants and animals were domesticated? The most important
grass was wheat which has the highest nutritive levels of the cereals.
The original domesticated wheats – einkorn and emmer – are no
longer cultivated (although they were still grown in medieval England),
having been replaced by bread wheat and hard (durum) wheat. Einkorn
still grows wild across south-west Asia though emmer, the most
important of the early wheats, was restricted to the Levant and
northern Iraq and Iran. Barley, despite the fact that it can withstand
harsher conditions than wheat, was little used for bread and was kept
for making beer. Rye is rarely found and oats, which began as a weed
in cultivated fields, was mainly used later as agriculture spread to
Europe. The wild forms of lentils, peas, broad beans and chickpeas

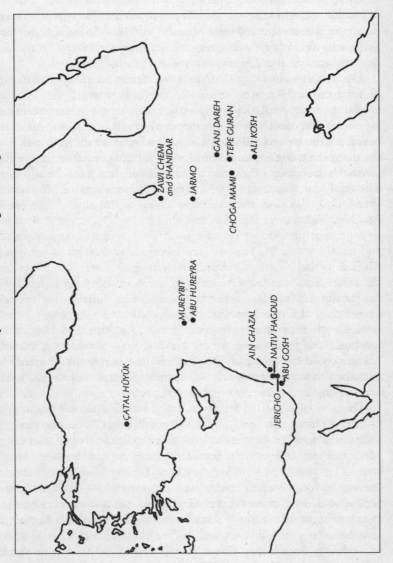

South-west Asia: key sites for the origins of agriculture

(the legumes essential for a balanced diet alongside cereals) grew wild across south-west Asia and were domesticated at about the same time as einkorn and emmer. The first domesticated animals were the Armenian variety of the west Asiatic mouflon (the ancestor of domestic sheep), first domesticated somewhere in southern Anatolia or northern Syria, and the Persian wild goat (the ancestor of modern forms), first domesticated in the Zagros mountains.

The earliest indications of the long transition to agriculture can be found in the Kebaran culture of the Levant which can be dated to about 18,000 BCE (about the same time as the advanced groups in ice-age Europe). The Kebaran people lived a semi-sedentary life based on the caves of the area. They gathered seeds from nearly all the plants that were later domesticated but concentrated their efforts on the large herds of gazelle in the region that were clearly being managed and selectively culled. This long-lived way of life changed about 10,000 BCE as the climate warmed at the end of the last ice age. The wild grasses (emmer, einkorn and barley), together with oak, almond and pistachio trees (all good food sources), spread rapidly and widely across the region. The changes this brought about can be studied in the Natufian culture that emerged along the coastal and hill zones from southern Anatolia to the Nile valley. In parallel with the herding of gazelle there was extensive exploitation of the wild grasses and the development of a sophisticated technology – bone sickles with flint blades, querns, grinding stones and pestles and mortars – to process the seeds. Food was so abundant that small villages could be sustained. The best studied of these is Abu Hureyra near the Euphrates in Syria. It was a small village with 300–400 inhabitants living in pit dwellings with reed roofs.

This way of life lasted for about 2,000 years before it was disrupted by further climate change. Continued warming after the ice age produced a climate across the Levant that was much closer to that of the modern Mediterranean and the wild grasses became restricted to much smaller areas. This produced a subsistence crisis for the Natufian groups and the way in which some of them reacted produced the earliest forms of agriculture. Many of the villages were abandoned and some groups may even have reverted to a mobile life of gathering and hunting. The majority, long adapted to a sedentary style of life, started to plant wild grasses in land near the villages. This enabled them to maintain their existing culture and continue to use all their seed processing and storage equipment. No doubt it hardly seemed a momentous step to take but the long-term consequences were profound.

The first signs of the change are apparent in the villages around the edge of the shrinking lakes of the Jordan valley about 8000 BCE although the process can be traced in about 150 sites spread across the Levant. In the first few hundred years rapid domestication was confined to the edge of the Mediterranean vegetation zone but then spread rapidly. The old idea that farming communities were 'colonisers' taking over new agricultural land has been abandoned as too redolent of European notions of the early twentieth century about imperialism and 'advanced' societies taking over those at a lower level. Instead there seems to have been a process of diffusion and adaptation as different groups adopted some 'agricultural' practices while retaining, often for considerable periods, some from their gathering and hunting past. This was certainly the case at Jericho and the nearby site of Nativ Hagdud which flourished around 7800–7500 BCE. Jericho had a population of about 300; Nativ Hagdud was much smaller with about twenty or thirty families. Hunting continued at both sites as did collection of seeds from the wild but, crucially, domesticated plants were grown in small fields. A relatively sharp transition took place at Abu Hureyra which was reoccupied, after a gap of three centuries, in 7700 BCE. The new village of one-storey, mud-brick houses grew to over seven hectares. At first the inhabitants lived as before – herding gazelle and gathering a wide range of plants. Then, within a couple of generations, there was a rapid switch to agriculture. The wide range of gathered plants was reduced to half a dozen culti-vated species (mainly einkorn and pulses) and the gazelle were abandoned in favour of domesticated sheep and goats.

The switch to sheep and goats in the Levant was almost certainly the result of their introduction from outside following their domes-tication in the Zagros mountains and along the edge of the Mesopotamian plain. This area has been less well studied than the Levant and it is likely that true farming, with its mix of domesticated plants and animals, developed here first. In the period around 9000–8500 BCE in places such as Zawi Chemi and Shanidar in the mountains of Kurdistan and Ganj Dareh near Kermanshah there were seasonal camps in the mountains which became villages as goats and sheep were kept in a semi-domesticated environment and local wild cereals were gathered before being cultivated. On the edge of the Mesopotamian plain the village of Ali Kosh was settled in about 8000 BCE. Sheep and goats were herded and taken up into the mountains in the summer. Emmer, einkorn, barley and lentils were cultivated from the start and fish and wildfowl were hunted in the nearby marsh.

It was from here that domesticated sheep and goats moved westwards into the Levant.

Further north in Anatolia at the site of Hacilar in around 6700 BCE the inhabitants cultivated barley and emmer while gathering other seeds from the wild and herded sheep and goats although the animals were not fully domesticated. The people still used baskets and animal skin containers rather than the pottery that was beginning to be made at other sites in the region. A few hundred years later a major settlement at Çatal Hüyük covered about twelve hectares and had a population of about 5,000. The small town depended on its control of a source of obsidian and a group of specialised craftsmen had emerged to process the stone before it was traded. Although most of the developments in agriculture were independent there were clearly exchanges of ideas, techniques and materials across south-west Asia between the different farming communities. Boats were first developed about 10,000 BCE when the highly prized obsidian on the island of Melos began to be traded widely.

By about 6000 BCE fully agricultural societies had emerged across the region. The original cereals became more productive through human selection – the first bread wheats emerged in northern Iran about this time when domesticated emmer was crossed with a wild wheat. Sheep were selected for their wool. Cattle, a much more difficult animal to tame, were probably domesticated between 7000 and 6000 BCE, although it was another several thousand years before they were selected to produce extra milk and humans began to consume dairy products. Pigs were first domesticated about 6500 BCE. As farming was adopted the population grew very slowly. Although the changes were hardly noticeable to each generation the long-term consequences were considerable. Villages sprang up everywhere and small towns emerged in some places. Although these societies seem to have been fairly egalitarian there was a much greater degree of social organisation. Societies developed complex rituals especially involving fertility now that their life depended upon the crops they planted and the animals they tended.

The spread of farming

South-west Asia was the first area in the world where humans produced artificial ecosystems on which they depended for their survival. From this core region farming spread first eastwards into north-east Iran and southern Turkmenistan near the Caspian Sea by about 6000 BCE. The spread of agriculture westwards from Anatolia into the Aegean area is unsurprising because their climates are similar and few changes were needed in farming methods. Generally there was a slow intensification of agriculture but in some cases such as Knossos on Crete there does seem to have been deliberate colonisation (about 6000 BCE). By this date there were farming communities in Greece and the Balkans and a thousand years later farming was being practised in south-east Italy. It was in southern Europe that olives, vines and figs were cultivated from about 4000 BCE.

The adoption of farming further west and north in Europe required a major series of modifications to the practices that had emerged in south-west Asia. The climate was wetter with year-round rainfall and the growing season was shorter. Planting had to shift from the autumn to the spring and winter feed was usually needed for the animals. Barley and oats became far more important crops. The limited range of tools available – mainly wooden hoes and digging sticks – were only of limited use in the heavy soils that predominated in central and north-west Europe. Shortly after 5000 BCE the farmers of the 'Banderkeramik' culture (named after their distinctive pottery style) spread up the Danube valley and then west to the Netherlands and east as far as the Vistula. They grew barley, einkorn, emmer and flax, adopted crop rotation and enclosed their cattle, sheep and goats in hedged fields. The spread of farming must have been anything but a smooth process and many sites must have failed as people slowly learnt how to adapt to the new and more difficult conditions. In general the first farmers settled on the lightest, most easily cultivated soils, especially gravel river terraces. The original settlements were small and widely scattered with gathering and hunting groups surviving and adopting some techniques from the farmers such as pottery for storage. Only as the population grew slowly and techniques improved did farming communities develop more widely.

Agriculture also spread south from the Levant into the Nile valley, though later than its adoption in south-east Europe. The valley was not settled on any major scale until about 6000 BCE because of the problems in coping with the annual flood. As numbers rose the pressure to adopt more intensive techniques increased and people began to shift to cultivating rather than gathering wild cereals. Exactly when this happened is not known because this period coincided with very low Nile floods and the sites of the first agricultural villages are now under water. The best estimate is that the first villages emerged about 4300 BCE and that agriculture was fully established by about 3500 BCE.

China

The second area of the world to adopt agriculture was China and it did so independently. Within the area of modern China two separate and distinctive types of farming emerged based on millet in the north and rice further south. The earliest took place in the south in the Hupei basin of the Yangtze and the coastal plain of the Hang-chou Bay in the Yangtze delta south of Shanghai. It was here, about 6500 BCE, that rice was first domesticated. Wild rice germinates in dry soil but grows in seasonally flooded shallow water. The exact ancestor of the domesticated variety is unknown but it was domesticated by creating artificial areas where it could be planted and then grow. This could easily be done in the middle and lower Yangtze valley which had vast areas of flooded lakes and lowlands. The companion legume for rice was the soybean. As in south-west Asia some of the earliest farming villages that developed continued hunting small birds and animals to supplement their diet.

In the north of China where the Yellow river comes down from the highlands onto the plains of east China and is joined by the Wei river, agriculture developed about 5500 BCE. In this area winters were harsh and the summer rain was unreliable. Farming was therefore based on two types of millet – broom-corn and foxtail – which were originally gathered from the wild in the nearby highlands. Millets have the lowest water requirements of any of the cultivated cereals. The wind-blown loess soils in the area were easily worked with the only tools available – simple wooden digging sticks and hoes. Animals were also important – the pig was domesticated by 5500 BCE (independently of south-west Asia) and the chicken (for the first time anywhere in the world) by about 5000 BCE. Five hundred years later the water buffalo was being used as a draught animal.

As in south-west Asia larger villages, small towns, trading networks and an increasingly sophisticated culture gradually developed as agriculture expanded. Agriculture also spread from China. Rice cultivation moved to the foothills of the Himalayas, upper Burma and northern Thailand and reached Taiwan by about 3500 BCE, East Timor by 2100 BCE and the Philippines about 400 years later. It did not reach Japan until about 400 BCE. Crops and animals were also transmitted eastwards into China, indirectly from south-west Asia via various groups in central Asia. Wheat and barley arrived about 2500 BCE and sheep and goats a little later, although the latter were never very important.

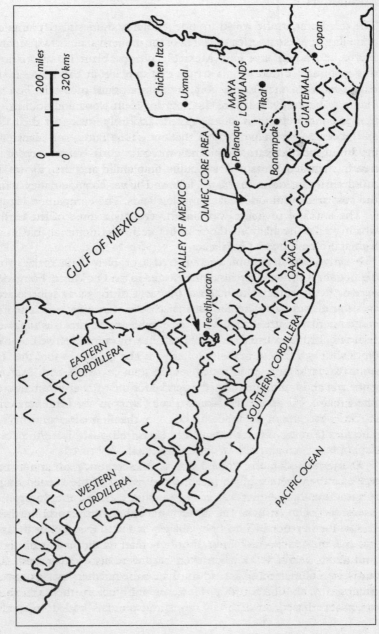

Mesoamerica

Mesoamerica

The third area in the world to independently domesticate plants and animals was Mesoamerica (the area of the modern states of Guatemala, Belize and south and east Mexico). The problem in studying the origins of agriculture in this area is that only about half a dozen sites have been excavated using modern archaeological methods. The first plants to be cultivated were the gourds – pumpkins and squash – at first as containers, then for their seeds and only finally for their flesh as it became sweet with domestication. This family of plants also produced the courgette and marrow and other early, but minor, domesticates included tomatoes, avocados and chilli peppers. By far the most important plant was maize, followed by its accompanying legume, the common bean.

The ancestor of maize was almost certainly a grass called teosinte which was gathered from the wild before it was domesticated in the highland area around Gudalajara in western Mexico about 3500 BCE. The early cultivators soon encountered a problem with maize which did not affect the wild cereals domesticated in the Old World. For genetic reasons it was difficult to increase the yield of maize by selection and crossing of varieties. The earliest maize cobs were very different from contemporary maize – they were only about six centimetres long. The earliest cultivated varieties were only slightly more productive than wild types although they were easier to harvest. The result was that the transition to agriculture took much longer than in south-west Asia and gathering and hunting remained essential for the early groups that cultivated maize. For the earliest known cultivators in the Tehuacan valley around 2700 BCE maize made up no more than a quarter of their diet. The first farming villages emerged in about 2000 BCE but the pace of development remained slow until maize yields improved.

Domesticated maize spread across Mesoamerica by about 1500 BCE and then further north into the south-west of the United States. Here, about 1200 BCE it was crossed with a local teosinte to produce 'maize de ocho' which was better adapted to the shorter growing season further north. The first villages in the area developed about 300 BCE but maize cultivation did not spread much further north for another thousand years when even hardier varieties were bred. The cultivation of maize also spread south into the northern parts of South America by about 1000 BCE. However, long before then a separate agriculture based on different crops and animals had developed in the Andes.

The Andes and the rest of the world

Gathering and hunting groups in the highlands of the Andes were herding wild guanaco and vicuna (the ancestors of the domesticated llama and alpaca) by about 7000 BCE – like sheep and goats they are docile and easily managed. Both animals forage on quinua which grows above 1,500 metres. The seeds pass through the animals undamaged and would grow when their dung was used as fertiliser. Quinua was certainly domesticated by 3000 BCE at about the same time as the llama and alpaca. The only other animal to be domesticated was the guinea pig, which had been used as a source of food since 9000 BCE. However, far more important was the domestication of four wild tubers. Three (oca, mashua and ullucu) remained confined to the highlands but the fourth – the potato (in a number of varieties) – was cultivated at lower levels and eventually became one of the world's major food crops. These tubers, which had long been gathered from the wild, were probably first domesticated around Lake Titicaca on the Peru–Bolivia border about 2500 BCE. An intriguing but still largely unresolved question is why and how potatoes were first eaten and then domesticated because nearly all wild potatoes are potentially toxic.

The evidence for domestication of plants in other parts of the world is sketchy in the extreme. Sorghum, currently one of the world's most important cereals, was first domesticated in the Sudan–Chad area. This must have been done at a fairly early date because it is known in India by 2000 BCE, having spread via the Arabian peninsula. Pearl millet was domesticated near the Sahara and when combined with local pulses (cowpea and groundnut) provided a balanced diet across much of the Sahel. The African variety of rice was domesticated about 1500 BCE and Old World cotton originated in the Sudan–Nubia area but was being grown in the Indus valley by 1800 BCE. Even less is known about tropical crops. Yams, found across Africa and southeast Asia, can be cultivated by replanting the stalk and this is done by numerous gathering and hunting groups. Taro and breadfruit originated in south-east Asia and spread into Oceania. Manioc and the sweet potato come from tropical South America.

By shortly after 2000 BCE all the major crops and animals in the world had been domesticated. (The horse was one of the last, probably after 3000 BCE somewhere in Ukraine.) However, for thousands of years there were separate streams of agricultural development because of the lack of contact between Eurasia and the Americas and the limited contacts between the different parts of Eurasia. Then, in two waves, the various systems were brought together and crops and animals diffused around the world. From the late seventh century CE the expansion of Islam created a great trading world that brought many of the semi-tropical crops of south-east Asia to the Near East and the Mediterranean. Then, almost a thousand years later, crops and animals from the Americas were brought to Europe (and slightly later to Asia) and European crops and animals were taken to the Americas and Australasia.

The impact of farming

The adoption of agriculture was the most fundamental change in human history. Not only did it produce settled societies for the first time, it also radically changed society itself. Gathering and hunting groups had few possessions and were largely egalitarian. A few people might be experts at making particular tools or finding types of food, and one or two among the elders might be responsible for the beliefs and traditions of the group and therefore have a special status. However, land was not owned and neither were its resources – they were available for whoever took them and food was shared within the group. The move to growing crops in fields and herding and breeding animals opened the way to seeing land, resources and food as 'property'. The far greater amount of time and effort required by farming reinforced this trend. Perhaps at first land and food was owned by the community but rapidly individuals did so and access to land and food became unequal.

The greater amount of effort involved in farming did mean that in most circumstances more food was produced and often this was more than the individual farmer and his family could eat. It was this food surplus that was the foundation for all later social and political change. It could be used to support people in occupations other than farming – craftsmen, bureaucrats, religious functionaries and eventually political and military leaders. In the broadest framework human history over the last 8,000 years has been about the acquisition and the distribution of this surplus food and the uses to which it has been put. The size of that surplus has largely determined how many people can be sustained outside of agriculture. Until the last few hundred years the surplus was small and unreliable – usually no more than a tenth of the population (sometimes less) could be anything other than peasants. This severely restricted the size and scope of the functions that the earliest states and empires could sustain. Only recently, with the huge increases in agricultural productivity, could industrial and post-industrial societies be sustained where now often less than five per cent of the population is engaged in agriculture. These linkages may have been more obvious in earlier, simpler societies. In medieval Europe and other feudal and quasi-feudal societies there was a direct relationship between land owned and the provision of military service and the church obtained its food through the direct ownership of land and the imposition of tithes. Rulers often had to move around their territories to eat the food surplus on the spot because there was little transport to move it about. Often the food surplus was extracted by force (it still was in the Soviet Union in the 1920s and 1930s) but now this is generally achieved by market forces (and huge subsidies in the industrialised states).

In the earliest agricultural communities it seems likely that the food surplus was given up voluntarily to support activities (primarily craft specialisations such as pottery and toolmaking, especially when metalworking developed, and religious functions) on which the community was agreed. However, even relatively small-scale agricultural societies quickly developed hierarchies that were headed by chiefs and clan leaders with authority to take and redistribute the food surplus on which the non-farmers depended. These functions were important because of the vulnerabilities produced by the adoption of agriculture. It is far from clear that *on average* the amount of food per person increased with the adoption of farming. Total output may have been larger but the population increased as, crucially, did the amount redistributed. The early farmers depended on a limited range of plants and were therefore vulnerable to crop failures due to poor weather, crop diseases and wastage during storage especially when in many cases the harvest was barely sufficient to last through until the next crop. A couple of bad harvests in succession could be catastrophic. Storage of food made it vulnerable to rodents, theft and conflict between groups. Unless redistribution of food could be ensured the non-producers (the leaders and craftsmen) would be the most vulnerable in the community.

In these circumstances it is easy to see why chiefs, clan leaders and religious authorities came to control the food surplus and the redistribution mechanisms in these early farming societies. It was essential for the smooth functioning of the community and it also provided leadership when these groups came into conflict with other groups, usually over land and access to water. However, as the leaders carried out these functions they took much greater control over the rest of the community. Societies at this level of development existed everywhere across the world for thousands of years. They could be found in much of Europe, especially central and eastern Europe, until the last thousand years. They lasted in North America until the destruction of the native American way of life by the Europeans after 1600 CE. In much of Africa similar societies survived into the early twentieth century.

The emergence of civilisation

In a handful of areas some societies, without any external influence, went much further, became coercive states and created the organisations, institutions and culture which we call civilisation. This happened at most six times in human history. Three took place in areas central to the emergence of agriculture – China, Mesoamerica and the central Andes – but two did not – Mesopotamia and the Indus valley. Egypt, the first of the early civilisations to be studied intensively by Europeans, is usually included in this category for traditional reasons, although it was influenced by developments in Mesopotamia. These societies were distinguished by a number of common features – they supported an elite of thousands of non-producers (priests, rulers, bureaucrats, craftsmen and warriors) who lived in cities and who exercised power over the rest of the population through forms of taxation, tribute and forced labour. In the cities there were complexes of public buildings such as temples, palaces and granaries, often on a grand scale. These societies were far more complex than the earliest farming communities, they were strongly territorial, warfare was almost constant and apart from those in the Andes they all developed a written script for record keeping and the bureaucratic control of society.

These changes were not an automatic process that took place in every society once it adopted agriculture and it is difficult to isolate the key factors that produced civilisation. Technology was not central – major developments in the use of metal followed the development of civilisation and those in the Americas managed almost entirely without it. Similarly writing followed from the emergence of civilisation and did not create it. The key element seems to have been the environment within which these early states emerged. It was the environmental problems they encountered and the levels of organisation this demanded that was crucial. Agriculture in these areas was not significantly more productive than elsewhere – given the technology available no major increase in food output by farmers was possible. What was significant was the way in which these societies were organised – the levels of power and authority within them increased dramatically over a relatively short time and this was combined with much greater inequality. As with the emergence of farming there was clearly a 'ratchet effect' in operation – once a step was taken it was difficult to reverse and changes in one area of society had major impacts elsewhere, were magnified and produced more changes in a positive feedback spiral. There were many similar patterns in the emergence of civilisation and so it is best to consider the changes involved in the first area to initiate them – Mesopotamia – and then consider the key differences elsewhere.

Mesopotamia

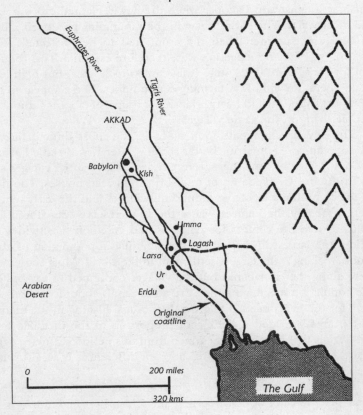

Mesopotamia

The major developments in early farming took place in the foothills of the mountains surrounding the plain of the twin rivers of the Tigris and Euphrates. The area was not settled by the first farmers until about 5700 BCE because of the difficult conditions found here. The summers were long and hot and the winters harsh and cold. There was little water available away from the rivers although the soil was productive when irrigated. The area also lacked nearly every key resource – stone, copper and timber. At first glance therefore this area seems an unlikely candidate for the first emergence of civilisation by about 3000 BCE. On the other hand it was probably the difficult circumstances that demanded a greater degree of social organisation from the beginning. The first farmers in the valley found that rainfall was almost non-existent in the south and the rivers at their lowest between August and October when the newly planted crops needed the most water. The spring thaw in the mountains produced a flood in the early summer when the crops had to be protected. Water control, storage and irrigation were therefore essential. This required not a strong, centralised state (that came much later) but high levels of co-operation within the farming communities. These appear to have been organised around large kinship groups of about a thousand people living in the same village.

The key developments took place in the most difficult area of Mesopotamia – Sumer in the far south – after it was settled around 5000 BCE. At this time there were a series of substantial villages such as Eridu and Ur made up of mud-brick and reed houses. The fields of the community clustered around the villages and the early temples were probably the centres where the food reserve was stored. The villages were the centres for religious and ceremonial activities and Eridu, the largest in the region, probably had a population of about 5,000, making it the largest settlement in the world. Most settlements were along the river banks but the villages continued to expand and the population was maintaining irrigation canals about five kilometres long for the fields needed to grow their crops. Fishing in the marshes of the area provided a significant proportion of the diet. However, the burials that have been excavated from this period strongly suggest that there was little social stratification or differences in wealth within the community.

The crucial period for the development of civilisation was the thousand years leading up to 3000 BCE and it can best be studied in the largest settlement in the area – Uruk – which became the first city in the world. About 3600 BCE a huge temple mound known as a ziggurat was constructed and then steadily expanded. Eventually the temples and other public buildings took up a third of the total area of Uruk which expanded to cover about 250 hectares (twice the size of ancient Athens at its peak). The city was surrounded by a wall almost ten kilometres long and the population rose to about 40,000 by 3000 BCE. The amount of labour needed for these tasks was obviously considerable and appears to have been organised by the temple authorities and the emerging secular ruler. The first primitive writing was developed about 3400 BCE to manage the various activities of the city. In the surrounding area there was a series of dependent, small towns and villages, each with their own irrigation system. The deities of Uruk attracted pilgrimages and offerings (voluntary and forced) from the surrounding population but 'tribute' and 'taxation' (the appropriation of the food surplus) for defence and state projects was enforced through the military and political authority of the city. Although Uruk was the largest city in Sumer a similar process of increasing control and authority over society can be traced in other towns and cities across the region. In the period between 3000 and about 2300 BCE (known as the Early Dynastic) the city states of Sumer were engaged in almost constant warfare over the resources vital for their societies – land and water. Lagash and Umma were at war for over 150 years around 2500 BCE over a boundary dispute about the control of water resources.

Until recently it was thought that Sumer consisted of a series of 'temple-states' in which the temples owned all the land, the inhabitants were regarded as temple servants of various ranks and the secular ruler was a deputy of the temple. This naturally reinforced arguments that it was religious functions that led to the requisitioning of the food surplus and the development of civilisation. This interpretation is now known to be a major misunderstanding based on a serious misreading in the 1930s of a single archive from the temple of Girsu. The accepted picture of how Sumerian society was organised is now very different and provides a very different perspective on the processes leading to the emergence of civilisation. In Sumer much of the land was owned by family groups and the 'temple' estates were the property of the city ruler and his family and he was able to distribute them to his followers. The ruler was the protector of the city in the name of the city's god; he maintained the temples and gained status through his relationship with the god, exemplified by his role in the ceremonials alongside the priests. The royal family held a number of key posts within the temple hierarchy. There was therefore no separation of the religious and secular spheres. In theory the temple controlled large tracts of land. Some estates were directly cultivated by peasants who owed a community obligation to work the land. From this land daily deliveries of food were made to support the staff of the temple. However, other land was assigned to office holders within the temple (usually members of the ruling family or the elite). Many of the temple positions were hereditary sinecures and the estates that went with them became hereditary too. The rest of the temple land was rented out to families and individuals. Temples also had extensive workshops employing specialist craftsmen who were fed by produce brought from the temple estates.

Soon estates were owned not by families but by individuals and this land could be bought and sold (with sales recorded on clay tablets complete with field maps). Agricultural labourers who owned no land became common – they were employed on contracts setting out the length of time they were to work, their wages and whether they were paid in silver or barley. However, they were outnumbered by the workers on the palace, temple and elite-owned estates. These people were dependent landless labourers tied to the estates so that they were effectively serfs or near slaves. In times of bad harvests whole families sold themselves, sometimes remaining on their land as dependent debt-tenants, sometimes becoming quasi-slaves supported by the people who bought them. Society had come a long way from the small villages of roughly equal peasant farmers.

The palace was the residence of the ruler, his family and their entourage of servants. It had its own workshops and was the centre for administration and the location of the state treasury. It also administered the extensive estates of the ruling family. The chief function of the ruler was as a war leader and this role seems to have been crucial in the emergence of secular authority. Rulers may have been temporary war leaders at first (perhaps even elected by the assembly of elders reflecting an earlier kinship structure) but by 2600 BCE at the latest rulers were hereditary monarchs. Warfare was central to the emergence of the Sumerian city states. The earliest cylinder seals from 4000 BCE show battles and prisoners of war, and warfare became more intense as the cities developed. Armies were made up of conscripts who served as part of their obligation to the state and they fought with axes, adzes, spears and leather shields. The ruler travelled to battle in a donkey-drawn chariot with solid wheels but fought on foot like the infantry. Cities were very vulnerable to the occupation of the land on which they depended for their food and sieges were brutal affairs. When a city was captured its walls were normally destroyed, the male population killed or enslaved and blinded, and the women and children enslaved.

By the late Early Dynastic period (about 2400 BCE) the endless warfare between the Sumerian cities had produced a growing concentration of power. Eventually the ruler of Uruk, Ur and Umma extended his control over nearly all of Sumer. In his turn he was defeated by Sargon from the northern area of Kish and the first 'empire' in human history had been established. Under Sargon and his successors the empire stretched from Sumer in the south to northern Syria and western Iran, and military campaigns were being mounted in Anatolia. Sargon's grandson, Naram-Sin, was so impressed by his conquests that he declared himself to be a god. The pattern of human history for thousands of years had been set.

Egypt

The processes that led to the emergence of civilisation in the Nile valley (the second to develop in the world) began later than in Mesopotamia but were much more concentrated. As late as 3500 BCE (when the first cities, extensive warfare and literacy could be found in Mesopotamia) the Nile valley had only just been fully settled and was characterised by small villages, a few irrigation works, some craft specialisation and a few 'chiefs' – rulers of a couple of villages. The civilisation that emerged in the next five centuries or so differed from that in Mesopotamia in two important respects. First, the amount of agricultural land was limited by the level of the annual Nile flood and therefore population densities were far lower than in Sumer. Cities did develop but they were ceremonial centres for the elite rather than the large residential cities of Sumer. Second, the Nile valley was unified at an early date and separate city-states did not develop. Despite frequent periods of disunity throughout its history the tradition remained that the great rulers and dynasties of Egypt were characterised by their ability to unify the valley. There was, however, one characteristic that was common to both Mesopotamia and Egypt – they were based on extensive social and political coercion of the mass of the population by a small religious and political elite.

Environmental factors seem to have played a key role in the emergence of the Egyptian state. Around 3300 BCE there was a drastic reduction in the level of the Nile flood which severely affected all the settlements along the valley. There was an increased need for social organisation to cope with the poor harvests and also reallocate land now that large areas were no longer flooded. The chiefs and leaders along the valley organised these activities and gained power and prestige. One of the key areas was around Hierakonopolis, a settlement of about 10,000 people in Upper Egypt. The town extended its control over its local rival, Nagada, and then over much of the upper Nile valley. The exact process of unification cannot now be established but was almost certainly characterised by warfare between the different chiefs and towns along the valley. However, the joining of Upper Egypt and Lower Egypt (the area of the Nile delta) around 3050 BCE under a ruler from Hierakonopolis, usually called Narmer, was always regarded as the key event in producing the Egyptian state.

The unification of the valley to create what is now seen as the quintessentially 'Egyptian' culture took much longer. All of the elements of this culture came from the court of the Pharaoh. The Pharaoh always wore a double crown – white for Upper Egypt and red for Lower Egypt. A sophisticated ideology, imagery and ceremonial were developed to emphasise the divine nature of the Pharaoh. The highly distinctive writing system known as hieroglyphic was developed as part of this process and extensive record-keeping (using a simpler hieratic script) became central to the state. By the time of the 'Old Kingdom' period (normally dated from about 2700 to about 2200 BCE) a strong, unified and highly complex state had developed in the Nile valley. The tombs and stelae erected by the rulers show scenes of constant warfare together with prisoners of war and booty being brought back from Nubia and the Sinai. About 2700 BCE the first pyramid and ceremonial courtyard was erected at Sakkara. These buildings reached a climax with the Great Pyramid built out of 2,300,000 blocks of stone weighing on average two and a half tonnes (some are fifteen tonnes and none is less than one and a half tonnes). The top was originally nearly 150 metres above the ground, although the gold-covered capstone is now missing. The pyramids were a symbol of the sun and the stars (in particular the circumpolar stars which, because they never set, were regarded as 'immortal') and the Pharaoh's relationship with them rather than the land he ruled. They symbolised the transformation of the divine Pharaoh after his death.

The pyramids were also symbols of the nature of the Egyptian state and the power of the elite over the rest of the population. The agricultural surplus was appropriated to support the rulers, both secular and religious, and their supporting servants, craftsmen and bureaucrats. In addition large amounts of labour were conscripted over long periods of time to work on these massive elite projects. There may have been some degree of popular support for these enterprises but many must have been forced to labour on them. A major task for the bureaucrats was to organise these drafts of labour from among the agricultural labourers and then ensure that they were fed while they were working.

The Indus valley

The civilisation which emerged in the Indus valley about 2300 BCE was the third to develop in the world, yet it is the least known of all the early civilisations. Its script is still undeciphered so little is known about how it emerged or its internal structure. Few of its achievements were transmitted to later states and it was the shortest-lived of the early civilisations – the peak of its prosperity lasted little more than three centuries and it declined rapidly after 1750 BCE.

Farming in the valley began about 6000 BCE based on wheat and barley, almost certainly diffused from the earliest villages in southwest Asia. A key crop was cotton – the first evidence of its cultivation anywhere in the world – and the animals that were domesticated (humped cattle, buffalo and pigs) came from local herds and sheep and goats were of little importance. From about 4000 BCE, as the population increased, substantial villages built from mud bricks proliferated and the culture across the valley became increasingly uniform. From about 3000 BCE extensive irrigation works were constructed to contain the annual flood of the Indus. This suggests considerable levels of control and direction and the increasing food surplus from the irrigated land reinforced these trends.

By 2300 BCE a complex society and state had emerged but the names of the rulers and even the names of the cities are unknown. There were only two major towns – the sites of Mohenjo-daro in the south and Harappa in the north. At their height they may have had a population of 30–50,000 (about the size of Uruk). Both cities were laid out to a similar plan, with a citadel of major public buildings and a lower residential area arranged to a regular street plan, and the buildings all used bricks of a similar size. The valley had a common set of weights and measures and a single script which was, no doubt, used for record-keeping. All of these factors suggest a high degree of central control and direction over society.

China

The way by which the early agricultural villages of the central plain of China around the Wei and Yellow rivers evolved into a complex civilisation is largely unknown. There are no written records and little archaeological work has been undertaken. However, the processes involved are unlikely to be very different from those in Mesopotamia, Egypt and the Indus valley. What is clear is that by about 1800 BCE an almost fully fledged complex culture and civilisation known as the Shang had developed. Central to this culture were the distinctive bronzes made by the unique piece-moulding process unknown elsewhere. Although the Shang emerged more than a millenium after the first civilisation in Sumer there is no doubt that it was not influenced from outside. The Shang period, which lasted until either 1122 or 1027 BCE, laid the foundations for the highly distinctive Chinese civilisation in culture, script and institutions.

The Shang are largely known through the site of An-yang, the first capital. It was not a city in the conventional sense and was more like a ceremonial centre with a wide area stretching almost 250 kilometres in every direction containing dependent villages. The Shang was certainly not the only 'state' in China – there are references to between eight and thirty-three other states linked by marriage and clan connections but where they were located is unknown.

The Shang ruled through a complex mixture of clan and royal elements and different members of the royal clan set up their own cities. These were planned and not natural creations and the city took its name from the ruler who controlled the surrounding agricultural land and owed food and service to the king in An-yang. Whether this could be enforced is doubtful. Indeed it is far from clear that Shang was ever a unified state rather than a loose confederation owing allegiance to a king to whom the other rulers were related either actually or nominally. In fact the cities were often at war and the mass of the population was organised into units of 100 households as the basis for providing the conscript army. As in other early civilisations the ruler had key religious functions which reinforced his authority. The high god *Ti* provided the two elements essential to society and the state – the harvest and military victory. The king's ancestors were able to intervene with him and therefore their worship was essential. He understood their will through divination from the cracks in heated animal bones on which questions were inscribed. It is these bones that provide the first examples of the fully mature and unique Chinese script that must have evolved over at least the previous millennium.

The Americas

The civilisations of the Americas developed relatively late in world terms. This stemmed from the comparatively late human settlement of the area (about 12,000 BCE) and the difficulty of increasing the yield of maize. The first farming villages did not develop until about 2500–1500 BCE (at least six thousand years later than Eurasia). The first relatively complex societies emerged shortly after 1000 BCE and the first cities and incipient states date to about the BCE/CE divide (a little later in the Andes). All of these societies had characteristics that distinguished them from those of the Old World, mainly stemming from the unique crops and animals of the region. Apart from the distinctive crops the Americas had no sheep, goats, pigs and, even more important, no cattle, horses or asses capable of being turned into draught animals. The llama and the alpaca which were domesticated in the Andes could not be used as draught animals. This severely restricted what these societies could achieve – the wheel was known but could not be adapted for transport. In addition metal was rarely used and agriculture remained dependent on stone tools.

The development of civilisation in Peru (which was isolated from Mesoamerica) was shaped by its environment. Much of the Pacific coastal area is dry, inhospitable desert but it is divided by about forty river valleys which descend from the Andes. Agriculture and irrigation were possible in the valleys but the societies remained relatively isolated from each other because of the intervening desert. The first villages emerged about 2500 BCE and pottery and woven textiles developed about 700 years later. By this time there were large ceremonial centres with monumental architecture in most of the valleys. It seems clear that these were irrigation-based agricultural socities where the elite were able to wield considerable power and mobilise large amounts of labour. Warfare was widespread. However, it was not until about 600 CE that the first true cities, similar to those found in Mesopotamia four thousand years earlier, began to emerge.

Although the first agricultural villages in Mesoamerica date to about 2000 BCE (slightly later than in Peru) the pace of development here was, eventually, faster. What is surprising is that the first complex society emerged around 1200 BCE in the semi-tropical jungles of the Gulf coast of Mexico – at first glance an unlikely environment for a sophisticated culture and unlike any of those found elsewhere in the world at this time. The culture is known as Olmec although that is the name of the people who lived in the area at the time of the Spanish conquest over two thousand years later, and the name of the people who created these ceremonial centres is unknown. The Olmec culture relied on the fact that with the rich soil of the region and heavy rainfall (nearly 300 centimetres per annum) it was possible to grow two crops of maize a year. This created a significant agricultural surplus to sustain the elite, establish the huge earth mounds of the ceremonial centres and carve the giant stone heads that are the distinctive feature of the Olmec culture. Olmec influence spread across Mesoamerica in the period until about 400 BCE and provided the foundations for all later civilisations. In particular they created the double calendar of 365 and 260 days that coincided once every fifty-two years and the method of counting using a base of twenty.

After the collapse of the Olmec there was a hiatus of about four centuries before the emergence, shortly after the BCE/CE divide, of the first major city and empire in Mesoamerica at the site of Teotihuacan in the north-east part of the valley of Mexico. The city covered a vast area and included temples, two huge pyramids and about 2,000 residential compounds, each of which seem to have contained about sixty people probably organised on a clan basis. At its height around 400 CE the population was probably about 100,000, making it one of the largest cities in the world (Rome was about half this size at that time). The city was clearly a ceremonial centre because of the numerous astronomical alignments found within it but its foundations were based on the irrigated agriculture of the valley. It was dominated by a religious and military elite (although the city had no defensive wall) with extensive powers of coercion and control over the mass of the population. The exact nature of the society is unknown because no real script has been found. Teotihuacan collapsed about 750 CE, almost certainly following a series of major revolts by major popular uprisings.

The early civilisations

By about 3000 BCE in Mesopotamia and Egypt, a few hundred years later in the Indus valley, a millennium or so later still in China and another two millennia later in the Americas, hierarchical, militaristic societies ruled by religious and political elites with immense powers of control over their populations were established. Societies that were broadly egalitarian were replaced by ones with distinct classes and huge differences in wealth. All of these changes depended on the development of farming and the use to which the food surplus was put.

These changes had two further consequences of great significance. The ability to support people not engaged in food production was the basis for all subsequent human, cultural and scientific advances. The demands of the elite produced the great temples, palaces, state buildings and other structures that make up the great ancient sites of the world and are now the main memorials of past societies. At the same time craftsmen produced exquisite works of art that are still admired. Settled societies also made possible the great developments in religious, spiritual and philosophical thought. The development of writing, once it had evolved from the demands of simple record-keeping, was fundamental for all later advances in knowledge. The earliest societies also developed extensive astronomical knowledge, notably among the Babylonians and the Maya, and many sites from Mesoamerica to China and bronze age Britain are aligned towards significant solar, stellar and lunar positions. The other side of the coin was the parallel development of increasing coercion within society and large-scale warfare. The great monuments could only be built by using large amounts of human labour. This may, in the early years of these more complex societies, have been provided on a voluntary or semi-voluntary basis but it was soon replaced by coercion. One of the reasons why internal control and discipline grew so much stronger in these societies was the external threat and increasing warfare. Many of the early cities had defensive walls, and warfare was almost constant. As the early empires developed armies grew bigger. By 1285 BCE Egypt was able to deploy an army 20,000 strong (larger than many in Europe before the seventeenth century) at the battle of Kadesh. Warfare was highly destructive and the civilian population was treated with great barbarity. Death or slavery was the usual consequence of defeat and the highly militaristic Assyrian empire of the early first millennium BCE forcibly deported over four and a half million people.

After the first states and empires were established there was no fundamental change in the way humans lived for several thousand years. The overwhelming mass of the population remained peasants, landless labourers or slaves, subject to extensive expropriation of their food, forced labour and the risks of extensive warfare. They lived constantly on the margin of subsistence, suffering hunger and the ever-present threat of famine. Only a very small minority could be supported in a more affluent or a more intellectually rewarding style of life.

Various states and empires rose and fell (usually as a result of the unexpected deaths of rulers, internal revolts and the fortunes of war) without fundamentally altering this way of life. Nevertheless their impact on their immediate environment was often far-reaching. They provide the first examples of intensive human alteration of the environment and its destructive impact. They also provide the first examples of societies that so damaged their environment that they brought about their own collapse.

5

DESTRUCTION AND SURVIVAL

I met a traveller in an antique land
Who said: 'Two vast and trunkless legs of stone
Stand in the desert . . . Near them, on the sand,
Half-sunk, a shattered visage lies, whose frown,
And wrinkled lip, and sneer of cold command,
Tell that its sculptor well those passions read
Which yet survive, stamped on these lifeless things,
The hand that mocked them and the heart that fed:
And on the pedestal these words appear:
"My name is Ozymandias, king of kings:
Look on my works, ye mighty, and despair!"
Nothing beside remains. Round the decay
Of that colossal wreck, boundless and bare
The lone and level sands stretch far away.'

(Shelley: 'Ozymandias')

The adoption of agriculture, combined with its two major consequences – settled communities and steadily rising population – placed
an increasing strain on the environment. That strain was localised at
first but as agriculture spread so did its effects. Some areas such as
the temperate forest ecosystems of north and west Europe with their
moderate temperatures, high rainfall and rich soils, were able to withstand the strain reasonably well. Other areas with more easily damaged
ecosystems and with a higher population density began to be affected
within a thousand years of the adoption of agriculture and a settled
way of life.

Agriculture involves clearing the natural ecosystem in order to
create an artificial habitat where humans can grow the plants and

stock the animals they want. The natural balances and inherent stability of the original ecosystem are thereby destroyed. Instead of a variety of plants and permanent natural ground cover a small number of crops make only a part-time use of the space available. The soil is exposed to the wind and rain to a far greater extent than before, particularly where fields are left bare for part of the year, leading to much higher rates of soil erosion. Nutrient recycling is also disrupted and extra inputs in the form of manures or fertilisers are required if soil fertility is to be maintained. The adoption of irrigation is even more disruptive since it creates an environment that is even more artificial than dry farming, which relies on rainfall. Adding large amounts of water to a poor soil may allow the farmer to grow his preferred crop but it can have catastrophic longer-term effects. The extra water drains into the water table and will eventually cause water levels to rise until the soil becomes waterlogged. The additional water also alters the mineral content of the soil: it increases the amount of salt and may eventually, especially in hot areas with high evaporation rates, produce a thick layer of salt on the surface which makes agriculture impossible. The only way in which this process can be avoided is very careful use of irrigation, not over-watering, and leaving the ground fallow for long periods.

The emergence of villages and towns (and an increasing population) meant that the demand for resources was now more concentrated, and efforts to increase supply would inevitably impose significantly greater strains on smaller areas. Forests suffered most as the requirement for wood to build houses, heat homes and cook food rose steadily. Local deforestation leading to increased soil erosion became a problem around settled areas. Recent evidence from central Jordan suggests that as early as 6000 BCE, within about a thousand years of the emergence of settled communities, villages were being abandoned as soil erosion caused by deforestation resulted in a badly damaged landscape, declining crop yields and eventually inability to grow enough food.

The creation of artificial environments to grow food and the rise of communities not only concentrated the environmental impact of human activities but also meant that it was far more difficult for human societies to escape the consequences of their actions. In sensitive ecosystems the foundations of society could be so damaged as to cause its collapse. These early societies were dependent on the production of a food surplus in order to feed and support the growing numbers of priests, rulers, bureaucrats, soldiers and craftsmen. If food production became more difficult and crop yields fell then the very basis of society was undermined. It is perhaps not surprising that the first signs of widespread damage emerged in Mesopotamia, the area where the most extensive modifications to the natural environment had first been made.

The decline and fall of Sumer

When, in 1936, one of the excavators of the earliest cities of Sumer, Leonard Woolley, wrote a book about his work entitled *Ur of the Chaldees* he was puzzled by the desolate, largely treeless landscape of contemporary southern Mesopotamia, similar to that imagined by Shelley.

> Only to those who have seen the Mesopotamian desert will the evocation of the ancient world seem well-nigh incredible, so complete is the contrast between past and present . . . it is yet more difficult to realise, that the blank waste ever blossomed, bore fruit for the sustainance [sic] of a busy world. Why, if Ur was an empire's capital, if Sumer was once a vast granary, has the population dwindled to nothing, the very soil lost its virtue?

The answer to Woolley's question is that the Sumerians themselves destroyed the world they had created so painstakingly out of the difficult environment of southern Mesopotamia.

The valley of the twin rivers, the Tigris and Euphrates, posed major problems for any society, especially in the south. The rivers were at their highest in the spring following the melting of the winter snows near their sources and at their lowest between August and October, the time when the newly planted crops needed the most water. In the north of Mesopotamia the problem was eased by the late autumn and winter rains but rainfall was very low and often non-existent further south. This meant that in the Sumerian region water storage and irrigation were essential if crops were to be grown. At first the advantages outweighed the disadvantages but slowly a series of major problems became apparent. In summer, temperatures were high, often up to about 40°C, which increased evaporation from the surface and as a consequence the amount of salt in the soil. Water retention in the deeper layers of the soil and hence the risk of waterlogging was increased by two factors. The soil itself had very low permeability. This was exacerbated by the slow rate of drainage caused by the very flat land, itself made worse by the amount of silt coming down in the rivers, probably caused by deforestation in the highlands, which added about 150 cm of silt every millennium and caused the delta of the two rivers to extend by about 25 kilometres a millennium. As the land became more waterlogged and the water table rose, more salt was brought to the surface where the high evaporation rates produced a thick layer. Modern agricultural knowledge suggests that the only way to avoid the worst of these problems is to leave the land fallow and unwatered for long periods to allow the level of the water table to fall. The internal pressures within Sumerian society made this impossible and brought about disaster. The limited amount of land that could be irrigated, rising population, the need to feed more bureaucrats and soldiers and the mounting competition between the city states all increased the pressure to intensify the agricultural system. The overwhelming requirement to grow more food meant that it was impossible to leave land fallow for long periods. Short-term demands outweighed any considerations of the need for long-term stability and the maintenance of a sustainable agricultural system.

About 3000 BCE Sumerian society became the first literate society in the world. The detailed administrative records kept by the temples of the city states provide a record of the changes in the agricultural system and an insight into the development of major problems. About 3500 BCE roughly equal amounts of wheat and barley were grown in southern Mesopotamia. But wheat can only tolerate a salt level of half of one per cent in the soil whereas barley can still grow in twice this amount. The increasing salinisation of the soil can be deduced from the declining amount of wheat cultivated and its replacement by the more salt-tolerant barley. By 2500 BCE wheat had fallen to only 15 per cent of the crop; by 2100 Ur had abandoned wheat production and overall it had declined to just 2 per cent of the crops grown in the Sumerian region.

Even more important than the replacement of wheat by barley was the declining yield of crops throughout the region. In the earliest phases of Sumerian society when areas went out of production because of salinisation they were replaced by newly cultivated fields. Rising population, and the demand for a greater food surplus to maintain the army as warfare became more frequent, reinforced the demand for new land. But the amount of new land that could be cultivated, even with the more extensive and complex irrigation works that were becoming common, was limited. Until about 2400 BCE crop yields remained high, in some areas at least as high as in medieval Europe and possibly even higher. Then, as the limit of cultivatable land was reached and salinisation took an increasing toll, the food surplus began to fall rapidly. Crop yields fell by 40 per cent between 2400 and 2100 BCE and by two-thirds by 1700 BCE. From 2000 BCE there are contemporaneous reports that 'the earth turned white', a clear reference to the drastic impact of salinisation. The consequences for a society so dependent on a food surplus were predictable. The bureaucracy, and perhaps even more importantly the army, could not be maintained. As the size of the army fell the state became very vulnerable to external conquest. What is remarkable is the way that the political history of Sumer and its city states so closely follows the decline of the agricultural base. The independent city states survived until 2370 BCE when the first external conqueror of the region – Sargon of Akkad – established the Akkadian empire. That conquest took place following the first serious decline in crop yields resulting from widespread salinisation. For the next six hundred years the region saw the Akkadian empire conquered by the Guti nomads from the Zagros mountains, a brief revival of the region under the Third Dynasty of Ur between 2113 and 2000 BCE, its collapse under pressure from the Elamites in the west and the Amorites in the east, and in about 1800 BCE the conquest of the area by the Babylonian kingdom centred on northern Mesopotamia. Throughout this period, from the end of the once flourishing and powerful city states to the Babylonian conquest, crop yields continued to fall, making it very difficult to sustain a viable state. By 1800 BCE, when yields were only about a third of the level obtained during the Early Dynastic period, the agricultural base of Sumer had effectively collapsed and the focus of Mesopotamian society shifted permanently to the north, where a succession of imperial states controlled the region, and Sumer declined into insignificance as an underpopulated, impoverished backwater of an empire.

The artificial system that was the foundation of Sumerian civilisation was very fragile and in the end brought about its downfall. The later history of the region reinforces the point that all human interventions tend to degrade ecosystems and shows how easy it is to tip the balance towards destruction. It also suggests that it is very difficult to redress the balance or reverse the process once it has started. Centuries later, when the city states of Sumer were no longer even a memory, the same processes were at work elsewhere in Mesopotamia. Between 1300 and 900 BCE there was an agricultural collapse in the central area following salinisation as a result of too much irrigation. Around Baghdad in the seventh and eighth centuries CE, both before and after the Arab conquest, the area was flourishing with high crop yields from irrigated fields supporting a wealthy and sophisticated society. But the same pressures seem to have been apparent as in Sumer over 3,000 years earlier. To boost food production four major new irrigation canals were dug between the Tigris and Euphrates, which, in turn, led to waterlogging, a rapidly rising water table and salinisation. At this time the population of Mesopotamia was probably about one-and-a-half million but the agricultural collapse brought about through intensive irrigation and the Mongol conquest in the thirteenth century caused a massive decline in population to about 150,000 by 1500 and brought about the end of the sophisticated society that had survived in the area for centuries.

The Indus valley

Many of the same forces that brought about the collapse of the first settled societies in Mesopotamia seem to have been at work in the Indus valley too, although it is not possible to identify the trends in such detail because the script used by the inhabitants has not been deciphered. As in Mesopotamia what was once a flourishing society and a rich and productive area was turned into a desolate region through over-exploitation of a delicate environment. The complex, highly centralised society which emerged about 2300 BCE lasted less than 500 years. The settlers in the Indus valley faced one major problem – the tendency of the river to flood over wide areas and change its course. Extensive works were built to contain the river and irrigate the fields to produce the food that supported and fed the ruling elite, priests and army. In the hot climate of the valley, irrigation had the same effect as in Sumer – a rising water table and increasing waterlogging. This produced progressive salinisation of the soil and eventually a layer of salt on the surface, which led to the gradual decline of crop production.

The other factor undermining the environment of the Indus valley was deforestation. The area that attracted the first settlers was richly forested with a plentiful supply of wildlife. Part of this forest was cleared to provide fields for agriculture. More important though was the construction technique used by the inhabitants. To build their huge temples and palaces Mesopotamian societies used mud bricks dried by the sun. The people of the Indus valley also used mud bricks but they dried them in ovens, and the process required enormous quantities of wood. Very rapidly the trees in the area were cut down. This exposed the soil and caused rapid erosion and decline in soil quality. About 1900 BCE the Indus valley society came to a sudden end. The immediate cause was probably external conquest following a period of internal decline. The extent of the environmental degradation through salinisation and deforestation suggests a substantial reduction in the available food surplus may have led to a reduction in the size of the army and increased vulnerability to external conquest – in much the same way that Sumer collapsed.

Deforestation

Extensive deforestation has been a problem for a number of societies throughout history. A larger human population put an increasing strain on local environments generally and in particular supplies of wood – the one readily available resource for heating, cooking and in many cases construction. Forests were cleared to make room for fields to feed the steadily growing number of people. It is the slow, steady, and in many ways unspectacular, destruction of the woods and forests around all settled communities that forms the background to the development of human societies. These early societies lacked the technology for large-scale clearance but between them the metal axe, ring barking and fire would have been very effective ways of obtaining wood or clearing a section of forest. No one generation would have been conscious of making any dramatic changes. Generally people seem to have accepted the process as a natural way of obtaining the resources they needed. As settlements moved into new areas the process of steady destruction would start again. (There is little evidence of any attempt at major planting or replanting, although coppicing was practised in many parts of Europe as a way of obtaining a sustainable crop of wood.) Over hundreds of generations the scale of this steady destruction could be massive. Huge areas that had once been dense forests would become treeless or retain only isolated pockets of woodland in the least accessible places.

In China the development of agriculture and the rise of the first settled societies had been based on the cultivation of millet on the easily worked loess soils in the north of the country. Although the soil was rich it was easily eroded once the natural grass cover had been removed in order to make way for fields of millet. Very rapidly huge gullies and canyons developed as the soil was blown away by the wind or washed away by the rain. At the same time hillsides were cleared of trees for fuel and construction. Steadily the deforested area increased until, by about two hundred years ago, nearly all the original forests of China had been cleared. The wholesale loss of trees in the highlands of China was one of the main causes of the often disastrous flooding of the Yellow river (so-called because of the amount of soil it carried from erosion upstream), which regularly resulted in major changes in its course in the lowlands with huge losses of life. The same sequence of events can be seen in Japan. The scale of forest destruction, particularly following the great rebuilding of castles and towns at the beginning of the Tokugawa period (shortly after 1600 CE), was such that it led to the imposition of strict government controls in the form of licences for further tree felling. The same problems can be identified in the great medieval Christian kingdom of Ethiopia. The original centre of the state was in the northern area – Tigre and Eritrea. Continual deforestation produced a badly degraded environment of poor soils and eroded hillsides, some in such a ruined state that they could no longer support shrubs or even grass. By about 1000 CE the damage was so great that the state had to shift to the south and a new capital in the central highlands. However, the same process was repeated, again leading to major environmental damage in this area. Just how quickly and completely people could transform the area surrounding a new or expanding settlement is illustrated by what happened around Addis Ababa after it became the capital of Ethiopia in 1883. Within twenty years a zone stretching for 150 kilometres around the town had been devastated – stripped of its trees by charcoal burners producing fuel for the capital.

The effects of the steady and continual cutting down of trees can be seen at their clearest in the Mediterranean region. Modern visitors regard the landscape of olive trees, vines, low bushes and strongly scented herbs as one of the main attractions of the region. It is, however, the result of massive environmental degradation brought about not by the creation of artificial systems such as irrigation but by the relentless pressure of long-term settlement and growing population. The natural vegetation of the Mediterranean area was a mixed evergreen and deciduous forest of oaks, beech, pines and cedars. This forest was cleared bit by bit for a variety of reasons – to provide land for agriculture, fuel for cooking and heating, and construction materials for houses and ships. Other activities ensured that it did not regenerate. Overgrazing by sheep, cattle and particularly goats meant that the young trees and shrubs were eaten before they could grow and mature. Gradually the flocks of animals reduced the vegetation to a low scrub of largely inedible plants. Removal of the tree cover, especially on steep slopes, led to large scale soil erosion which ruined agricultural land (already short of manure because the farmers practised transhumance, moving flocks of animals to different areas for summer and winter) and the large amount of silt brought down in the rivers blocked water courses and caused large deltas and marshes to form at river mouths.

The process of long-term environmental decline can be traced in every area around the Mediterranean and the Near East. Overall it is now estimated that no more than 10 per cent of the original forests that once stretched from Morocco to Afghanistan even as late as 2000 BCE still exist. One of the first areas to suffer was the hills of Lebanon and Syria. The natural climax forests here were particularly rich in cedars, and the cedars of Lebanon became famous throughout the ancient Near East for their height and straightness. They were prized by the states and empires of Mesopotamia as building materials and control of the area or trade with its rulers was a high priority. Later the cedars became one of the mainstays of Phoenician commerce and were traded over a wide area. Gradually the trees were cut down until the renowned cedars of Lebanon were reduced to a pathetic remnant in a few places – there are now just four small groves of cedars left in the region, maintained as a symbol of former glory.

In Greece the first signs of large-scale destruction began to appear about 650 BCE as the population rose and settlements expanded. The root of the problem was overgrazing on the 80 per cent of the land that was unsuitable for cultivation. Although the Greeks were well aware of techniques for preserving soil such as manuring to maintain the structure of the soil and terracing to limit the erosion on hillsides, the pressure from a continually rising population proved too great. The hills of Attica were stripped bare of trees within a couple of generations and by 590 in Athens the great reformer of the constitution, Solon, was arguing that cultivation on steep slopes should be banned because of the amount of soil being lost. A few decades later the tyrant of Athens, Peisistratus, introduced a bounty for farmers to plant olives, the only tree that would grow on badly eroded land because it had roots strong enough to penetrate the underlying limestone rock. Many of the writers of ancient Greece such as Herodotus, Xenophon and Aristotle were aware of the problem but the most graphic description of the effects of deforestation and soil erosion was left by Plato in his *Critias*:

What now remains compared with what then existed is like the skeleton of a sick man, all the fat and the soft earth having wasted away, and only the bare framework of the land being left . . . there are some mountains which now have nothing but food for bees, but they had trees not very long ago . . . there were many lofty trees of cultivated species and . . . boundless pasturage for flocks. Moreover, it was enriched by the yearly rains from Zeus, which were not lost to it, as now, by flowing from bare land into the sea; but the soil it had was deep, and therein it received the water, storing it up in the retentive loamy soil, and . . . provided all the various districts with abundant supplies of spring water and streams, whereof the shrines still remain even now, at the spots where the fountains formerly existed.

The same problems can be identified in Italy a few centuries later as population rose and Rome grew from a small city into the centre of an empire encompassing the Mediterranean and much of the Near East. About 300 BCE Italy and Sicily were still well forested but the increasing demand for land and timber resulted in rapid deforestation. The inevitable consequence was much higher levels of soil erosion, and as the earth was carried down in the rivers, the gradual silting up of ports in the estuaries. The port of Paestum in southern Italy silted up completely and the town decayed, while Ravenna lost its access to the sea. Ostia, the port of Rome, only survived by constructing new docks. Elsewhere large marshes developed around river mouths built with the soil eroded in the hills. The Pontine marshes were created about 200 BCE in an area which had supported sixteen Volscian towns four hundred years earlier.

The creation of the Roman empire increased the pressure on the environment in other areas of the Mediterranean as the demand for food increased. Many of the provinces of the empire were turned into granaries to feed the population of Italy, particularly after 58 BCE when the citizens of Rome started to receive free grain for political reasons. North Africa, for example, contains a whole series of impressive Roman remains, such as the great city of Leptis Magna in Libya, from what were once some of the most flourishing and highly productive provinces of the empire. But now they lie surrounded by vast deserts, a memorial to widespread environmental degradation brought about by human actions. The area continued to flourish even after the final destruction of Carthage in 146 BCE but the growing Roman demand for grain pushed cultivation further into the hills and onto vulnerable soils that were easily eroded when deforested. There is no single date that marks the decline of the north African provinces – it was a long drawn-out process of increasing strain and deterioration in the environment as soils eroded and the desert slowly encroached from the south. The process was intensified after the fall of Rome as tribes such as the Berbers moved into the cultivated areas, bringing with them their large flocks of grazing animals, which completed the work of removing the remaining vegetation cover. Similar pressures can be identified in Asia Minor where the interior of the old Roman province Phrygia was completely deforested by the first century CE. A few decades later the emperor Hadrian had to restrict all access to the remaining forests of Syria because of the amount of deforestation. Some regions in the area were less seriously affected and continued to prosper as food exports to the main imperial cities and towns such as Antioch and Baalbek flourished until the early Byzantine period. But both are now ruins, some of the limestone hills of the area have lost up to two metres of soil and Antioch is under more than eight metres of water-borne silt from hillsides ravaged by deforestation.

The causes of the decline and fall of the Roman empire in the west are still subject for debate by historians. However, most would agree that it was the result of the interaction of a number of factors causing internal political decay and vulnerability to external pressure. It would, therefore, be too simple to see environmental degradation as the single or even the main cause of the decline and fall. But there is no doubt that it was an important contributory factor and that the difficulties in extracting the food surplus needed to feed both the population of Rome and large standing armies was one of the causes of internal weakness in the empire.

The deterioration of the environment of the Mediterranean region did not end with the fall of Rome. Freed from some of the demands of the imperial system some areas may have recovered, enabling secondary forests to develop as population fell. The recovery in population levels by around 1000 CE followed by a steady rise meant that deforestation continued until the present day. As more forests were cleared, more soil was eroded away. The same trends can be identified in Spain where overgrazing by the huge flocks of sheep kept by the *Mesta*, the most powerful of the medieval guilds, permanently degraded the environment of large parts of central Spain, in particular La Mancha and Extremadura, producing vast expanses of poor-quality grass and scrub.

The decline and fall of the Maya

The development of settled societies in the Americas produced the same sequence of events as in Eurasia – the clearance of land for agriculture, deforestation and soil erosion. There is a strong suspicion that the collapse of the great city of Teotihuacan in the valley of Mexico and some of the early city states in the coastal area of Peru in the first centuries CE were linked to problems arising from the overuse of irrigation and the consequent failure of the agricultural base leading to an inability to maintain the superstructure of the state. But the clearest case of environmental collapse leading to the demise of a society comes from the Maya, who once flourished in what are now parts of Mexico, Guatemala, Belize and Honduras. They were one of the most extraordinary societies of their type found anywhere in the world. Some of the first explorers to find the 'lost cities', the Americans John Stephens and Frederick Catherwood in the late 1830s, were as much at a loss as Leonard Woolley in Mesopotamia to explain what had happened. At Copan and Palenque they mused on a vanished society:

> We . . . strove in vain to penetrate the mystery by which we were surrounded. Who were the people that built this city? . . . architecture, sculpture, painting, all the arts which embellish life, had flourished in this overgrown forest; orators, warriors and statesmen, beauty, ambition, and glory, had lived and passed away, and none knew that such things had been or could tell of their past existence . . . In the romance of the world's history nothing ever impressed me more forcibly than the spectacle of this once great and lovely city, overturned, destroyed and lost; discovered by accident, overgrown with trees for miles around, and without even a name to distinguish it.

The main obstacle to understanding what really happened is that the Mayan script has only recently been deciphered. Archaeological research since the sites were rediscovered and the use of increasingly sophisticated techniques in the last five decades or so have also brought new insights.

Mayan society was a remarkable achievement in that it developed in dense lowland tropical jungle. The earliest settlements in this area date from about 2500 BCE. Population rose slowly and settlements grew in size and complexity so that by about 450 BCE it is possible to identify separate ceremonial areas and buildings within the settlements. Two hundred years later at Tikal in Guatemala a complex, hierarchical society had emerged (easily identified by the large differences in status between the burials of different groups). Steep pyramids over thirty metres high with temples on their summits were being built out of the local limestone rock in the north acropolis of this major centre. Over the next two or three centuries this process was repeated throughout the area until a whole series of major settlements had developed with a remarkably uniform culture displayed in the architectural styles and common script. The considerable intellectual achievements of the Maya were reflected in their astronomy (where they made detailed and accurate calculations not just about the phases and positions of the sun and the moon but also planets such as Venus) and in their highly complex and extremely accurate calendar, based on a fifty-two-year cycle counting from a fixed date in the past equivalent to 3114 BCE (although the significance of this date remains unknown). All the Mayan sites have a large number of stone stelae inscribed with a series of dates and texts. The main phases of Mayan history are clear and much of the political history can now be reconstructed from the recently deciphered texts. By the first centuries CE a large number of elaborate ceremonial centres had developed throughout the region. For a couple of centuries after 400 there was a strong influence from the city of Teotihuacan in central Mexico but when that declined after 600 the Maya entered their most spectacular period. Huge pyramids, often aligned towards significant astronomical points, were built at all the centres and large numbers of stelae erected. Then, within a few decades after 800, the whole society began to disintegrate. No stelae were erected, the ceremonial centres were abandoned, population levels fell abruptly and the cities were soon covered by the encroaching jungle.

Until the 1960s it was widely believed that the Maya were virtually unique in their world in that they were peaceful and governed not by secular rulers and a military elite but by a religious caste obsessed by the intricacies of their calendar and astronomical observations. Since only the dates on the stelae could be understood it was assumed that these recorded various events associated with astronomical and calendrical cycles. The way in which the Maya obtained their food and supported the priestly elite in a lowland jungle environment remained a puzzle. Twentieth-century studies of the Maya suggested that the only viable strategy would have been a swidden system, involving clearing a patch of jungle with stone axes during the dry season between December and March and then setting fire to the area just before the start of the rainy season when the maize and beans would have been planted with a digging stick to be harvested in the autumn. The cultivated patch would have been abandoned after a couple of years as weeds re-invaded and made clearance too difficult. This agricultural system is widely used in tropical areas and is highly stable in the long term but it can only support a small population in any area because of the need to have a large amount of land for each farmer – the cleared patches cannot be reused for about twenty years or more until the jungle has regrown (the task of clearing jungle is far less laborious than that of clearing grass and scrub). It was, therefore, assumed that the Maya lived in small, shifting settlements scattered throughout the jungle and only coming together at the ceremonial centres, where the small priestly caste lived permanently, for part of the year.

In the last fifty years these assumptions about Mayan society have been abandoned and a radically different picture of the Maya adopted which helps to explain why the society collapsed so abruptly. The most important change has stemmed from a new understanding of the texts engraved on the stelae. It is now clear that these do not represent religious dates but rather they are monuments to the different secular rulers of the cities, setting out the dates of their birth, accession and death together with the major events of their reign. The picture of a peaceful, religious society has been replaced by one which sees Mayan society as dominated, in the same way as other early societies, by a secular elite supported by armies and engaged in fairly continuous warfare between different cities. Recent archaeological work has also made the nature of these cities much clearer. They were not merely ceremonial centres occupied by a small elite but true cities with a large permanent population. At the centre were the huge ceremonial areas with magnificent temples and palaces built around a plaza. Beyond were complexes of thatched huts on platforms grouped around courtyards where most of the people lived in extended family groups. They provided the labour force that constructed the public buildings and residences for the elite. Excavations in the outer areas of Tikal suggests that, at its height, the population was at least 30,000 and possibly as high as 50,000 (of the same order as the cities of Sumer). Other cities, though not quite so large, would have followed the pattern and it seems likely that the total population in the Maya an at its peak might have been near to five million in an area that now supports only a few tens of thousands.

This new knowledge about the nature of Mayan society has been complemented by new information about the way the Mayans obtained their food. Obviously a swidden system could not be productive enough to support such a large population. Not enough land was available between the cities, which in some cases were no more than about sixteen kilometres apart, to make this feasible. Hunting and fishing would have provided no more than useful supplements and although the ramon, or breadnut tree, whose nuts can be ground to make flour, grows in profusion in the Mayan area, studies of the current-day Maya suggest that it would only be used as a food of last resort. Archaeological fieldwork in the 1970s discovered that a much more intensive agricultural system was in fact used by the ancient Maya. On the hillsides they would clear jungle and make fields using extensive terracing to contain the inevitable soil erosion. Equally important though was the construction of raised fields in swampy areas. Grids of drainage ditches were dug into swamps and the material from the ditches was used to form raised fields. Traces of the huge areas once covered with these fields have now been found in the jungle from Guatemala across to Belize. In the fields crops such as maize and beans were grown for food together with others such as cotton and cacao.

This intensive cultivation system was the foundation for all the achievements of the Maya. However, when too much was demanded of it, it could not withstand the strain. The crucial period came after the waning of Teotihuacan influence about 600. It was marked by increasing warfare between the Mayan cities and a much greater emphasis by the elite on the construction of more and larger ceremonial buildings taking up huge amounts of labour. Population continued to rise steadily and a higher proportion lived in cities where they were available to man the armies and work on the construction projects. Cultivation became more intensive. However the ecological basis to support such a massive superstructure was simply not there. The soils in tropical forests are easily eroded once the tree cover is removed. Mayan settlements clustered, not surprisingly, around the areas of fertile soil but three-quarters of the fertile soil in the area occupied by the ancient Maya is today classified as highly susceptible to soil erosion. Around Tikal for example about three-quarters of the soil is classified as highly fertile but nearly two-thirds is highly vulnerable to erosion once it is cleared of trees. Clearing the forest, therefore, ran the risk of bringing about soil deterioration and declining crop yields. This problem would be exacerbated by the lack of domesticated animals to provide manures to maintain soil structure and fertility. The forest was cleared not just to provide land for agriculture but also for fuel, construction materials and for making the huge amounts of lime plaster that coated the ceremonial buildings. Population pressure pushed fields and terraces into ever more marginal areas which were even more vulnerable to erosion. Across the Maya area the vulnerable soils were increasingly exposed to the wind and the rain and eroded away.

Soil erosion caused by deforestation would have reduced crop yields in the affected areas. The associated higher levels of silt in the rivers would have seriously damaged the extensive raised fields in the marshy areas by altering the delicate balance between water levels and the fields and by making the ditches much more difficult to keep clear. The first signs of declining food production are evident in the period before 800, when the skeletons from burials show higher infant and female mortality and increasing levels of deficiency diseases brought about by falling nutritional standards. A reduction in the food surplus on which the ruling elite together with the priestly class and army depended would have had major social consequences. Attempts were made to increase the amount of food taken from the peasant cultivators, leading to internal revolt. Conflict between the cities over the declining resources would have intensified, leading to more warfare. The fall in food supplies and the increasing competition for what was available led to very high death rates and a catastrophic fall in population, making it impossible to sustain the elaborate superstructure the Maya had built upon their limited environmental base. Within a few decades the cities were abandoned and no more stelae were erected to commemorate rulers. Only a small number of peasants continued to live in the area. The deserted fields and cities, buried under dense jungle, were not found again until the nineteenth century.

The Nile valley

The most striking example of a society establishing a sustainable balance between the natural environment and its demand for food is Egypt. For about five thousand years after the emergence of settled societies in the Nile valley the Egyptians were able to exploit the annual flood of the river as the basis of a succession of states from the various dynasties of the Pharaonic era, through the Ptolemies to the period of the Roman empire, and under the Arabs and the Mamluks, until the use of new technology in the nineteenth century began to undermine the system.

Each year the Nile would flood the vast length of its lower valley, depositing huge quantities of silt from its twin sources in Ethiopia and Uganda. To some extent the Egyptians happened to be the beneficiaries of somebody else's environmental problems. Much of the silt was the result of deforestation and soil erosion in the highlands and although, at present, it is estimated that the Nile carries about a hundred million tonnes of silt a year, it was probably less in earlier periods. The heaviest rainfall in the highlands occurred in June and the flood reached Egypt, nearly 3200 kilometres away, in September. Here it spread out over the narrow valley (no more than twenty kilometres wide in places) through natural overflow channels into basins to produce the rich, permanently renewed soil of Egypt. The flood ended in November but this short period was exactly right for the sowing of autumn crops.

The great stability of the agricultural system originally adopted by the ancient Egyptians and used (without major changes) by their successors lay in the fact that what they did was to exploit a natural process, with only minimal human interference, and only then at a low technological level. The water control system was built upon regulating the natural flow of the river to provide the right amount of water at the right time together with the silt to fertilise the land, rather than producing an artificial environment. In some places natural levees were breached to ensure that the water reached the maximum possible area and in others artificial banks were built to provide semi-natural basins to enable water to be retained for longer. The natural irrigation of the flooded area made it both impractical and unnecessary to build artificial canals. Because of its underlying geological structure and essentially natural irrigation system the Nile valley suffered none of the problems and unwanted and damaging side-effects of Mesopotamia's artificial irrigation system. Within a month of the flood the water table was more than three metres below the surface so waterlogging was not a problem and there was no build-up of salts in the surface layers. This eliminated the need for the long fallow periods essential, but not adopted, in Mesopotamia if the twin disasters of waterlogging and salinisation were to be avoided. The silt, rich in nutrients and regularly renewed, also eliminated the need for heavy manuring and ensured the continuing fertility of the soil. The level of this fertility can be judged by the fact that in the eighteenth century CE crop yields in the Nile valley were about twice as high as in France. The absence of salinisation is demonstrated by the increasing importance in Egyptian agricultural output of wheat, a crop more sensitive to salt levels, compared with barley – the exact opposite of what happened in Mesopotamia.

Although there were some changes in the agricultural system over the centuries, it showed a remarkable degree of continuity. In the Pharaonic period Egyptian farmers planted wheat, barley, beans and chick peas in the flood basins. Vegetables, particularly onions, lentils and also fodder crops, were planted in those areas where water was still available in the spring. On the higher land, date palms flourished. Sorghum, which was resistant to semi-drought conditions, was introduced as a main summer crop in the Ptolemaic period and crops such as sugar, cotton and rice were cultivated for the first time in the Islamic period. The technology involved in utilising the annual flood was, like the rest of the system, simple. For the first two thousand years or so the fields were watered from the channels by buckets moved by hand. About 1340 BCE the shaduf (or bucket and pole system) was brought into use and its greater efficiency increased the amount of cultivated land by about 10 per cent. A similar increase was again achieved about 300 BCE when the animal-drawn water wheel was introduced. After that there were no technological changes of importance until the nineteenth century.

The long-term stability of the Egyptian agricultural system was maintained as long as there were only limited modifications to the natural flood regime. However, there were problems stemming from the fact that there could be major fluctuations from year to year in the flood level of the Nile and from a tendency for there to be sustained periods of especially high or low floods. In the short term the results for both Egyptian farming and society could be disastrous and these fluctuations had a major influence on the course of Egyptian history, particularly under the Pharaohs. Very high floods destroyed many settlements and the artificial works for containing water. Very low floods left some areas dry and lacking in silt. For example, although the 1877 flood was only about 180 centimetres below normal, it still left about a third of the valley without water and silt. Either extreme reduced crop yields with severe social implications.

In general the overall trend has been towards a fall in flood levels, probably caused by declining rainfall in the highlands that are the source of the Nile. However, there were major fluctuations within that trend. After about 3000 BCE flood levels fell by about a quarter but it was the long series of extremely low floods in the period 2250–1950 BCE that brought about major social disruption and the end of the Old Kingdom in Egypt. These low floods drastically reduced crop yields, leading to mass starvation, death of livestock, lack of seeds for the next planting season and abandonment of marginal land. This happened at a time when the demands of the state (huge construction projects and support of a growing bureaucracy, army and priesthood) were imposing an ever heavier burden on a weakened agricultural base. The result was the political and social collapse of the Pharaonic state in the wake of a full-scale revolt by the peasantry. There followed two centuries of disunity before Egypt was reunited at the beginning of the so-called Middle Kingdom. This period was marked by a series of very high floods (those between 1840 and 1870 BCE were nearly nine metres higher in places than the modern average) which, although destructive, at least ensured adequate water and silt for crop production. A substantial decline in flood levels after about 1150 BCE again caused major political and social problems. The fall in food production made it more difficult to support a large number of non-producers, especially the army, at a time when the Egyptians were under considerable external pressure from a wave of what their records refer to as 'sea-peoples', attempting to settle in the Nile delta. The powerful Egyptian state under the command of the Ramesid dynasty that had built the great temples of Abu Simnel collapsed. Egypt was once again disunited and was only reunited about two hundred years later under external conquerors.

Despite these vicissitudes that had such a major impact on Egyptian history, the Nile flood continued to provide a stable way of supporting a large population and a complex society in Egypt until the nineteenth century. Only then were major changes made which, within a relatively short space of time, began to have widespread effects. In the 1840s the first artificial irrigation systems were built to provide more irrigated land to grow extra crops, not for food but other items such as cotton for sale in Europe. Within decades permanent irrigation had produced widespread salinisation and waterlogging in the newly cultivated areas. In 1882 the British agricultural expert Mackenzie Wallace described the 'white nitrous salts covering the soil and glistening in the sun like untrodden snow'.

Many of the earliest settled societies were unable to strike a balance between the need for food for the mass of the population (as well as for the rulers, bureaucrats, priests and soldiers) and the ability of their environment to sustain intensive agriculture over a long period. Some overreached themselves in the end, although for a considerable period, often many centuries, they appeared to be highly successful. In Mesopotamia, the Indus valley, the jungles of

Mesoamerica and other areas too, a fragile environment collapsed under pressure. The demands of an increasingly complex society began to overstretch the capacity of the agricultural base of the society to support the large superstructure that had been erected. In the end, the unwanted, and unexpected, side effects of what at first appeared to be solutions to environmental difficulties became problems themselves. The result was falling food production and increasing difficulty in supporting a large number of non-producers. Attempts to increase the proportion of the declining amount of crops going to the elite often produced internal revolt. The inability to feed a sufficiently large army often led to conquest by enemies. Only in a few cases, like the Maya, was there a dramatic collapse of an entire society or wholesale abandonment of the land no longer fit for agriculture. In the Mediterranean and China there was long-term degradation which seriously damaged the resource base of these societies. The struggle to provide enough food was to be one of the central features of nearly all the rest of human history. It remains acute for the majority of people in the world.

6

THE LONG STRUGGLE

Agriculture did not solve the problem of producing enough food to meet the needs of the world's population. Until about the last two centuries in every part of the world nearly everyone lived on the edge of starvation. Throughout this period, the rise and fall of empires and the emergence of new states and their decline, the underlying economic and social conditions remained broadly unchanged for thousands of years. Although in areas less environmentally sensitive than Mesopotamia, the Indus valley and the tropical jungles of Mesoamerica a complete collapse of society did not take place, there was still a high price to pay in the wholesale loss of life. All but about 5 per cent of the people in the world were peasants, directly dependent on the land and living a life characterised by high infant mortality, low life expectancy and chronic undernourishment, and with the ever-present threat of famine and the outbreak of virulent epidemics. The food they ate was almost entirely vegetable in origin (especially in Asia, Africa and the Americas) and the dietary staples were the three major crops of the world – rice in Asia, maize in the Americas and wheat (supplemented by oats and rye) in Europe. Because of the overwhelming dependence of these societies on agriculture there were severe limits on the scale of other activities and the numbers of soldiers, priests and craftsmen that could be supported by the peasantry.

Food and population

The human condition around the world varied from place to place and period to period depending on the balance between agricultural output and the population level. Although the number of people in the world was far lower than at present, there was the persistent threat of overpopulation and starvation because of the inefficiency of the agricultural system. Agriculture allowed far more people to be supported than in gathering and hunting societies but reliance on a small number of crops grown in a specialised environment increased vulnerability to crop failure. Continuous cultivation of the same area lowered soil fertility. A vicious circle was established between the need to use as much land as possible to grow crops for direct human consumption, the limited amount of land available for animals and the lack of animal manures to maintain soil fertility. The limitations on agricultural output were exacerbated by problems in distributing food. The total amount that could be stored was limited and losses were high because of inadequate facilities. A primitive transportation system meant that anything more than local distribution of food, except by water, was extremely difficult. The market for food was therefore limited and often crop failure in one area could not be alleviated by moving in supplies from elsewhere because they were not available or, even if they were, they could not be transported. These problems were exacerbated by the religious and secular elite who enforced the collection of food through taxes, tithes and other forms of direct expropriation, often leaving the peasantry with insufficient food for survival. Armies moving through the countryside looting the food they needed and destroying crops and animals made the situation worse.

Only very slowly did changes in agricultural systems leading to higher productivity, together with better storage and distribution facilities, begin to alleviate these problems. For most societies until the nineteenth century population size and the amount of food available were often out of balance, both in the short term and the long term. In the short term, annual fluctuations in supply as a result of bad harvests or outbreaks of warfare could bring disaster. In the long term, population could increase to a level where it was almost impossible for a large part of the population to obtain an adequate diet. It was difficult to increase food production as quickly as population was increasing, and although many of these early agricultural societies around the world used fairly crude methods for restricting the growth in numbers (infanticide or a tradition of late marriage), food supply and population were rarely in balance. Over the very long term it is clear that slow improvements to the agricultural system meant that more people could be fed. But overall population growth rates remained very low. Until about three hundred years ago the world's population never increased by more than about 0.1 per cent a year – a twentieth of the current rate. Within this upward trend there was no steady growth in population or in food production. Instead there were bouts of rapid growth in numbers followed by sudden halts and declines caused by population levels outstripping food supplies or by the consequences of war and disease. More often than not, increasing population put ever greater pressure on a limited agricultural system, producing more poverty and malnutrition. The conflict was normally only resolved in a drastic fashion through mass starvation and death until the population was more in balance with the output from farming.

There are no reliable population statistics until the last two hundred years. Nevertheless, working from partial censuses, estimates and returns produced for other purposes such as taxation, demographers have been able to make calculations which, although not accurate in detail, give a broad picture of the number of people in the world, their distribution and the main periods of growth and decline. About the time agriculture was adopted in the three core areas the world's population was about four million (the same as one large city today). The spread of agriculture enabled more people to be fed and human numbers rose steadily, doubling every thousand years to reach about fifty million by 1000 BCE (roughly equivalent to the current population of England and Wales). That figure doubled in only 500 years to reach 250 million at the peak of the Han and Roman empires about 200 CE.

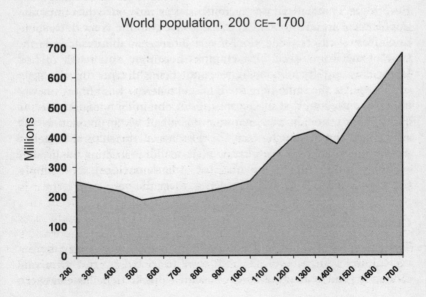

World population, 200 CE–1700

With the decline of those empires a widespread increase in instability, warfare and destruction meant that there was little further growth anywhere in the world until about the year 1000. Then, in both China and Europe the numbers rose to reach a temporary peak of about 350 million people in the world by 1200. For a century population increased only slowly to some 400 million as the limits of food supply were reached. After 1300 starvation and plague sharply reduced numbers so that by 1400, when some recovery had taken place, there were still only about 350 million people in the world. Numbers rose sharply in the next two centuries to reach about 550 million by 1600. Then for the next century a deteriorating climate affected food production and restricted the growth in numbers so that the world's population in 1700 was just under 700 million. The eighteenth century saw the most rapid growth in history to that date, bringing the total to 900 million by 1825.

The distribution of people in the world also changed markedly over this long period. Before the spread of agriculture the world's population was fairly evenly distributed across the globe but the rise of settled communities reduced the share of Africa, the Americas and Oceania from about 40 per cent to less than 15 per cent of the total as the great empires of the Near East, Mediterranean, India and China became the main centres of human society. Within this general picture there have nearly always been more Chinese and Indians than Europeans. That pattern was established very early once the temporary importance, around 3000 BCE, of the societies in the Near East, stemming from their early adoption of agriculture, had diminished. By the time of the Roman and Han empires there were about fifty million Europeans but fifty million Chinese and about the same number again of Indians. At the same time the total population of the Americas was about five million, that of Oceania about one million and Africa had about twenty million inhabitants, over half of whom lived in North Africa near the Mediterranean. These broad relationships still applied a millennium later between Europe (sixty million), China (eighty-five million), India (ninety million), the Americas (fourteen million), Oceania (two million) and Africa (forty-six million).

China

The examples of China and Europe show that although very different agricultural systems evolved, both were restricted by environmental limitations and neither was able to sustain a long-term balance between

China population, 200 CE–1700

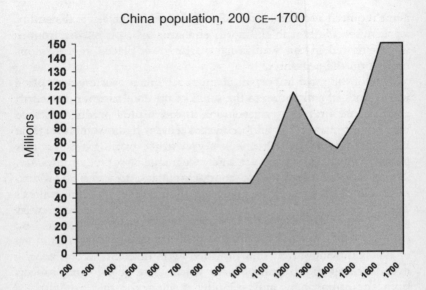

population and food supplies. In China the development and emergence of a settled society took place in the north, where it was based on the dry farming of millet. Until the end of the Han empire in 220 CE the centre of the Chinese state remained in the north. It was in this period that one of the most distinctive features of Chinese society developed – the almost complete dichotomy between the ruling elite and the bulk of the population who were peasants living in small villages. Throughout Chinese history there have been long periods of political unity followed by disunity but the way of life of the overwhelming majority of the people continued basically unchanged in the village world. The main task of the elite was to ensure that enough food was obtained from the peasants to maintain themselves and the army that was largely stationed in the north to provide protection against attacks from the nomads of Central Asia. The collapse of the Han empire under the pressure of barbarian attacks pushed the centre of gravity of the Chinese state slightly south towards the Yangtze river, into an area which became one of the main grain-producing areas. The later reunification of China, after 589, made it necessary to transport the food surplus northwards to the military centre of the empire. The huge Grand Canal, which extended for almost 2,000 kilometres, was built in the early seventh century to move food supplies from the Yangtze valley to the north for the army and to the capital, which alone required 400,000 tonnes of grain a year. By the eleventh century an army of 300,000 in the north and over 750,000 on the frontier was being fed in this way. Not surprisingly it placed an enormous burden on the peasantry.

One of the most important changes in Chinese society took place with a wave of emigration to the south of the country after the fourth century, encouraged by a revolution in agricultural production. Rice had been domesticated and cultivated widely across south-east Asia for thousands of years but it was grown like other grains such as millet or wheat. About 500 BCE a new technique of wet-rice production in paddy fields developed in south-east Asia and spread slowly to reach China, Korea, Japan, India and Java in the next millennium. The inherently poor nutrient content of tropical soils was circumvented by growing the rice partially covered in water in special fields fed by complex water management techniques to produce large quantities of slowly moving water. This provided extra nutrients in two ways – it encouraged the growth of algae which were able to fix nitrogen from the atmosphere and secondly it allowed larger amounts of organic matter – vegetable waste and both human and animal manures – to rot in the water. The continual trampling involved in working the fields made the soil impervious and therefore able to retain the nutrients. This system produced huge increases in crop yields but required enormous amounts of labour not only for cultivating the crop but, more importantly, for constructing and maintaining the fields and the water control systems. In the centuries after about 400 CE there was a steady wave of people moving south to open up and settle new lands using the paddy field systems. The main rice-growing provinces of the Yangtze delta, the Hsaing valley, Szechwan and Kwantung were colonised, and steady if unspectacular improvements in techniques further increased production. The most important of these was the introduction, in the eleventh century, of new varieties of quicker growing rice from Vietnam that enabled two crops a year to be grown in the most favoured areas in the south and both a rice and a wheat crop further north.

The Chinese developed the most sophisticated agriculture in the world (based on techniques such as crop rotation that were still largely unused in Europe), producing very high yields from intensively farmed land. By about 1200 China was the largest, most literate and most advanced country in the world. The expansion of settlement into the new rice-growing areas in the south allowed population to grow from about fifty million under the Han (a level that, with some fluctuations, was maintained for several hundred years afterwards) to reach about 115 million in the early thirteenth century. However there were a number of structural problems that ensured that the balance between food supply and population was never satisfactory. Improvements in techniques or farming of new land brought only temporary increases in food supplies which were soon counterbalanced by increases in population.

The Chinese were unable to make the structural changes in the agricultural system that might have increased food supply on a large scale. Yields were about as high as was possible before the introduction of modern artificial fertilisers. The most fertile areas were already densely populated and inputs could not be increased because there was not enough land to keep more animals which might have provided the extra manure. Social mores emphasised equal land division within the family, which meant a large number of very small farms, each of which could only produce a very small surplus, if any. Output could only be increased by small improvements in productivity and by cultivating new land, although this was often of marginal quality and therefore produced lower yields. The Chinese agricultural system was certainly impressive in its total output and its very intensive production methods. However, the high level of population and the difficulty of making any significant qualitative changes meant that the mass of the population were dependent on a system that could produce only a low level of food for each individual. Disasters like the Mongol invasion which resulted in the deaths of about 35 million Chinese or the massive epidemics in 1586–89 and 1639–44, which killed about a fifth of the people on each occasion, reduced the pressure of population for a while. But after about 1700, as the population rose dramtaically, there is no evidence of any significant increases in yields. The area under cultivation rose but the amount of food available per person was about the same in 1850 as it had been three hundred years earlier. The result of this high intensity system (dependent on a large amount of labour to produce high levels of food from the land used) was that the overwhelming majority of the population lived permanently on the verge of starvation.

Europe

Chinese agriculture was able to maintain a large number of people on the brink of starvation. Medieval European agriculture was a low productivity system that kept a smaller number of people in the same condition. Europe too found it difficult to expand output on a sustainable basis. The overriding problem here was that soil fertility was steadily reduced through continual cropping, leaching out of nutrients in rain and low level soil erosion, and there was a lack of replacement nutrients. Fertility in the arable fields could only be maintained through use of animal manures, but the number of animals that could be fed throughout the year was very small because of the lack of fodder crops. Many animals had to be slaughtered in the autumn because of an acute shortage of winter feed. Keeping animals on pasture was difficult in the long term when the manure was removed to fertilise arable fields, because the yields of hay and grass for the animals would then fall. A vicious circle was therefore established. As the number of animals that could be fed declined so did the manure available for arable fields and with it crop yields. Most manure was also stored in the open for long periods which severely lowered its nutrient value. Crops were reduced by the use of only a two-field rotation system under which a field alternated between being planted in the autumn with winter grain and being left fallow for a year. This resulted in an inadequate amount of time to restore nutrients and remove any diseases or pests in the soil that might reduce crop yields. A shortage of marls (or liming materials) also meant that it was difficult to reduce the acidity of the soils and this kept yields low. The consequence was that in medieval Europe the agricultural system was only able to sustain a low level of productivity. In the short term it was sometimes possible to increase production. Extra inputs could be introduced by keeping more animals in new fields or new arable fields could be brought into use but these improvements could not be sustained in the long term because of the draining of nutrients from the pastures and fields. There was therefore a tendency for the fertility of the soil to fall back to a level where the risk of crop failure was very high.

Changes to the European agricultural system came gradually. About 800, in north-eastern France a new three-field rotation was adopted. Under this system one field was planted in the autumn with winter wheat or rye and a second the following spring with oats, barley or perhaps peas whilst a third was left fallow. This not only increased fallow periods and the variety of crops but also helped to some extent to raise the fertility levels and distributed work more evenly throughout the year. The three-field system spread across Europe – but only slowly. Its use was not widespread in England until after 1250 and even later in other areas. However, the output of food was still constrained by the amount of inputs available. An important contribution to raising output was the invention of the heavy plough which was probably developed in the sixth century and its use spread across Europe during the next four hundred years. The earliest lightweight ploughs were simply an enlarged digging stick dragged by a single animal or a pair of oxen; they did not turn over the soil and they left a wedge of undisturbed earth beneath each furrow. This plough was not well suited to the heavier soils of northern Europe. The heavy plough needed a team of eight oxen but it did enable a much larger area to be cultivated than before, though it could not solve the problem of low nutrient levels. Improvement in crops came very slowly with the increasing use of legumes capable of fixing nitrogen and therefore improving soil fertility and of fodder crops to provide winter feed for animals. It was not until about 1300, and then only in a small area of Europe (principally Flanders), that their use became at all widespread.

For thousands of years European agriculture remained at a low level of productivity and the majority of people lived on the edge of starvation. There is evidence that European societies were more successful than those elsewhere in the world in limiting their population. The most important factors were late marriages or fewer marriages. The age of marriage and the number not marrying tended to rise when the population was higher and so nearer the limits of food production and fell when the population pressure eased following famine or disease. These methods could never be more than partially successful and there were still many occasions when even at the relatively low levels common in the medieval and early modern periods Europe was 'overpopulated' in relation to the supply of food.

In 1000, the population of Europe was about forty million. In the next three centuries, numbers doubled to reach eighty million in 1300. This rise in numbers was sustained mainly by a major increase in the amount of land under cultivation rather than any fundamental changes in productivity. Nevertheless, many parts of Europe became severely overcrowded – the population in northern Italy, Flanders, Brabant and the Paris area was probably as high as in the early nineteenth century despite the lower level of agricultural productivity. The supply of new land was virtually exhausted in the late thirteenth century, and yields were falling as more land was put under crops as a short-term measure to try to increase food production. This reduced the amount of pasture and therefore the number of animals and the amount of manure produced. The shortage of land, combined with a rising population, caused cereal prices to rise and many people suffered from under-

Europe population, 200 CE–1700

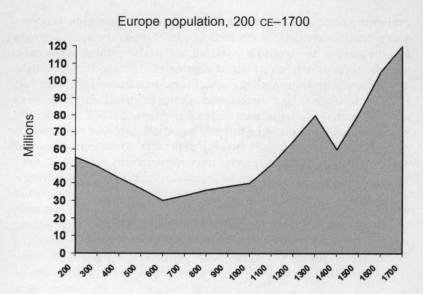

employment in an overcrowded labour market and lived at a very low level of subsistence. The growing shortage of food was exacerbated by the fact that the nobility and clergy were taking about half of the peasants' output of food once tithes, rents, taxes and the time they were forced to spend working on the lord's estate are taken into account. By 1300 European population was beyond the optimum that could be supported by the technology and institutions of the time. At the beginning of the fourteenth century there is clear evidence of falling population brought about through permanent malnutrition and near starvation in many areas of Europe from Tuscany and Provence to Normandy and south-east England. The major famine of 1316–17 added to the number of casualties but the pressure of population on resources was not removed until after the outbreak of the Black Death in 1346 and the subsequent recurrences of the plague for the rest of the century.

The period from the late fourteenth century until the mid-fifteenth was one of comparative prosperity as the population remained below the peak of 1300 for about two hundred years. However, by 1600 it was near 100 million, higher than in 1300, even though there had been little improvement in agricultural productivity. The signs of 'over-population' and an imbalance between food supply and numbers appeared again. Settlements expanded into new areas but usually the soils were poor and output low. In England agricultural prices were rising from about 1500 as shortages started to develop and real wages fell by a half in the period between 1500 and 1620, causing immense distress to those already on the margins of society and unable to find any form of regular work. By the 1620s population growth slowed as malnutrition and higher mortality caused by inadequate food supplies took their toll. The same symptoms can be found in France. By 1570 most usable agricultural land was in production and, although numbers were kept in check by the continual civil wars of the period, a crisis was reached early in the seventeenth century. Food prices continued to rise, land holdings got smaller and real wages fell drastically. Population rose on occasions to about 20 million but fell back rapidly since at that level it was out of equilibrium with the number that could be fed in the long term. A series of severe famines between 1690 and 1710 demonstrated that population was still higher than the agricultural system could regularly feed.

Climate, food and population

In all the agricultural societies around the world the outcome of the harvest was crucial. A bad harvest was a calamity but two in succession could bring disaster – not just for the poor who were the first casualties of reduced supplies and increased prices but also for the peasants and eventually for the whole of society. The temptation, especially for the peasants, was to consume what little food was available, leaving inadequate quantities of seeds to plant for the next crop, thus increasing the chance of disaster the next year. The elite in each society were normally in a position to be able either to force the peasantry to hand over enough food or to buy it even at highly inflated prices. Starvation affected those unable to retain enough of their crop to live on until the next harvest and those, mainly in the towns, who could not pay high food prices. Under such circumstances large numbers, already badly nourished, easily succumbed to the outbreaks of disease that normally followed a period of bad harvests.

The most important influence on the outcome of the harvest and therefore the state of society was the weather. Any one of a number of possibilities – a wet, cool period that reduced the chances of seeds germinating, a dry spell when crops were growing or wet weather during harvesting – could severely reduce the harvest. Nearly every government paid great attention to detailed accounts of the weather from around the country, the prospects for the harvest, and the price of grain and bread as main indicators of the level of social unrest. Crops were affected not only by annual variations in the weather but also by the longer-term cycles in the earth's climate, which could cause widespread dislocation of agricultural systems. Colder periods shortened growing seasons, made some areas marginal and, in general, reduced yields. While they lasted there would be an increased risk of a poor harvest in any particular year and a succession of poor harvests resulted in the build-up of internal strains within society. Warmer periods increased the areas where crops could be grown, improved food supply and reduced the pressure stemming from the number of people who had to be fed. Most of the available evidence about these long-term climatic cycles comes from Europe and the detailed effects of the changes were different in other parts of the world – for example, one of the coldest spells in Japan coincided with one of the warmest in Europe.

Since the end of the last ice age there have been alternating periods of warmer and colder weather in Europe. After a steady increase in average temperatures from about 10,000 BCE, which marked the end of the last ice age, there was a warm period in the two thousand years after 5000 BCE. Vegetation zones moved northwards and this period coincided with the development and spread of agriculture across Europe. A general decline in temperatures then set in, reaching a low point between 900 and 300 BCE, a time of very high rainfall too. A rise in temperatures was noticeable by around 100 BCE when vines spread further north, but it petered out around 400 CE with a cool spell that lasted for around four hundred years. Then there was a warm period that was shorter than the first peak and reached its height about 1200. A steady decline followed, reaching a long low between 1430 and 1850, the 'Little Ice Age', when average temperatures were between one and two degrees centigrade lower than the present. The main impact was felt in the very severe winters, with summer temperatures not much different from current levels. These climatic changes have been established through a multitude of different techniques – pollen records from mud cores, cores from the Greenland ice cap, isotope analysis of foraminifera in ocean cores, the recording of changes in lakes, glaciers and tree lines plus some historical records. Although the broad trends have been firmly established, it is only for the last two periods – the warm spell that ended around 1200 and the 'Little Ice Age' – that detailed analysis of the effects of climate are possible.

At the height of the warm period which lasted for about four hundred years before 1200 the tree line in central Europe was about 150 metres higher than today, vines grew in England as far north as the Severn and farming was possible on Dartmoor as high as about 400 metres. Large parts of the uplands of southern Scotland were arable land and in 1280 the sheep farmers of Northumbria were complaining about the continual encroachment of arable fields on their upland pastures. One of the most important effects of the milder climate was on the Viking voyages and settlements. Iceland was colonised from Norway in 874 at the beginning of the warmer period and Greenland from 986. Both of these societies were on the climatic margins of Europe and their existence was largely dictated by the weather. The Greenland settlement flourished during the warm period with a population of about 3,000, almost 300 farms, sixteen churches and even a cathedral in the main village. But it remained a marginal and highly vulnerable society, dependent on the mild weather for its very existence.

The gradual cooling after 1200 caused a steady decline in Greenland. The hay-growing season gradually became shorter and shorter, yet the Viking settlers tried to retain their way of life based on cattle instead of shifting to the more readily available marine resources. As the climate deteriorated the Inuit moved further south and the Viking western settlement of Godthaab Fjord was destroyed shortly after 1350. The severe climate meant that pack ice remained in the seas around Greenland throughout the summer and contact with the rest of Europe was lost after 1408. The eastern settlement at Julianehaab died out, probably under Inuit attack, about 1500. Iceland, too, became a much more marginal society under the impact of a worsening climate. Wheat growing died out (a 1°C fall in annual temperatures in Iceland reduces the growing season by almost a third) and marine resources became overwhelmingly important in the economy. In the harsher climate the numbers that could be supported were much less and the population fell from about 77,000 at the height of the warm period around 1100 to 38,000 in the late eighteenth century.

The increasingly colder climate affected the rest of Europe too. The uplands of southern Scotland reverted to pasture and the growing of vines for wine-making died out in England about 1400. But the real impact of a much worse climate was felt after the middle of the sixteenth century – a series of severe winters and a period of much greater climatic instability began that was to last for almost three hundred years. After 1580 the glaciers in the Alps, Iceland and Russia advanced, in many places almost two kilometres, and did not begin to retreat until after 1850. Between 1564 and 1814 the Thames froze in the winter at least twenty times, as did the Rhone three times between 1590 and 1603, and even the Guadalquivir at Seville froze in the winter of 1602–03. At Marseilles the sea froze in 1595 and in 1684 there was pack ice off the coast of England. From the 1580s the Denmark Strait between Iceland and Greenland was regularly blocked by pack ice even in summer. Across Europe the lower temperatures reduced the growing season by about a month and lowered the height at which crops could be grown by about 180 metres, with consequent adjustments to the cultivation areas of nearly all crops. There were outbreaks of even more severe weather within this overall pattern. A series of cold mistral winds destroyed many of the olive groves of Provence between 1599 and 1603 and very heavy frosts around Valencia in the same period ruined many of the fruit trees.

The effects varied in different parts of Europe. There was no simple relationship between the temperature, the amount of rain and the size of the crop since the most important factor was how these influences were distributed through the seasons. The overall decline in temperatures had its greatest impact in Scandinavia, where the reduced growing season made many areas extremely marginal for growing crops. Further south a very cold winter might have some beneficial effects by killing a higher than normal number of pests. But even here the consequences of a deteriorating climate can be detected. In England there was a shift towards spring rather than autumn sown crops in order to try to avoid damage from a harsh winter. In the Netherlands, buckwheat, which is hardy and has a short growing season, but which was hardly grown in Europe before 1550, became increasingly important in the next hundred years. Other evidence from the Netherlands suggests that a cold, late spring reduced grass growth so that pastures were late to develop, reducing dairy output and increasing prices. It also led to a slaughter of cattle the following year if the hay crop was not sufficient to provide enough feed till the new grass had grown. In other areas increased rain could be most damaging, especially in the winter, by reducing arable yields because of waterlogged soils. This long period of poor climate came at a time when European population was already at the limit that the agricultural system could support. The worsened growing conditions meant a significant reduction in food production, leading to increased malnutrition, widespread famine and death. One consequence was a period of much greater internal instability within the

European states, which was particularly acute in the early seventeenth century.

Food and famine

For the overwhelming majority of people, food meant vegetables. Nearly all the available land was needed to feed humans rather than provide pasture and there was not enough surplus grain to feed many animals. Where animals were kept their output was low. A medieval cow in Europe produced one-sixth of the milk and one-quarter of the meat of a modern animal. In China all but 2 per cent of the calorific value of the diet came from vegetables, primarily rice. In Europe most people survived on a monotonous diet of vegetable and grain gruels and bread; meat and fish were rare items except for the upper classes. As late as 1870, 70 per cent of the French diet consisted of bread and potatoes and in 1900 only about a fifth of the calories came from animal products. Throughout Europe the majority of people lived on a maximum of about 2,000 calories a day (about the level of modern India), slightly higher in more prosperous countries such as England and Holland, but everywhere there were gross inequalities within society that meant that most lived on far less than this. In the early nineteenth century, in Norway, France and Germany, the average food consumption was still lower than contemporary Latin America and North Africa. The poorer regions of Europe had a particularly meagre diet. In some areas of France in the eighteenth century, for example in the Auvergne and the foothills of the Pyrenees, large parts of the population were still dependent on chestnuts for two or three months a year together with slops of maize and buckwheat, with some milk from a cow fed on weeds from the side of the road. These people lived on a diet that was far worse than that of their gathering and hunting ancestors.

There was no steady improvement in food supplies. For a century or more after the plagues of the fourteenth century people were reasonably well fed because of the reduction in numbers but standards fell dramatically between 1500 and 1800 – indeed the amount of meat consumption in Germany did not return to medieval levels until the mid-nineteenth century. Supplies also fluctuated during the year. The best time of year would be the harvest (assuming it was at a reasonable level) – it was a time for celebration and for eating more food than usual. Winter was a bad time but the worst period was the early spring – before the first crops were ready. Food problems were exacerbated by any outbreak of animal disease (which could spread rapidly in what were often undernourished herds) such as the epidemic of rinderpest which spread from Russia into western Europe between 1709 and 1714 and which killed one and a half million cattle. For human beings a permanent state of poor diet led to constant malnutrition, poor health, susceptibility to disease and a continuing high level of mortality. For example, in France in the late seventeenth century between a fifth and a quarter of the population died before their first birthday, half before they were twenty and only one in ten survived until they were sixty.

The endemic level of inadequate diet and malnutrition for most of the people in the world was frequently turned into disaster by the outbreak of famine, which usually stemmed from crop failure following bad weather. In China in the two thousand years between 108 BCE and 1910 there were 1,828 years (over 90 per cent of the total) in which famines involved at least one province in the country. In France between 970 and 1100 there were sixty years of famine at a time of expanding agricultural output and in Tuscany between 1351 and 1767 there were 111 years of famine but only sixteen with good harvests (less than 4 per cent of the total). In France the rate of general famine affecting the whole country was high between the tenth and eighteenth centuries. In that period by far the worst century was the eleventh, with twenty-six famines affecting the whole country, but the lack of any overall improvement in the situation is demonstrated by the fact that the eighteenth century had the second worst record with sixteen famines. The best, or least bad, century was the twelfth (when new land was being brought into production) with only two famines, followed by the fourteenth which had four, an improved performance largely due to the fact that the population was severely reduced by the plague after 1346. When food supplies did fail the scale of the subsequent tragedy could be immense. In 1696–97 in Finland between a quarter and a third of the population died as the result of famine. About the same proportion died in Bengal in 1769–70, when the total number of deaths was around ten million, and Ethiopia suffered a similar loss of its population between 1888 and 1892.

The origins and effects of widespread famine can be illustrated by the events of 1315–17 when medieval Europe experienced its worst ever food shortages at a time when the population was at the very limit that the agricultural system could support. In 1314 the harvest was reasonable but the weather in 1315 was dreadful, being wet in every season. The spring sowing failed in most areas because of waterlogged fields, ploughs stuck in the mud and the hay crop was not properly ripe or dry when cut and stored. Crop yields were about half the normal level and what was available was of low quality. By early 1316 food was already in short supply across the whole of Europe and seed for the next crop was being eaten. The winter and spring were again very wet and the rain continued through the summer, producing another harvest at about half the average level. The resulting food shortage brought catastrophe to most of Europe. Wheat prices rose to three times their normal level and in some places of acute shortage they were over eight times higher. This meant that many of the poor could no longer buy food but even people with money could not buy simply because there was no food available – as Edward II discovered for himself when the court arrived in St Albans in August 1316. The King of Bohemia lost thousands of sheep because he could not buy feed for them. All over Europe animals were killed in large numbers as feed supplies ran out. The poor were dying in large numbers or turned to robbery in an attempt to get food; huge bands of starving peasants swarmed across the countryside. The food that was available was often of very low quality – bread was mixed with pigeon and pig droppings, and animals that had died of disease were eaten, causing outbreaks of disease in the human population. Some people were driven to even more desperate measures, as the many reports of widespread cannibalism in an area stretching from England to Livonia on the Baltic coast bear witness. In Ireland in 1318 bodies were dug up from graves to provide food and in Silesia executed criminals were eaten. There were still numerous cases of cannibalism as late as 1319. Animal diseases, increased by the lack of feed, added to the carnage, killing about 70 per cent of the sheep in some areas and in the four years 1319–22 some two-thirds of Europe's population of oxen died. Only slowly did better weather and improved harvests bring some relief from the catastrophe.

The conditions that agricultural societies, utterly dependent on fragile food supplies, faced during a time of harvest failure is illustrated in graphic and harrowing detail by an account, contained in the parish register of Orslosa in western Sweden, of the terrible famine at the end of the sixteenth century:

In 1596 at midsummer-tide the land was abundantly covered with splendid grass and much corn, so that everybody thought that there would be sufficient corn in the country. But . . . when the people were at Skara market [June], there came so much rain and flood that all the bridges floated away. And with that same flood . . . the water went over all the fields and pastures, so that the corn and the grass were ruined, and thus there was little of both grain and hay . . . In the winter the cattle fell ill from the rotten hay and straw which was taken out of the water . . . It went the same way with the cows and the calves, and the dogs which ate their dead bodies also died. The soil was sick for three years, so that even those who had good farms turned their young people away, and many even their own children, because they were not able to watch the misery of them starving to death in the homes of their fathers and mothers. Afterwards the parents left their house and home going whither they were able, till they lay dead of hunger and starvation . . . People ground and chopped many unsuitable things into bread, such as mash, chaff, bark, buds; many people were so weak and their bodies so swollen that innumerable people died. Many widows, too, were found dead on the ground with red hummock grass, seeds which grew in the fields, and other kinds of grass in their mouths . . . Children starved to death at their mothers' breasts, for they had nothing to give them to suck. Many people, men and women, young and old, were compelled in their hunger to take to stealing . . . At times these and other inflictions came and also the bloody flux [dysentery] which put people in such a plight that countless people died of it.

Europe emerged from the constant threat of famine very gradually. The 1594–97 famine affected the whole of the continent after a succession of four bad harvests that again produced widespread cannibalism and the eating of cats and dogs. That period marked the last severe famine in southern England although the north of the country was still seriously affected in 1623 when, for example, one in eight of the population of Penrith died. France was still badly affected by famine throughout the seventeenth century and the first decade of the eighteenth. In 1693–94 about 10 per cent of the population of northern France died: in the Auvergne it was twice this figure and the total number of deaths may have been about two million. In 1709–10 a famine of similar proportions occurred and also affected Prussia on a large scale.

The last severe crisis to affect the whole of Europe came in 1816–17. Although partly a result of the dislocation caused by the Napoleonic wars, its main origins lay in appalling weather across the continent (it affected the United States too). This was the result of the large amount of volcanic dust in the atmosphere following the eruption of the Tamboro volcano in Indonesia in 1815. Throughout the summer the weather was cool (July was the coldest on record) and rainfall was 50 per cent above average. In England two-thirds of the year's rain fell in the summer months of July, August and September. As a result the harvest was six weeks late and the wine harvest was the latest ever known – extending into November in some places. Crop failure was widespread, wheat prices doubled from an already high level and the real wages of peasants and workers fell drastically. The result was widespread food riots in England, France and Belgium in 1816 and across most of the continent the following year. Death rates rose, though they were not as high as in earlier famines, and the number of epidemics increased, especially in southern Europe.

The last major famine to affect part of Europe came thirty years later in Ireland. The social and economic conditions and fragile agricultural base of the country were in many ways similar to those in the rest of Europe centuries earlier. The root cause of the famine was population pressure on the available land. The population of Ireland rose ten-fold from about 800,000 in 1500 to eight and a half million in 1846 and inheritance practices had produced a large number of very small holdings. There were about 650,000 landless labourers living in permanent destitution and most of the rural population lived in squalid, one-room cabins. The need to provide food from tiny plots of land encouraged adoption of the potato, universally regarded throughout Europe as the food of the poor. If the small plots of land grew potatoes they could provide a family with a monotonous diet but one which supplied a minimal level of nutrition. By the early nineteenth century potatoes took up about 40 per cent of the total crop area in Ireland and constituted the sole food of nearly half the population. However, the potato was not well adapted to growing in the wet climate of Ireland. Disease and poor weather brought about widespread crop failure in 1739–41 when about 500,000 people died. By the 1830s poor harvests were becoming the norm, which meant that even in good years a high proportion of the population would be on the edge of starvation, particularly during the early summer before the new crop was ready.

Catastrophe was triggered by the arrival in June 1845 from America of potato blight, a type of fungal disease which causes rapid deterioration in healthy plants and spreads quickly given the right weather conditions. It can wipe out a whole crop and cause tubers to rot in storage. By August the disease had spread throughout Europe and for the next two years virtually no potatoes were on sale anywhere. The failure of the 1845 crop in Ireland was only partial but that of 1846 was almost total. The human consequences of that failure in Ireland were largely the result of the policies adopted by the British government and their determination not to interfere with the workings of the free market in food. The Corn Laws were repealed to allow the import of grain (partly because the English harvest had been bad). Although a large quantity of grain was imported, the problem was that the impoverished Irish peasantry had little money with which to buy wheat or maize and often no utensils to cook it. At the same time a large part of the Irish grain harvest was exported, often under armed guard. At the height of the famine in the summer of 1846 the government closed all public relief works, such as road building, in order to stop the people becoming dependent on government welfare. The only works that were allowed had to be paid for by the local districts which, because of the famine, had little revenue to fund such projects. The provision of relief food was left to private enterprise or voluntary efforts. Later in the year some works were reinstated but only for about 500,000 people – when even in a good year about two or three million would have been destitute. Government food stocks were only sold at market prices so as not to undercut private traders. Not until the winter of 1847 were soup kitchens provided, although in parallel all relief works were stopped as were any government imports of food. The consequence was that, overall, about one million people died either directly from lack of food or from the subsequent outbreak of diseases that affected the under-nourished population. Another million people emigrated during the famine period and immediately afterwards, often in wretched conditions. By the end of the nineteenth century a further three million people had left Ireland and the population of the island was four and a half million, almost half what it had been in the mid-1840s.

The Irish famine illustrates two important aspects of the problem of food supplies. First, that it was still possible even in a supposedly advanced area of the world such as Europe in the nineteenth century for a million people to starve to death. Second, that famine is not a simple matter of food shortage. There was plenty of food in Ireland – those that died could not afford to buy it and the British government was not prepared to give it to them. The question of who is entitled to obtain food (either through purchase or gift) has formed a central part of contemporary analysis of recent famines in the poorest countries of the world. In cases such as the late-twentieth-century famines in Ethiopia, the Horn of Africa and the Sahel there was plenty of food in the country and exports continued. What happened was that certain groups of people were unable to obtain food, either because their own crops had failed or because they could not afford to buy food at the high prices which were a consequence of the famine. For example in British-controlled Bengal in 1943 (when about three million starved to death) people died even though there were well-stocked food shops that were protected from looting by the police and army. In Russia in 1911–12 there was a major famine affecting sixty regions in the country, but amidst widespread starvation and death a fifth of its grain production was exported to the West (it made up about a quarter of the world's trade in grain). In the Soviet Union, in the early 1930s, peasants died because the government expropriated the harvest to feed the cities and provide exports to fund imports required for the industrialisation drive.

In its deepest sense the problem of famine stems from the change of attitude towards food that goes back to the emergence of agriculture. Gathering and hunting groups do not regard food as something to be traded but as available to all within the group. The problem of entitlement arose once ownership of land and food became the norm when settled, agricultural societies emerged. The dependence of these societies on a limited range of crops increased the risk of failure and when this happened the poorest members of society found themselves unable to obtain food. The problem of access is evident in the earliest accounts of famine in ancient societies although in some cases, and the great medieval famine of 1315–17 is probably an example of this, there was an absolute shortage of food.

The problem of access to food emerges plainly throughout Europe once contemporaneous accounts are available. The frequent response of those suffering from an inability to buy food at prices they could afford was to turn on the merchants, who were accused of withholding supplies from the market or moving them out of the region to areas where they could be sold at even higher prices. Fearing social unrest governments often intervened to try and provide food, particularly in important towns. Occasionally they tried to buy food or force merchants to sell it but the normal response to shortage from ancient Greece and Rome through medieval to early modern Europe was to try and fix prices. It was rarely successful and often counter-productive by ensuring that food was withheld from the market. The scale and frequency of outbreaks of popular discontent is illustrated by the series of riots in various towns across France at the turn of the fifteenth and sixteenth centuries; there were violent protests against merchants moving food out of the area in Bayonne (1488), Montauban and Moissac (1493), Paris (1500), Agen (1514) and Lyons (1517). For at least another three centuries these events were constantly repeated across France (and other parts of Europe). The fears and perceptions of the people about the supply of food and those they believed were withholding it from the market were a central factor in influencing the actions of crowds during a number of crucial episodes during the French Revolution. The same reactions to food shortage were still apparent as late as the agricultural crisis that affected Europe in 1816–17. For example, in 1816 at Dumfries in Scotland, a crowd seized oatmeal at the docks when it was being exported and took it to the town where they forced it to be sold to the public at what they thought were reasonable prices. A year later in Toulouse crowds stopped the export of grain and forced its sale in the town.

New crops, improved farming

An existence under the constant threat of starvation and in the face of the daily reality of an inadequate diet and malnutrition has been the common lot of most of humanity since the development of agriculture. Only slowly, in a few areas of the world, did some societies (principally Western Europe and its colonies in North America and Australasia) emerge from this long struggle to survive. They were able to do so as a result of a combination of developments which made larger quantities of food available to them. Over the centuries a number of small-scale improvements slowly raised agricultural output and productivity. It is possible to trace a slow improvement in European output and efficiency in the six hundred years after 1200: by 1800 yields were about two and a half times higher. This was the result of a wide variety of changes. The range of fodder crops was increased, legumes were more widely used to improve fertility, better breeding of animals and more cross-breeding enhanced output, rotations became more complex and manuring more widespread as more animals could be fed during the winter months. Just as important though was the introduction of new crops and animals, which widened the agricultural base, provided greater stability against failure and improved food output. Some of these changes were the result of the slow diffusion of animals within a limited area – particularly from southern Europe to the northern parts. For example the Romans introduced the chicken from the Mediterranean region to north-west Europe and the white grape to the Moselle valley. Rabbits were introduced into Britain from southern Europe as a domesticated animal in the twelfth century and only subsequently did they escape into the wild. Pheasants and fallow deer were also introduced about this time. But the major changes in distribution of plants and animals in the world took place in two phases – in the Islamic world from the seventh to the tenth century and then following the first European contacts with the Americas after 1492.

The worldwide diffusion of crops and animals

Chief centres for diffusion			
South-east Asia	Europe	Americas	Africa
Sugar cane	Wheat	Maize	Hard Wheat
Rice	Barley	Tobacco	Sorghum
Orange	Oats	Potato	Coffee
Lemon	Sheep	Tomato	
Lime	Cattle	Manioc	
Spinach	Horse	Cocoa	
Aubergine	Pig	Rubber	
Banana	Bee	Pineapple	
	Rabbit	Avocado	
		Peppers	
		Squash	
		Sisal	
		Turkey	

There was no contact between the agricultural system of Europe and the Americas until the sixteenth century CE. The various empires that dominated Mesopotamia had numerous contacts with western India (and often controlled parts of the area) but very little with states further to the east. Even at the time when both the Roman and Han empires were at their height there was only a small amount of contact between them (little of it direct) and they continued to develop in their own way.

The rapid rise of Islam after the death of Muhammad in 632 and the conquest of most of the Near East, north Africa, Spain, Armenia, Georgia, Iran, Afghanistan and north-west India by the early eighth century did not result in a long-lasting unified empire, but it did produce a vast area of considerable cultural uniformity which extended its influence further into south-east Asia and along the east coast of Africa through trading links. The creation of the Islamic empire with its trading networks brought about a substantial diffusion of crops from south-east Asia and India westwards to the Near East, the Mediterranean region and eventually parts of southern Europe. India was a major centre for the transmission and Oman was an important area where the sub-tropical crops of south-east Asia were gradually acclimatised to new growing conditions. From these centres the new crops spread into north Africa, with some reaching as far west as Spain and others penetrating west Africa, and progressing along the trade routes of the east coast of Africa to reach Zanzibar and later Madagascar.

Some of the crops such as the coconut palm could only grow in sub-tropical climates and therefore did not spread beyond the Gulf area and east Africa but most were gradually acclimatised to new growing conditions over a wide area. The most important of these crops for the future history of world agriculture was sugar cane. It was brought from India to Mesopotamia in the seventh century and then it spread westwards to the Levant, Egypt and the islands of the eastern Mediterranean, especially Cyprus, by the tenth century. Almost as important as sugar cane was the slow spread of hard wheat from Ethiopia to the Mediterranean, where it became a staple part of the diet in north Africa (couscous) and in Italy (pasta) after the thirteenth century. Rice also spread from the Near East to large areas of Africa and reached the Po valley in northern Italy by the late fifteenth century. Citrus trees, the sour orange, lemon and lime were brought by Islamic traders from south-east Asia (though they originated in eastern India) to the Mediterranean, where their culti-vation spread quickly, reaching Seville in southern Spain by the tenth century. Sorghum, which had originally spread from Africa to India about 2000 BCE, moved back westwards to north Africa and Spain. Vegetables such as spinach and aubergines were brought from

Iran and India to north Africa and eventually to Spain by the eleventh century.

The new crops which diffused through the Islamic world and beyond did not transform agriculture or provide more than a few mainly supplementary crops. A much more radical change took place after the Spanish conquered the Caribbean islands and the Aztec and Inca empires in the early sixteenth century. The Europeans who went to settle the Americas took with them their own crops and animals (wheat, sugar cane, cattle, sheep and horses). In the process they significantly altered the environment (a story that will be examined in later chapters). Equally important were the consequences of the diffusion of previously unknown American crops, which influenced the agriculture of not just Europe but also the Near East, India, Africa and China. The two most important introductions from the 'New World' were the dietary staples of Mesoamerica and the Andes – maize and potatoes.

Maize was a highly productive crop (the yield was about twice that of wheat) but it took a long time to spread, particularly in Europe, even after the development of new varieties, possibly because of the poor climate in the area during the 'Little Ice Age'. The central point for the dispersal of maize was the Mediterranean area, where it could be grown without difficulty. It proved particularly well suited to conditions in Egypt, where it became a staple crop by the seventeenth century. In Europe it did not reach the Balkans until the eighteenth century and its movement further north had to await both an improvement in the climate and the development of types able to thrive in cooler conditions and a shorter growing season. Maize was not common in India until the early nineteenth century but then spread rapidly. China however adopted maize very early in the sixteenth century and it was soon a primary food crop in the upland areas of the south-west of the country but did not reach the north for another three centuries. The great attraction of maize was its high yield, enabling more people to be fed from the same amount of land. In China this was important at a time when rice-growing had begun to reach its natural limits. In the seventeenth century rice constituted about 70 per cent of the national food output, but this had fallen to less than 40 per cent by the early twentieth century as crops originally of American origin became more important. Maize also arrived in west Africa from Brazil in the sixteenth century and rapidly replaced millet and sorghum to become a central part of the diet because of its higher yields.

Adoption of the potato as a major crop was also a long-drawn-out process. It reached Spain by 1570, England and Germany by the end of the sixteenth century and Scandinavia about a hundred years later, and was introduced into North America from Europe in 1718. It seems that most people did not take readily to eating potatoes and originally they were grown as a fodder crop rather than for human consumption. Only in Ireland and in parts of the Balkans did the potato become a staple food before the nineteenth century. Its chief advantage – the ability to provide a large amount of food from a small area – was widely recognised but the potato was normally only adopted after the failure of other crops.

The other major crop of the Americas to be widely adopted was a tropical plant – manioc. In the early seventeenth century it was taken from Brazil to the mainland of Africa, where its very high yields combined with drought and pest resistance were rapidly appreciated. Once the problems associated with processing it into an edible food by removing the poisons were understood, it formed a vital part of the diet in the tropical areas, becoming particularly important in the nineteenth century. In this period farmers also adopted it in the southern parts of India. Apart from the major crops of maize, potato and manioc, the Americas also provided important supplementary crops. Particularly important was the tomato, adopted initially in the Mediterranean area (and also in India and the Near East) and later further north as varieties capable of growing in a cooler climate and shorter season were developed. (Europe now produces about 40 per cent of the world's tomato crop.) A wide variety of beans (an important source of protein) were rapidly adopted as were flavourings and spices such as chillies, to such an extent that they are now seen as integral parts of the 'local' cuisine of many areas of the world such as India.

One major beneficial effect of the spread of new crops was that the subsistence base of many societies, which was often narrow and therefore highly vulnerable, became wider and this reduced the risk of catastrophic crop failure and famine. Another important gain was nutritional. Not only was more variety available in the poor and very limited diets of the majority of people but many of the foods, particularly plants such as tomatoes and chillies, were rich in vitamins and could help to reduce the risk of some deficiency diseases. However in some areas over-reliance on maize (especially without adopting the American way of preparing and cooking it) produced the deficiency disease pellagra. Despite improvements in reliability, quality and variety associated with the introduction of new crops the basic problem remained that of quantity. Some of the more productive crops, especially maize and potatoes, improved food output but in nearly every society these new crops did not solve the age-old problem of keeping a balance between human numbers and the amount of food that could be produced. In many cases the new higher-yielding crops, rather than providing more calories per head, had the effect of allowing the population to grow more quickly until it reached a point at which it was out of balance with food supplies, as happened among the potato-growing peasantry of Ireland before the disaster of the famine.

The limited escape from the long struggle

Only slowly, and in the relatively recent past, did a few societies begin to escape from a situation where a large part of the population lived on a poor diet barely adequate for minimum subsistence and under the constant threat of starvation. The first to move along this path was the Netherlands in the sixteenth and seventeenth centuries. Here the population doubled from around one million in 1500 to two million in 1650, which required a major adjustment in the agricultural system. Food output was increased as new land was brought into cultivation (some of it was completely new land created from drained marshes or reclaimed from the sea). More intensive farming systems were introduced (using clover, legumes and fodder crops together with increased amounts of manure from the extra animals that could be kept). In this period Dutch agriculture was certainly the most productive in Europe with yields about two-thirds higher than in England. However, much of the extra food required by the rising population was imported from the semi-colonial grain growing areas around the Baltic through the Dutch domination of the trade of the area. Without this additional source of food it is doubtful whether the larger numbers could have been fed.

England gradually adopted many of the improved agricultural methods developed in the Netherlands so that by the middle of the eighteenth century, output was rising faster than population growth. But the gain proved only temporary as the country experienced an unprecedented growth in population between about 1780 and the end of the nineteenth century. Again population growth at about 1 per cent a year threatened to outstrip the capacity of even the improved English agricultural system. The population of England and Wales rose from 7.5 million in 1780 to 14 million in 1831 and 32.5 million in 1901. The response of English agriculture to this rise was much the same as in the past – new land was brought into cultivation (about a 50 per cent rise between 1700 and 1850), fallow land was reduced (equivalent to a further 40 per cent increase) and lower quality food was grown (the area under potatoes increased three-and-a-half-fold in the first half of the nineteenth century). Agricultural productivity rose, partly through the further spread of some of the new ideas introduced in the eighteenth century but also through the introduction of more machinery, new fertilisers and new feeds such as oilcake (which was becoming common in the 1820s). Another bonus (a side effect of the large amount of cheap labour available in the countryside) was achieved by more weeding. Nevertheless in the early nineteenth century the rising population of England was only fed by increasing food imports from Ireland.

The rural population rose rapidly in this period (almost doubling), producing immense social pressure through rising rents and prices. Poverty was on the increase, especially in the east and south of the country where new machinery was introduced, and the number dependent on the potato rose to about two million. It was only the drift of a large part of the rural population into the new industrial towns in the search of employment that avoided a major rural crisis in the early nineteenth century. Even so, by the 1840s, only minor improvements in subsistence levels had been achieved for most of the population. This period, marked by the triumph of the industrial over the agricultural interest with the repeal of the Corn Laws in 1846, saw the swing towards importing food. In the 1840s about 5 per cent of Britain's food was imported (from places other than Ireland). By the end of the century the situation had been transformed: 80 per cent of the grain for human consumption, 40 per cent of the meat and over 70 per cent of dairy products were imported. It was only this use of other countries to supply the food that Britain needed that was to provide a solution to the perennial problem of feeding a rapidly rising population and improving the amount of food available per head.

The real revolution in the European food situation came after about 1850 with large-scale emigration to the colonies together with North and Latin America, the equally large-scale importation of food from the rest of the world and the use of imported resources such as guano from South America and other fertilisers from colonial territories. The key to western Europe's success in breaking free from the long struggle to survive that had dominated the experience of nearly every society since the development of agriculture lay in its changing relationship with the rest of the world and, in particular, its ability to control an increasing share of the world's resources.

7
WAYS OF THOUGHT

Human actions have shaped the environment in which successive generations and different societies have lived. The driving force behind these actions has been simple – the need to feed, clothe and house the increasing population. However, the way in which human beings have thought about the world around them has been important in legitimising their treatment of it. The way of thinking about the world that has become dominant in the last few centuries originated in Europe. Other traditions, particularly those of eastern religions, have provided radically different interpretations, but they have been less influential.

One of the fundamental issues addressed by all traditions is the relationship between humans and the rest of nature. Are humans an integral part of nature or are they separate from it and in some way superior to it? The answer to this question is crucial in determining how different thinkers and religions decide which human actions can be regarded as legitimate or morally justified. From this flow other related questions about whether all the plants and animals in the world are there solely for the benefit of humans and whether humans have a responsibility to guard and take care of the rest of nature (or God's creation). In the last two hundred years or so these religious and philosophical questions have been overtaken by questions of economics – how scarce resources should be used and distributed. Although these may not seem at first glance to be philosophical questions, they have exercised an influence way beyond the sphere of economists and academics. They, too, have had a fundamental impact on the way humans view the world and justify their actions.

Classical thought

The origins of European thought about the relationship between humans and nature can be traced back, as in so many other areas, to the influence of the philosophers of ancient Greece and Rome and the ideas that the Christian Church inherited from its Jewish origins. The strong conviction running through both classical and Christian traditions has been that human beings have been put in a position of dominance over the rest of a subordinate nature. Although the idea that humans have a responsibility to preserve a natural world of which they are merely guardians can be traced, it has remained a minority tradition.

Many thinkers have looked at the world around them and seen that what is now recognised by ecologists as competition and co-operation between plants and animals in ecosystems has produced an ordered world in which every part seems to have a role and purpose within an overall plan. This has led them to argue that such a plan can only have been conceived by a God, or the gods, and they have gone on to speculate about the position of humans within this plan. One of the first to do so was Xenophon in his *Memorabilia*. He attributes to Socrates the argument that everything about humans (such as the eyes and hands) has a purpose and that the gods have also provided everything for the benefit of man. Euthydemus, one of the participants in the fictional debate, responds by saying, 'I begin to doubt whether after all the gods are occupied in any other work than the service of man.' But he is worried that 'the lower animals also enjoy these blessings' until Socrates reassures him that within the overall plan it is clear that these animals are only produced and nourished for the sake of humans. This argument, based on a perceived plan and design within nature, makes frequent reappearances in Western thought until the nineteenth century when developments in scientific theory, notably Darwin's ideas on the origins of species, natural selection and adaptation, served to undermine it.

Over the centuries the argument about design and purpose developed mainly in the sense that the new thinkers proposed new evidence from within nature to illustrate the diverse way in which plants and animals were so well adapted to their particular roles. This tended to reinforce the idea that, since everything had been so well provided for humans, then they must indeed be the most important creatures on earth and so entitled to use nature as they saw fit. Another early expression of this ultimately anthropocentric view of the world is to be found in Aristotle. In *Politics* he argues that plants are made for animals and concludes with the statement, 'Now if Nature makes nothing incomplete, and nothing in vain, the inference must be that she has made all animals for the sake of man.'

With the Stoics, especially Panaetius, and also Cicero, some more subtle arguments were added to this general approach, by emphasising both the aesthetic and utilitarian aspects. To them the world is seen as both beautiful and useful. Beauty is pleasant to look upon and therefore ought to be preserved, but human beings, by meeting their demands for food and goods, improve upon nature. Cicero, for example, makes little distinction between the untouched, natural world and that modified by human action – the two are assumed to be identical. Although the Epicurean thinkers emphasised the harsher aspects of nature – wild beasts, natural disasters, crop failures – that went with the beauty, classical thought was generally characterised by an idea of humans as orderers of nature. Humans were placed on a higher plane than other animals: their ability to create their own world implied a superior position as a finisher of raw creation. The classical thinkers were well aware that human actions were changing the world around them (as Plato makes clear in the passage from *Critias* about deforestation and soil erosion cited earlier on p.75) but apart from those who, like Xenophon and Hesiod, saw human history as a story of decline from a past golden age, they generally regarded human actions in modifying the environment as perfectly natural and beneficial.

Christian thought

The rise of Christianity and its adoption as the state religion of the late Roman empire in the fourth century introduced a new element – Jewish thought – which had until then been confined to a small and previously uninfluential people on the very margins of what contemporaries regarded as the civilised world. The Christians incorporated into their sacred texts earlier Jewish religious books. Genesis, which became the first book of the Christian Bible, contains two different, incompatible creation myths (neither very different from many others in Near Eastern religions) but both give essentially the same view of the relationship between God, human beings and the natural world. In the first of these (Genesis Chapter 1) God creates humans as the climax of his previous five days' work. With divine blessing they are granted domination over the rest of creation:

Be fruitful and multiply, and fill the earth and subdue it; and have dominion over the fish of the sea and over the birds of the air and over every living thing that moves upon the earth . . . Be fruitful and multiply and replenish the earth and subdue it.

In the other myth (Genesis Chapter 2) man is created first and then the Garden of Eden with all the plants and animals and then finally woman. But in this myth the animals are still created for the benefit of humanity and it is Adam who gives them their names. Then God nearly destroys the world in a great flood but in a new contract with Noah and his family as the only human survivors, God once again, this time in even starker terms, gives them and their descendants dominion over the world:

Every moving thing that lives shall be food for you; and as I gave you the green plants, I give you everything . . . The fear of you and the dread of you shall be upon every beast of the earth, and upon every fowl of the air, upon all that moveth upon the earth, and upon all the fishes of the sea; into your hand are they delivered.

This theme appears in many of the other Jewish sacred books incorporated into the Bible. Psalm 8 for example says, 'Thou hast given him [man] dominion over the works of thy hands,' and Psalm 115 has the same message: 'The heavens are the Lord's heavens, but the earth he has given to the sons of men.'

The early and medieval Christian thinkers accepted, almost without demur, the view inherited from Jewish writing that God had given humans the right to exploit plants, animals and the whole world for their benefit. Nature is not seen as sacred and therefore it is open to exploitation by humans without any moral qualms – indeed humans have the right to use it in whatever way they think best. God is typically portrayed as above and separate from the world and what matters above all is the relationship of the individual with God and not with the natural world. Indeed in this way of thought human beings are not seen as part of the natural world since they are unique and have been placed by God on a pedestal above all other living things.

In medieval Europe this view, increasingly influenced after the twelfth and thirteenth centuries by the rediscovery of ancient Greek writers such as Aristotle, provided a widely accepted framework for understanding the world and the human position within it. Although different writers might emphasise different aspects, typically the world was seen as a planned and ordered creation, made by God in his goodness, that would survive until the day of judgement. The ordering of the natural world was evidence of the work of a benevolent creator. Humans enjoyed a wholly different status from other animals as the only creatures with a soul and a life after death. The sixth century writer Cosmas Indicopleustus proclaimed in his *Christian Topography*, that man was 'the king of all things on earth and reigns along with the Lord Christ in the heavens'. Thomas Aquinas encapsulated these views, and provided, as he did for much of medieval Christian thought, their most coherent and logical expression. He argued that there was a hierarchy of beings from the most insignificant up to God but, although there was a reason for the existence of each, the whole plan was known only to God. Humans took their unique place above the animals and their dominion over nature was part of the logical divine plan – rational creatures should rule over irrational ones (animals) and this was well illustrated by the human ability to domesticate animals. Similarly the work of human society in altering nature by extending cultivated areas and using the resources of the world was seen as part of the divine plan to tame the wild and as part of a continuous process of improvement on nature. The Reformation in the sixteenth century brought no fundamental change in this point of view, indeed by re-emphasising the importance of biblical texts it tended to reinforce it. Calvin, one of the leaders of the movement, stands firmly behind the view that it had taken God six days to make the world perfect for the arrival of humans and that God 'created all things for man's sake'. The unique position persistently attributed to humans in Jewish and, following from it, Christian theology produces a highly anthropocentric view of the world which was to have a profound and enduring impact on later European thought even when it was not specifically religious.

Within Judaism and Christianity there were exceptions: the special position of humans in the world was challenged by a few thinkers. In this minority tradition slightly different aspects of the Jewish creation myths were emphasised, particularly that of the Garden of Eden. Humans were depicted as the stewards of God's creation with the task of caring for it on his behalf. This viewpoint is well illustrated by the Jewish thinker Maimonides, who wrote:

It should not be believed that all beings exist for the sake of the existence of man. On the contrary, all other beings, too, have been intended for their own sakes, and not for the sake of something else.

This same note of dissent from the widely accepted tradition can also be found in the ideas of Francis of Assisi. His views reflect much of mainstream Christian thought, in particular the idea that nature, its purpose and order, and the way each part of it was so well adapted to its way of life, demonstrated the work of a benevolent creator. For him the natural world could, therefore, be seen as being, like the Bible though at a lower level, an illustration of the nature of God. Francis saw all creatures as equal parts of this creation, each a part of God's plan but not put there for the utilitarian purposes of humans. This idea, which was revolutionary in its implications and, like much of his thought, widely detested within the Church, was never more than a minority viewpoint within the overall structure of Christian belief, although, against a background of growing public concern about the environment, it has been emphasised more strongly in the last few decades.

The rise of secular thought

The increasingly rapid development of secular thinking in Europe from the sixteenth century produced little alteration in the assumptions and beliefs inherited from classical and medieval thought about the relationship between humans and the natural world. The basic anthropocentrism of Christianity continued, though in a slightly modified form. The world was still seen as part of an organised and rational divine plan. Writers used the increasing amount of biological knowledge to illustrate, with ever more examples, the wisdom of God in providing such a perfectly adapted world. This approach is perhaps best exemplified by the work of John Ray, an English writer at the end of the seventeenth century, and encapsulated in the title of his book *The Wisdom of God Manifested in the Works of the Creation*. Humans were still regarded as placed in a special position by God, above other creatures and able to use them and the natural world for their own benefit. Interventions in, and modifications to, the natural world could readily be interpreted as taking part in God's plan to improve upon creation. This concept of a special role led some to adopt more extreme positions. For example Marsilio Ficino, an Italian Renaissance author, wrote, 'Man not only makes use of the elements, but also adorns them ... man who provides generally for all things, both living and lifeless, is a kind of God.' A similar viewpoint is found in the writings of Sir Matthew Hale, a seventeenth century lawyer, who in *The Primitive Origination of Mankind*, wrote:

> this was the one End of the Creation of Man, namely To be Vicegerent [sic] of Almighty God, in the subordinate Regiment especially of the Animal and Vegetable Provinces.

For these thinkers the unique position of humans and their right to exercise control over nature stemmed from their position as the highest link of what was widely seen at the time as a great chain of being, stretching from the most insignificant to the most important. The idea that humans needed to interfere with, or add the finishing touches to, nature in order to maintain civilisation was a widely held assumption, as was the idea that nature was at its best not in its primitive or rude state but when it was controlled and shaped by humans. From such views it was a small and logical step to welcome the way that increasing human knowledge brought the prospect of greater control over the natural world and to believe that this would be pleasing to God because humans were taking full advantage of the wonders of his creation.

One of the major themes of seventeenth-century writings was the emphasis placed on human domination over nature and the role of humans in completing God's work. Human actions directed towards this end were seen as being beneficial and harmless. At this time, a slowly developing scientific method and a growing body of scientific knowledge were working in the same direction. René Descartes in his *Discourse on Method* emphasised the importance of scientific method through the use of mathematics to measure and quantify, together with a process of analysis designed to reduce wholes to their constituent parts. The widespread adoption of this reductionist approach to scientific enquiry was to have a profound impact on the shaping of European thought generally. It inevitably led to a fragmented view of the world – to a focus on the individual parts of a system rather than on the organic whole. It led to an emphasis on the way in which the constituent elements operated separately rather than the ways in which they interacted, both by competition and co-operation. This tendency was reinforced by a mechanistic approach to natural phenomena, which can again be traced back to Descartes who wrote, 'I do not recognise any difference between the machines made by craftsmen and the various bodies that nature alone composes.' Animals were therefore mere machines. Whatever new intellectual methods

Descartes introduced, God was still central to his view of the world and humans still occupied a special place in that scheme, set apart through the possession of minds and souls which enabled them to dominate nature. His mechanistic view of the world seemed to be vindicated by the spectacular success of Newton in the late seventeenth century in applying physical laws, such as that governing the force of gravity, to explain the workings of the universe. A popular new image of the relationship between God, humans and nature was launched. For the next two hundred years or more there are frequent references to the idea of God as the great designer of a machine the workings of which humans could, through their God-given intellectual faculties, seek to understand. Writing two hundred years after Descartes, the American economist H. C. Carey could still declare in 1848 that 'the earth is a great machine, given to man to be fashioned to his purpose'.

The idea that the application of science is a powerful aid to progress and a vital tool to enable humans to dominate the world is strongly expounded in the works of Francis Bacon. He started from the traditional view when he wrote that 'the world is made for man, not man for the world', and:

> Man, if we look to final causes, may be regarded as the centre of the world, insomuch that if man were taken away from the world, the rest would seem to be all astray, without aim or purpose.

He went on to urge that the whole point of scientific endeavour was to restore the dominion over the world that had been lost with the fall of Adam and Eve in the Garden of Eden. As he wrote in the *Novum Organum*: 'Let the human race recover that right over Nature which belongs to it by divine bequest.' The way in which that right could be regained increasingly used the language of violence. A contemporary poet, Abraham Crowley, wrote of 'the husbandman's innocent wars, on beasts and birds', and words such as 'mastery', 'conquest' and 'dominion' were common in describing human efforts to control nature. Bacon in his correspondence with John Beale said that the purpose of studying the natural world was that 'Nature being known, it may be master'd, managed and used in the service of human life'. Descartes too saw the purpose of science and increasing human knowledge as being part of a wider struggle so that 'we can . . . employ them in all those uses to which they are adapted, and thus render ourselves the masters and possessors of nature'.

During the second half of the eighteenth century the idea of a perfectly designed world came under attack, notably in Voltaire's satirical attack on Leibnizian optimism in *Candide*. This vein in the work of philosophers was assisted by later developments in scientific thought. Charles Darwin's *Origin of the Species*, published in 1859, opened up a debate about the origins of man, undermined the orthodox view of divine creation and put forward the idea of the natural selection of characteristics that helped survival in a highly competitive world. In the hands of philosophers such as Herbert Spencer, this theory about the origin of species was transmuted into a statement about the nature of society and the morality of human actions. Spencer in effect refurbished the old ideas about the special position of humans compared with the rest of the natural world but now in a Darwinian form. For Spencer all life was a struggle for 'the survival of the fittest'. Humans had to struggle against nature in order to survive and in doing so they had demonstrated their fitness to be on the topmost rung of the ladder. Variations on the old theme appear in many other nineteenth-century writings, evidence of the potent attraction of an idea which not only powerfully reinforced a sense of purpose and of superiority but also readily justified areas of traditional human interference with the natural world as well as new activities such as the vast increase of industrial output. The philosopher Immanuel Kant wrote, 'As the single being upon the earth that possesses understanding, he [man] is certainly titular lord of nature . . . he is born to be its ultimate end.' Given this position, Kant felt sure that the human relationship with nature could not be subject to any moral censure. Other familiar views also appear again in only slightly different guises in a number of modern thinkers. For example John Stuart Mill in his *Three Essays on Religion* wrote of nature: 'Her powers are often towards man in the position of enemies, from whom he must wrest, by force and ingenuity, what little he can for his own use.' The founder of psychoanalysis, Sigmund Freud, said in *Civilisation and its Discontents* that the human ideal was 'combining with the rest of the human community and taking up the attack on nature, thus forcing it to obey human will, under the guidance of science'.

The idea of progress

As well as the strong sense of continuity that emerges in European thinking about the relationship between humans and the natural world over many centuries it is also possible to trace the emergence of a powerful new element – the idea of *progress*. This idea is such a fundamental constituent of modern thought that it is very difficult to appreciate just how recent it is or how other societies viewed the world before the idea took hold. The ancient world had little concept of the idea of progress: history was usually seen as having no particular direction at all or, if it did have one, to be a story of a decline from a golden age. For writers such as Xenophon, Hesiod and Empedocles human society had been steadily decaying from an age of gold to one of silver to the age of iron in which they lived. Other societies, for example the Cherokee Indians, have had the same idea of a lost golden age as have the Chinese thinkers such as the Taoist Chuang Tsu. Both the early Christians and those of medieval Europe saw the history of the world as one of decline, of innocence lost in the Garden of Eden, never to be regained on earth. They also believed that a day of judgement would occur in the not very distant future that would mark, not just the final trial for both the living and the dead, but also the end of all earthly history. Many secular figures in the fifteenth and sixteenth centuries believed that, although they had rediscovered much of the culture of the ancient world, their own age was far inferior to the revered Greeks and Romans, not just in cultural terms, but also in civic virtue and valour. Within an intellectual framework of this sort it was almost impossible to interpret human history as one of uninterrupted progress.

It is not until the end of the seventeenth century that the continuing increase in scientific knowledge and the steady advance of technology (both areas in which western Europe was by then demonstrably in advance of ancient societies) began to convince some thinkers that history might be a chronicle of progress rather than of decay. Gradually it came to be commonly accepted among European intellectuals that history was the story of a series of irreversible changes in only one direction – continual improvement. The eighteenth century was marked by a wave of optimism about the future and the inevitability of progress in every field. In 1793 the English writer William Godwin in his *Political Justice* wrote in glowing terms of the prospects:

Three-fourths of the habitable globe, are now uncultivated. The improvements to be made in cultivation, and the augmentations the earth is capable of receiving in the article of productiveness, cannot, as yet, be reduced to any limits of calculations. Myriads of centuries of still increasing population may pass away, and the earth be yet found sufficient for the support of its inhabitants.

In the same year the Marquis de Condorcet published his *Sketch for a Historical Picture of the Progress of the Human Mind*. The book was a statement of his belief in human potential and the unlimited scope for human progress:

> The perfectibility of man is truly indefinite; and that the progress of this perfectibility, from now onwards independent of any power that might wish to halt it, has no limit than the duration of the globe upon which nature has cast us . . . this progress . . . will never be reversed as long as the earth occupies its present place in the system of the universe.

It is perhaps possible that if Condorcet had known that he was to die the next year in jail during the period of terror in the French Revolution he might have taken a less sanguine view of human nature and history.

His contemporary, the Reverend Thomas Malthus, did take a much darker view of human history. His *Essay on the Principle of Population*, published in 1798, argued that there was a permanent cycle in history. Human numbers increased until they were too great for the available food supply to support, at which point famine and disease would reduce the population until it was again in balance with the amount of food that could be produced. Malthus could see no way out of this terrible cycle. During the nineteenth century this Malthusian view of history was largely ignored and the idea of progress became almost universally accepted as the natural, unspoken assumption in Western thought. Such optimism seemed justified by the huge material progress made by western Europe and North America in the nineteenth century: their ability to feed an ever larger population, the unprecedented growth of cities, the development of industry, the steady emergence of new inventions and the ever rising standard of living. The acceptance of progress was at the heart of Western thought in men such as Saint-Simon, Comte, Spencer and John Stuart Mill. Its strongest manifestation is in the thought of Marx and Engels with their idea of the inevitable progress of human societies through different economic foundations and their related power structures. Human history was, they argued, the march of progress from tribal through feudal and capitalist societies until its climax in the inevitable victory of the proletariat and socialism. By the late nineteenth century the idea of progress had become a part of popular culture as well – something so ingrained in the unspoken assumptions of nearly all Western views of the world that almost any change could be equated with progress. Although severely battered by the events of the twentieth century, it remains the dominant view about the nature of human history and the prospects facing contemporary societies.

Other traditions

Although men like Francis of Assisi emphasised the stewardship of man over God's creation he still retained the idea that humans were placed above the rest of creation. Other religious traditions did not put humans in such a position. In the last fifty years or so, as the environmental impact of Western ways of thought has become more apparent, there has been an increasing interest in other ways of thinking about the world. Chinese Taoist thought emphasised the idea of a balance of forces, within both the individual and society. The aim of both should be to live in a balanced and harmonious way with the natural world. This way of thought was also holistic rather than the reductionist approach which had dominated Western thought after Descartes. The Indian tradition, exemplified in writings such as the *Upanishads* and religions such as Jainism and Buddhism, is built on a radically different view of the world from the Judaeo-Christian perspective (that also predominates in Islam). All creatures, including humans, are seen as part of a world characterised by suffering in which all need release from a continuing cycle of existence in which, through karma and reincarnation, actions in one life affect the next. Overwhelming importance is attached to the need for compassion for all creatures trapped in this cycle of existence. Humans are in a privileged position but not because they are rulers of the world on behalf of a god. Rather they are the only creatures capable of achieving enlightenment and therefore they should take the rare opportunity they have to escape from the cycle of suffering.

Although generalisation is dangerous it is clear that the world view of the 'Eastern' religions, developed centuries before the emergence of Christianity, does emphasise a less aggressive approach by humans to the world they find around them. It does not conceive of humans as placed above and beyond a natural world which is theirs to exploit. Humans are only a small part of a much greater whole. They are set apart by greater intellectual and spiritual capabilities but this should be directed towards the goal of enlightenment and as part of the path to achieving this they should not take life unnecessarily. The ideas of suffering and universal compassion are far more important than dominion over the natural world.

The available evidence about the beliefs of gathering and hunting groups suggests, not surprisingly given their close relationship with the natural world, a variety of beliefs about the interdependence of humans, plants and animals. In general they do not see a distinction between nature and society. Their world is seen as being composed of people, animals and inanimate things all of which are linked together in a single whole without distinct categories. Perhaps the most explicit statement of these beliefs comes from Chief Seattle of the Squamish tribe, who wrote to the president of the United States in 1854 to protest (without effect) on the way the white settlers were treating the native peoples and the environment of North America:

> What is man without the beasts? If all the beasts were gone, men would die from great loneliness of spirit, for whatever happens to the beasts also happens to man . . . Teach your children what we have taught our children: that the earth is their mother. Whatever befalls the earth befalls the sons of the earth. If men spit upon the ground, they spit upon themselves. The earth does not belong to man, man belongs to the earth. Man does not weave the web of life, he is merely a strand in it. Whatever he does to the web, he does to himself.

It is difficult to imagine a statement more at odds with the prevailing Western view about the relationship between humans and the world around them.

All of these views differ radically from the way in which modern Western thought has been built up over the last two thousand years or so. It is therefore obvious why they have been influential in shaping the developing environmental movement of the last half century or so. However, it is far from clear how influential they have been in creating any radically different ways in which human societies have acted. The great empires that dominated India at various times and the Chinese society and state have been just as environmentally destructive as Western societies. They too have cleared forests, ploughed up land and used resources as they saw fit. They too were driven by deeper economic forces.

The emergence of modern economic thought

While the influence of religious and philosophical thought continues to colour the way in which people view the world around them a new force has emerged in the last two centuries or so – the ideas of economics. This 'social science' is a relative newcomer among academic disciplines and is ridden with controversy among its practitioners. It has, however, come to exercise a profound influence over the way in which the world is seen and analysed and it is central to the way in which the environment is valued and treated. Not only the professed economic system of a society, but the hidden assumptions of economics and the value systems that it enshrines, are central to understanding the modern view of the relationship between humans and the natural world.

Modern economies and societies are dominated by market mechanisms to such an extent that it is difficult to appreciate that this is a comparatively recent phenomenon. Only in the last few centuries have societies developed that were controlled by the free (or relatively free) operation of markets for land, labour and capital with other considerations relegated to a subordinate position. For thousands of years societies were organised in very different ways. Gathering and hunting groups normally had few possessions, the idea of land ownership was completely alien to them (hence their inability to come to terms with European concepts) and goods vital to the group as a whole (for example food) were regarded as available to all and shared if necessary. Many of the early states and empires had systems for the redistribution of food between members of the community and the population of Rome was fed through the free distribution of grain by the state. In these societies, and in more modern ones until the early nineteenth century, most people lived in small units, self-contained for most purposes, engaged in semi-subsistence farming and using the small surplus they produced to buy or barter for items produced by local craftsmen. They had little or no contact with a money economy. In medieval Europe (and many other societies) land was rarely traded – it was held in return for military service or rent provided in labour or goods rather than money. Only very slowly over many centuries did a market in land emerge although most people did not have the wealth to participate in it. The market for labour was also far from free. It was largely organised in guilds (of both owners and workers) that regulated working hours, apprenticeships and output. The state also regulated large areas of activity, particularly in trade, and restricted the operation of markets in what were regarded as the wider national and strategic interest.

Only slowly, over many centuries, did markets emerge in Europe as trade expanded, wealth increased, land was bought and sold and labour became more of a commodity to be bought and sold. By the eighteenth century, particularly in the richer areas of western Europe, a relatively free market in land, labour and capital was becoming predominant. The first writer to provide a sytematic analysis and apologia for the revolution in behaviour and social organisation that was taking place was Adam Smith (now regarded as the founder of modern economics) in his *Inquiry into the Nature and Causes of the Wealth of Nations* published in 1776. He argued that individuals acting in their own self-interest, whether as producers or consumers, but regulated by competition between them, would produce the most beneficial outcome for society as a whole. He argued that the operation of a hidden, 'guiding hand' would ensure that the sum of the self-interest of every individual in society was equivalent to the best outcome for the whole of society. He shared the common eighteenth-century belief in the inevitability of progress and saw society as engaged in a process of continual improvement through investment, greater productivity and the accumulation of individual wealth. At the core of his ideas was the concept of competition and of a self-regulating market operating through prices set by a balance between supply and demand so as to produce the optimum allocation of resources through both the economy and society as a whole. Smith, together with other writers such as Ricardo and Mill (now categorised as 'classical' economists), placed the production of goods at the centre of economics. From the start, therefore, economists concentrated much of their effort on how the production of goods was organised and on the need for a free market in the various factors that went into production – land, labour and capital.

The assumptions of classical economics have, over the last two centuries, become deeply ingrained into modern industrialised societies. However, markets rarely operated in the manner described by Smith and others. Monopolies, oligopolies, price fixing and other forms of anti-competitive agreements were common. In some sectors markets might operate reasonably efficiently. In others, such as agriculture where there was a major time lag between supply and demand, they did not. More important, it was far from clear that when aggregated together the sum of individual markets did produce the optimal solution for society as a whole. Regular cycles of boom and slump also led to chronic social problems. The worst of these, the Great Depression of the 1930s, led to the emergence of a new type of economics, named after its main progenitor, John Maynard Keynes. It accepted the underlying values of liberal economics but argued that faults in the operation of market mechanisms should be compensated by government activity (through taxation and government spending) to regulate the overall level of demand in the economy and so avoid the worst of the boom and slump cycles. Keynesian economics brought in its train new methods for assessing the level of activity in an economy – gross domestic production (GDP) as a measure of amount of production, consumption and investment. The success of an economy is now generally judged by the rate at which GDP is increasing every year and it is still widely regarded as the best measure of economic and social progress.

Marxist economics

In some ways the most radical dissent from classical and liberal economics came from Marx and Engels in their theoretical works written in the mid-nineteenth century. This was carried further by their followers, in particular Lenin and the revolutionaries who took power in the Soviet Union after 1917. However, despite their antipathy to free-market capitalism, Marx and Engels, especially in their attitudes to the environment, shared many of the assumptions of classical economics and the idea of progress and took them to extremes not found elsewhere. They argued that the 'value' of any product came from the amount of human labour put into producing it and therefore ignored the value of the resources involved in its production. In some of his early works such as the *Economic and Philosophical Manuscripts* written in the early 1840s, Marx adopts a more idealistic view of nature than in his later work. But even in these works Marx adopted without question the common view that nature only had meaning in terms of human requirements. For example, he wrote, 'Nature taken abstractedly, for itself, and fixedly isolated from man, is nothing for man.' In his later works Marx takes this argument further and argues that the 'great civilising influence of capital is that it rejects the 'deification of nature' so that 'nature becomes, for the first time, simply an object for mankind, purely a matter of utility'. Engels argued in a similar way when he wrote that in the future humans will be able to 'learn and hence control even the more remote natural consequences of at least our most ordinary productive activities'.

Marx, Engels and, in particular, Lenin rejected any of the more libertarian socialist ideas that increased happiness might be obtained by limiting or reducing production and consumption and seeking a simpler, more harmonious life. Instead they saw the first aim of communism as raising the proletariat to the level of consumption achieved by the bourgeoisie in nineteenth-century Europe. Socialism was to be built on the productive capacity of an advanced industrial society organised through the factory system with a large degree of state power. Communism would be more efficient than capitalism in producing goods. The possibilities open to such a society were therefore limitless. As Engels declared, 'The productivity of land can be infinitely increased by the application of capital, labour and science.' Lenin and his successor Stalin were determined to give the development of industry the highest priority in the new state. The environmental consequences could be disregarded in the context of a materialist philosophy which saw the highest human achievement as the ability to alter the natural world as required. It focused on a glowing future of continual progress as evoked by the Soviet historian, M. N. Pokrovsky, in his *Brief History of Russia,* published in 1931:

It is easy to foresee that in future, when science and technique have attained to a perfection which we are as yet unable to visualise, nature will become soft wax in his [man's] hands which he will be able to cast into whatever form he chooses.

The strength of the Marxist view of the ability of humans to shape nature in any way that they wanted can be judged from the fact that Pokrovsky's book was later condemned for giving the role of the environment too much importance in human history!

The critique of economics

There is a fundamental flaw in classical economics and all the modern systems derived from it – Marxist, welfare, Keynesian and modern ultra-liberal economics. They all deal only with the secondary problem of the distribution of scarce resources between competing ends and how the optimal allocation of resources can be achieved. The resources of the earth are treated as capital – a set of assets to be turned into a source of profit. Trees, wildlife, minerals, water and soil are simply commodities to be sold or developed. More important, their price is simply the cost of extracting them and turning them into marketable commodities. Yet this view ignores the basic fact that the resources of the earth are not just *scarce*, they are *finite*. Economics is unable to incorporate this fact into its analysis and so the economic systems based upon it encourage both the producer and the consumer to use up resources at whatever rate current market conditions dictate. Economics assumes, in defiance of all logic, that resources, in terms of land, materials and energy, are inexhaustible and that growth in the overall level of the economy can continue for ever. It also assumes that as one material or source of energy becomes scarce its price will rise and this will encourage the development of substitutes. In the short term this is certainly possible but the process cannot be continued indefinitely into the future. There is no way in economics that current prices can take account of the problems that will have to be faced in the future. (Since there is in theory an almost infinite future ahead of human societies current prices would have to be infinite in order to conserve them for that future.) Instead, economics argues that since the most rational action for individuals and hence societies is to pursue immediate self-interest then there is no need to take account of posterity.

Another problem for economics is that markets and therefore prices do not reflect true costs. Some commodities such as air never enter a market mechanism and are therefore treated as free goods available to all. However, as early industrial societies found out, if commodities such as air and water are treated as free then levels of pollution will rise rapidly because firms do not have to pay a price for the smoke and gases they put into the air or for the waste products and effluent that they tip into rivers. The rest of society does, however, bear the cost of such activities – water not fit for drinking, air not fit to breathe or even simple problems such as people's washing becoming dirty through air pollution. These 'externalities' can only be controlled through government regulation.

Many of these problems are also reflected in the failure of GDP to measure the real state of society and the economy. GDP does not measure some sorts of activity and economic contributions, for example housework if it is done without wages being involved, subsistence agriculture and voluntary community work. The 'black economy' (which is very large in some countries) is by definition not measured. There are also problems with what GDP does measure. It includes many items that are not benefits to society as a whole. For example, the shorter the life of cars and the more frequently they break down, the greater will be the amount of activity in an economy and the level of GDP will rise. This ignores the fact that individuals (unless employed in a car factory or a garage) would be better off if they had more reliable, longer-lasting cars. GDP also takes no account of some of the effects of using cars – higher levels of pollution and greater traffic congestion and delays. Other social problems such as poor housing and health care are also not included. On the other hand some social problems actually lead to a rise in GDP. In the United States as levels of crime continue to rise one of the major growth industries has been in crime prevention and security. This is now worth $65 billion a year and makes its contribution to GDP but it is doubtful whether this measures an overall benefit to society. Similarly over-eating, obesity and an obsession with weight have created a diet industry worth $32 billion a year and the stresses of modern life have resulted in sales of Prozac worth $1.2 billion a year, both of which increase GDP. The growing doubts about the failure of GDP to reflect social costs and problems have led in the last couple of decades to attempts to create better measures of the real increase in the standard of living, taking account of factors such as pollution, traffic congestion, crime, health care and access to education. One of these measures is the Genuine Progress Indicator or GPI and it reflects a very different picture from GDP. In the second half of the twentieth century the GDP per head in the United States rose from $12,000 to $35,000, suggesting a near tripling in the standard of living. However, the GPI would suggest that the standard of living taking account of wider social problems increased by only 8 per cent.

Only a small minority of economists have tried to deal with these deeper questions and raised doubts about the ability of conventional economics to deal with environmental questions. E. F. Schumacher in his book *Small is Beautiful*, published in the early 1970s, argued for an approach to economics 'as if people mattered'. He never created a coherent economic theory but argued that efforts should be concentrated on questions about the appropriate size and scale for activities and technology and identifying the real needs of people rather than assuming they would be met by ever increasing levels of production and consumption. He characterised this approach as 'Buddhist Economics' and although his writings became international bestsellers his ideas have had little impact on policy makers or mainstream economics. Another trenchant analysis of the shortcomings of traditional economics was made by Hazel Henderson in her book *Creating Alternative Futures* published in 1978. Like Schumacher she criticised the fragmentation of economic thought, its failure to take into account humanity's dependence on the natural world and its even greater failure to recognise the values and unspoken assumptions that lie deeply embedded within it. The result, according to Henderson, is that:

> Economics has enthroned some of our most unattractive predispositions: material acquisitiveness, competition, gluttony, pride, selfishness, shortsightedness, and just plain greed.

Modern liberal economics

From the mid-1970s there has been a strong reaction against Keynesian economics and a return to ultra-liberal economic thinking. In part this has reflected changes in the real world. Governments, even in the industrialised world, have become increasingly less powerful in economic terms. In the 1970s they lost control over the exchange rate for their currencies and gradually gave up the attempt to manipulate and regulate economic activity. They could no longer attempt to control the free flow of finance and speculation around the globe. Privatisation of nationalised industries and major reductions in the power of trade unions followed and the one tool that could be used to influence the level of economic activity – interest rates – was handed over to central banks to determine. Increasingly governments became competitors and suitors for the favours of transnational corporations offering favours in return for investment.

The collapse of the Soviet Union and its satellite states in the late 1980s and early 1990s ended the world's brief and limited flirtation with state-controlled socialism. Unrestrained free-market capitalism became the only acceptable ideology. Governments had little choice but to go along with this trend and attempt to gain what benefits they could from an increasingly globalised economic system. International institutions such as the World Bank and the International Monetary Fund enthusiastically backed this process and used their power when they could to ensure that this ideology was accepted everywhere as the only route to economic success. In the mid-1990s a World Trade Organisation was set up with the aim of liberalising world trade to the maximum amount possible and it was given the power to enforce its decisions on governments. The change in the balance of power between the interests of transnational corporations and those of governments is illustrated by the fact that governments in the late 1940s had specifically rejected such an organisation because it would have too much power.

All of these developments had significant implications for the environment. The actors in the world economy, in particular transnational corporations, may all be acting logically within the confines of liberal, free-market capitalism in trying to maximise profits in whatever way they can through increasing production and consumption. However, to the outsider it was clear that the system was on autopilot and that there was no aim other than to continue growth as far into the future as possible. Yet the continued consumption of resources and the output of different types of pollution might not be sustainable. The problem was that there was a disjunction between governments and the economy. Although governments might, in some cases, agree to regulate and reduce certain types of pollution they had little or no power to ensure that the economic system followed suit. These problems were exacerbated by the nature of modern democratic politics in which governments were largely judged by their ability to sustain continued economic growth and not interfere with the 'right' of consumers to buy and use products such as cars in whatever way they saw fit.

The policies of international institutions such as the World Bank, IMF and WTO were also often environmentally damaging. The rules of the WTO specifically exclude environmental protection as a reason for discrimination in trade. The inevitable result has been a tendency to reduce environmental standards to the minimum. This can be seen by some the proposals put forward at the WTO meeting in Hong Kong in late 2005 to end what was seen as environmental 'discrimination' in trade. Argentina proposed that the WTO should require all countries to remove the 'recyclable logo' from cans and bottles. Thailand wanted there to be no labelling of products to identify that they contained genetically modified ingredients and they also wanted to end all 'excessive' safety testing for toxins in shellfish. China objected to the imposition of energy efficiency standards for hot-water boilers and air-conditioning units. By the early twenty-first century the pressures of free-market capitalism, enshrined as the dominant ideology of the world, took little account of environmental constraints and possible future problems.

8

THE RAPE OF THE WORLD

Over the last 10,000 years human activities have brought about major changes in the world's ecosystems. The expansion of settlements, the creation of fields and pastures for agriculture, the clearing of forests and the draining of marshes and wetlands have all reduced the habitats of almost every type of plant and animal. The deliberate hunting of animals for food, furs and other products (and in many cases for 'sport') has drastically reduced the numbers of many species and driven others to extinction. In addition humans have moved plants and animals around the world, often with unexpected and near catastrophic results. The scale of wildlife losses in earlier periods is difficult to assess. There is more evidence, though still very patchy, for the period after 1600 but it was not until the twentieth century that detailed research was undertaken, largely prompted by growing awareness of the scale of the losses. Overall there is no doubt that the scale of destruction has increased markedly in the last couple of centuries.

A reduction in wildlife habitats and the extinction of species on a local scale can be identified from the time of the first human settlements. In the Nile valley the extension of the cultivated area and the draining of marshland, combined with hunting, led to the elimination of many species originally native to the area. By the time of the Old Kingdom (2950–2350 BCE), animals such as elephants, rhinoceroses and giraffes had disappeared from the valley. The spread of settlement around the Mediterranean produced the same results with the animals at the top of the food chain being the most vulnerable. By about 200 BCE the lion and leopard were extinct in Greece and the coastal areas of Anatolia, and wolves and jackals were confined to the remote mountainous areas. The trapping of beavers in northern Greece had also driven them to extinction. The Roman addiction to killing wild animals in games and other spectacles added to the slaughter. The scale of destruction, which continued across several centuries, can be guessed from the fact that 9,000 captured wild animals were killed during the 100 day celebration to mark the dedication of the Colosseum in Rome and 11,000 were slaughtered to mark Trajan's conquest of the new province of Dacia. By the early centuries CE the elephant, rhinoceros and zebra were extinct in north Africa, the hippopotamus in the lower Nile and the tiger in northern Iran and Mesopotamia.

Europe

The great festivals of the Roman empire stopped in western Europe after the fifth century but the destruction of wildlife continued in other ways and eventually on a far larger scale. Early medieval Europe consisted of large areas of almost undisturbed natural ecosystems with a small population living in scattered settlements. The steady expansion of settlement, particularly after about 1000 CE, gradually reduced the habitats on which plants and animals depended for their survival. Whole species became extinct and others disappeared from large areas or were severely reduced in numbers. The aurochs (the wild ancestor of modern cattle) was a woodland animal that suffered badly from deforestation. It was extinct in Britain as early as 2000 BCE and slowly disappeared from the rest of the continent. The last known specimen died in the Jaktorowa forest in Poland in 1627. The European bison was still found in the early medieval period across a wide area of what is now Belgium and Germany. By the eighteenth century it was only found in eastern Europe and the last wild animal died in the Bialowieza forest in Poland in 1920. The great auk, a flightless seabird, was once found in huge colonies along the Atlantic coasts of Scotland and Iceland. It was highly vulnerable – in 1540 two ships were filled with freshly killed auks in just half an hour (producing five tonnes of salted birds) and the sailors killed more birds to eat fresh. Sailors also ate the eggs and because the auk only laid one egg a year its ability to breed was easily undermined. By the eighteenth century the bird was rare around the Scottish coast. The last pair were killed in Iceland in 1844.

Many more species which were once common in the whole of Europe have become extinct across large areas. Wolves were still seen in large numbers throughout western Europe until about four centuries ago. As late as 1420 and 1438 wolf packs were seen in the streets of Paris in daylight. A century later enough still survived for Francis I to organise official hunts. In 1640 there are accounts of wolves coming down from the hills of the Jura to terrorise the inhabitants of Besançon. In Britain there were still enough wolves to undertake full-scale, official hunts in Scotland during the sixteenth century. The last recorded sighting of a wolf in England came in 1486, in 1576 in Wales, in 1743 in Scotland and during the early nineteenth century in Ireland. The brown bear was also common across medieval western Europe (although it was extinct in Britain by the tenth century). Numbers declined steadily through hunting and habitat destruction and it now survives only in a few remote mountainous areas. The beaver was also common but was trapped for its fur and died out in Britain in the thirteenth century and slightly later across most of the rest of Europe.

Britain, as one of the first countries in Europe to be densely settled and industrialised, illustrates many of the forces that affected the whole of the continent. The crane became extinct in the sixteenth century and the sea eagle, which was still common in the 1870s, died out in the twentieth century. The conversion of grassland to arable farms, combined with extensive hunting, drove the great bustard to extinction in 1838. The osprey, which was so common in the eighteenth century that it was not a matter for comment, was, wrongly, believed to be a major predator on salmon. With the increase in salmon fishing in the nineteenth century it was ruthlessly hunted and driven to extinction. A few pairs returned in the twentieth century but it now only breeds under extensive protection. The game bird of the Scottish forests, the capercaillie, was once common across the whole country but the steady clearing of the forests reduced its numbers and extent. By the seventeenth century it was only found north of the river Tay and the last example was seen in Inverness-shire in 1762. It was reintroduced in 1837 but further forest clearance reduced its numbers to about 2,000, which may be too low for survival. The golden eagle was still found in Derbyshire in the early seventeenth century and in the Cheviots in the nineteenth century. It now only survives in remote areas of the Scottish highlands. The chough was common in inland areas of Scotland until the early nineteenth century but it is now found in only a few isolated areas along the coast. The red kite was once one of the commonest birds of prey – in the sixteenth century it was still found in the centre of London scavenging on rubbish in the streets. Over the next few centuries it was hunted and its habitats destroyed. By the early twentieth century only about five birds survived. However, its numbers slowly grew but it was confined to central Wales until the late twentieth century when it was deliberately introduced elsewhere.

It is clear from contemporaneous texts that the idea of conservation and the preservation of wildlife was noticeable by its absence until the twentieth century. The general attitude towards the natural world was well summed up by a seventeenth-century English cleric, Edmund Hickeringell, who wrote that: 'So noisome and offensive are some animals to human kind, that it concerns all mankind to get quit of the annoyance, with as speedy a riddance and despatch as may be, by any lawful means.' In 1668 John Worlidge published his *Systema Agriculturae* which contained a calendar for the year with the following tasks for animals regarded as 'harmful' to farming:

February: pick up all the snails you can find, and destroy frogs and their spawn
April: gather up worms and snails
June: destroy ants
July: kill . . . wasps and flies

Official policy was equally destructive. In 1533 the English parliament passed an act (the Scottish parliament had passed a similar act in 1424) requiring all parishes to have nets to catch rooks, crows and choughs. This was extended in 1566 so that church wardens were authorised to pay for the corpses of foxes, polecats, weasels, stoats, otters, hedgehogs, rats, mice, moles, hawks, buzzards, ospreys, jays, ravens and kingfishers. In every area of England large hunts were carried out to try to exterminate various animals. In 1732 at Prestbury in Cheshire 5,480 moles were killed, at Northill in Bedfordshire between 1764 and 1774, 14,000 sparrows were killed (and 3,500 eggs destroyed) and at Deeping St James in Lincolnshire in 1779, 4,152 sparrows were killed. On one estate in Sutherland in the early nineteenth century, 550 kingfishers were killed in just three years. Also on two estates in Sutherland 295 adult and 60 young golden eagles were killed (and an unknown number of eggs destroyed) in the seven years after 1819 in an attempt to preserve fish and game for 'sport'. During the First World War the government ordered the destruction of sparrows in an attempt to reduce crop losses. Special clubs were set up and their success can be judged from the fact that the one at Tring in Hertfordshire killed 39,000 in three years.

Hunting for 'sport' across Europe has been highly destructive but it is impossible to measure the scale of destruction across the centuries. A few isolated examples provide a clue to what happened year after year. At Wainfleet in the mid-eighteenth century, 31,200 ducks were killed in a single year. At one village in Lincolnshire an average of 3,000 wildfowl were killed every year between 1833 and 1868. Trapping and killing of migratory birds remains common across Europe – probably about one in six of them are killed every year. (In Italy about 200 million birds are shot every year.) Shooting birds for food has always been on a massive scale – in 1898 the Paris markets sold 270,000 wild quail. The range of birds eaten in the past was also much wider than today: it included curlews, plovers, blackbirds, larks, thrushes and even gannets. During the nineteenth century about 1,300 gannets a year were killed on the Bass Rock alone. Wild bird eggs were also considered a delicacy – by the 1870s the lapwing was almost extinct in the north of England because of the demand for its eggs. Other demands were equally destructive. The great crested grebe was driven to the point of extinction in nineteenth-century Britain because of the demand for its down to make ladies' muffs. In 1850 the large copper butterfly became extinct in England because people collected its caterpillars. The nineteenth- and early twentieth-century craze for birds in cages wreaked havoc in the bird population. In 1860, 14,000 goldfinches a year were being captured just from the area around Worthing in Sussex. In the first decade of the twentieth century 7,000 linnets a week were sold in London markets.

The impact of Europe on the rest of the world

Outside of their home continent the Europeans had an even greater impact and in a much shorter period. When the first Europeans reached the Americas, Australasia and the Pacific they were overwhelmed by the new and strange animals and plants that they found. One European arriving in Australia in the 1830s noted some of the striking differences:

> trees retained their leaves and shed their bark instead, the swans were black, the eagles white, the bees were stingless, some mammals had pockets, others laid eggs . . . even the blackberries were red.

Above all the first explorers and settlers were stunned by the sheer profusion of wildlife in areas that had seen little or no human settlement. Their accounts give some idea of the teeming mass of life that untouched ecosystems can support. In 1658 when the French explorer Pierre Radison reached Lake Superior, he reported that there were 'stores of fishes, sturgeons of vast bigness, and pikes seven feet long. A month's subsistence for a regiment could have been taken in a few hours.' Thirty years later one of the first settlers in Florida reported that 'quantities of wild pigeons, parrots and other birds were so numerous that boatloads of birds' eggs were taken.' In 1709 when an English sailor, Woods Rogers, landed on Mas Afuera, one of the Juan Fernandez Islands off the coast of Chile, he wrote that the seals and sea lions were 'so thick on the shore, that we are forced to drive them away, before we could land, being so numerous, that it is scarce credible to those, who have not seen them'. In the late eighteenth century Captain Cook arrived in Australia and found that the sea was so full of fish that they broke the fishing nets with their weight and flocks of thousands of birds could easily be shot because they had no fear of humans. On the Great Barrier Reef the botanist of the expedition, Joseph Banks, wrote in his journal that the butterflies were so numerous that:

> the air for the space of 3 or 4 acres was crowded with them to a wonderful degree; the eye could not be turned in any direction without seeing millions, and yet every branch and twig was almost covered with those that sat still.

A few years later Captain Thomas Melville was sailing into Sydney harbour when he saw more sperm whales in a day than he had seen in six years in the old whaling grounds off Brazil. He wrote that 'we sailed through different shoals of them from 12 o'clock in the day till sunset, all around the horizon, as far as I could see from the mast-head'.

This profusion of wildlife seemed to the early explorers and settlers as a convenient and readily available source of food which they proceeded to draw on regardless of the consequences. This unrestrained killing had a dramatic impact especially on islands where there were vulnerable flightless birds. On Mauritius the introduction of pigs and rats together with hunting by sailors drove the ground-nesting dodo to extinction by 1681. One of the greatest sets of extinctions was on Australia after the Europeans arrived in the late eighteenth century. The duck-billed platypus was still common in the Blue mountains in 1815 but had disappeared by 1850. An expedition to the Murray–Darling basin in 1856–57 recorded thirty-one species of native animal – twenty-two of them are now extinct. Others, especially the kangaroo and the emu, suffered badly from organised hunts. In 1850 one hunter, Captain Foster Fyans, lamented that in the colony of Victoria:

> Emus and kangaroos on our arrival were plentiful in all parts
> ... also bustards in large flocks of ... 40 or perhaps more. The
> bustards now are scarce, and only met with in distant places.
> The kangaroo and emu are nearly extinct in the district; the
> country is almost void of game.

By the end of the nineteenth century some of the rare marsupials such as the hare-wallaby and the banded hare-wallaby were already extinct and the last of the bilbies, which in the early nineteenth century had been the commonest form of native game, was shot in 1912.

The vogue for big game hunting in the nineteenth and early twentieth centuries in Africa and India (a 'sport' that is now returning for the super-rich and powerful) had a major impact on animal numbers, especially carnivores such as tigers and lions. Fashions of the period (some of which still continue), including crocodile skins for shoes and handbags, elephant tusks for ivory and rhinoceros horn for its supposed aphrodisiac properties, added to the slaughter. The demand for exotic feathers in hats meant that in 1869 Brazil alone exported 170,000 dead birds for their plumage. In 1913 the London salerooms were able to offer feathers from 77,000 herons, 48,000 condors and 162,000 kingfishers. Similar numbers could be found in Paris, Rome, New York and other major cities in the industrialised world. Plants were affected too – the craze for rare orchids in nineteenth-century Europe meant that Brazil was exporting over 100,000 plants a year from the tropical forests.

The European impact on the wildlife of North America was even greater than in Australia. When the first Europeans reached the Great Plains they found huge herds of bison roaming the area. There were probably between forty and sixty million animals. Some native Americans hunted the herds for food and hides. They probably killed about 300,000 a year which was below the natural replacement rate. The herds were therefore about their original size when the Europeans began exploiting them in the 1830s. About two million a year were killed for their meat – enough to start reducing the size of the herds. In 1871 bison hides were first made into leather and the rate of slaughter stepped up to unprecedented levels: about three million animals a year were being killed in the 1870s and 1880s. This was more than enough to drive the bison to the point of extinction by the 1890s. It now survives in a few carefully managed herds.

The passenger pigeon

Probably the most terrible example of mass slaughter was not the bison but the passenger pigeon – it is a story that almost defies belief. The early Europeans in North America frequently commented on the huge numbers of blue, long-tailed, fast and graceful pigeons in the country. One of the first settlers in Virginia wrote:

> there are wild pigeons in winter beyond number or imagination, myself have seen three or four hours together flocks in the air, so thick that even have they shadowed the sky from us.

Similar reports can be found from the Dutch on Manhatten Island in 1625, from Salem in Massachusetts in 1631 and from some of the first explorers in Louisiana in 1698. As late as 1854 in Wayne County, New York, a local resident wrote:

> There would be days and days when the air was alive with them, hardly a break occurring in the flocks for half a day at a time. Flocks stretched as far as a person could see, one tier above another.

On 8 April 1873 at Saginaw in Michigan there was a continuous stream of passenger pigeons overhead between 7.30 a.m. and 4 p.m. Other reports describe flocks a mile wide flying overhead for four or five hours at a time during their migration in the early spring from the south to their breeding grounds in New England, New York, Ohio and the southern Great Lakes area. The flocks were so thickly packed that a single shot could bring down thirty or forty birds and many were killed simply by hitting them with pieces of wood as they flew over hilltops. Their roosting sites were correspondingly enormous – some covered an area eight kilometres by twenty kilometres with up to ninety nests in a single tree. The weight of the birds, often standing on top of each other, was such that branches and even whole trees were broken. The exact number of passenger pigeons in North America when the Europeans arrived is unknown – the best guess would suggest about 5 billion, or a third of all the birds in the region and the same as the total number of birds found in the United States today.

One reason why the passenger pigeon existed in such huge numbers was the lack of natural predators apart from hawks and eagles. It was, however, surprisingly vulnerable to human actions. Each female only laid one egg a year so losses were not easily replaced. Only a flimsy nest was made and its habit of nesting in vast colonies and migrating in huge flocks made it very easy to attack. The birds fed on acorns, chestnuts and beechnuts and so when the woodlands were steadily cut down by the first European settlers habitats and food supplies were reduced. On its own this would have reduced numbers slightly – it was hunting that brought about a crisis. The native Americans captured the birds in large nets and by the 1630s the European settlers were doing the same. The young squabs were regarded as a great delicacy and the adults were killed for their feathers as well as their meat. In the first two hundred years of European settlement the number of birds fell but probably still numbered several billion in the midnineteenth century – the scale of the human attack was limited in this period.

The billions of passenger pigeons were driven to extinction in about fifty years. The onslaught began with the development of large-scale commercial hunting to supply the developing cities on the east coast of the United States with cheap meat. This depended on the development of railways which, by the early 1850s, linked the Great Lakes area with New York. In 1855 about 300,000 pigeons a year were being sent to New York. Killing on this scale was still small-scale compared with the slaughter of the 1860s and 1870s. The scale of the operation can be judged by figures that seem almost incredible but which were carefully recorded as part of a legal and highly profitable trade. On just one day (23 July 1860) 235,200 birds were sent east from Grand Rapids in Michigan. In one year (1874) Oceana County in Michigan sent 1,000,000 birds to markets in the east. Two years later it was sending 400,000 a week at the height of the season and 1,600,000 in the year. In 1869 Van Buren County in Michigan sent 7,500,000 birds to the east coast. By the end of the 1870s passenger pigeon numbers were severely reduced but in 1880 Michigan was still able to send 527,000 birds to the east. By the end of the 1880s the large flocks, which had once been so common, had disappeared and sightings of passenger pigeons were a matter for comment. The last known specimens died in most of the eastern states in the 1890s. The passenger pigeon became extinct in the wild in about 1900 when the last birds in Ohio died. The last pathetic survivor of a species that had once been numbered in its billions died in captivity in 1914.

The introduction of new species

The expansion of European settlement after 1500 drove many species to extinction and severely reduced the numbers of many others. However, this was not the total of the European impact on ecosystems around the world. They took with them the domesticated plants and animals (and many of the pests) they had known in Europe. It was the start of a major homogenisation of the world's plants and animals. Many of the animals escaped and went wild, and plants often replaced established native species. The effects of this movement were felt in every part of the globe but especially in the Americas and Australasia.

The European domesticated animals – pigs, cattle, sheep and horses – all had a major impact. Pigs were introduced everywhere. Once in the wild, they multiplied rapidly in the forests where they could find plenty of food. Australia now has over 20 million wild pigs. Cattle were first taken to the Americas by Columbus in 1493 and within fifty years were found in huge herds as far apart as Florida, Mexico and Peru. Not surprisingly they flourished on the pampas of South America – in 1700 there were probably 50 million animals there. By the middle of the nineteenth century they were so numerous that walls for fields were made from cattle skulls placed nine deep. In Australia the current wild herds of cattle can be traced back to just eight animals that escaped from a domesticated herd in 1788. Domesticated sheep did not flourish in the Americas until they were taken to the highlands of Mexico in the 1540s. Within thirty years there were immense migratory herds of sheep – 200,000 strong in the Michoacan area alone. By 1614 there were over 620,000 sheep in the area around Santiago in Chile. In Australia there were no hoofed animals before the Europeans arrived. Within a hundred years (the end of the nineteenth century) there were 100 million sheep and 8 million cattle in Australia. In New Zealand there were 9 million sheep within thirty years of its annexation by Britain. This massive increase in the number of grazing animals had an impact on the native grasses which were not adapted to intensive grazing – they were replaced by European varieties. Like cattle and sheep, the horses the Europeans took to the Americas also went wild. They migrated from Mexico to the Great Plains where they were domesticated by the native Americans. Many groups changed their way of life and shifted from small-scale agriculture to hunting of the bison. When the Europeans moved westwards over the Appalachians in the late eighteenth century they regarded the huge herds of wild horses as pests and shot them. Other animals were introduced too. The camel was introduced into the central desert of Australia in the late nineteenth century as a pack animal but it was not a success and the animals were allowed to go wild. Australia now has more camels than Arabia. The honeybee was taken to North America (there were no indigenous bees so the native Americans used maple syrup as a sweetener) and became naturalised in about 1800. They were also taken to Australia in 1822 where they soon outnumbered the local stingless bee.

The greatest ever environmental disaster following the deliberate introduction of a new animal came in Australia after Thomas Austin, a farmer near Geelong in Victoria, established a few rabbits for game in 1859. Rabbits are rapid breeders and with no natural predators their numbers rose rapidly and soon crops were being devastated over a wide area. By 1880 they had reached New South Wales and South Australia. In the mid-1880s large-scale eradication campaigns began: 1.8 million rabbits were killed in Victoria and nearly 7 million in New South Wales. This had little effect and their relentless spread continued. A decade later rabbits were moving across the Nullarbor desert towards Western Australia. New countermeasures were tried. A 1,600 kilometre-long fence was built between the north and south coasts in 1902–07 but it was breached in the 1920s. Within a century the rabbit population of Australia increased from a handful to about 500 million. Crop losses mounted and almost continual eradication campaigns had no significant effect. In 1950, as a desperate measure, the disease myxomatosis was deliberately introduced from Brazil. The death rate was very high but, as with nearly all forms of disease, some of the rabbits were naturally immune and were able to continue breeding. Within seven years the death rate from myxomatosis had dropped to about 25 per cent. Despite periodic outbreaks of the disease the rabbit population of Australia has survived and numbers are once again rising rapidly.

If Thomas Austin had known about what happened on the island of Porto Santo in the Madeiras in the fifteenth century he might have thought again about introducing rabbits. When the first Portuguese settlers arrived in the 1420s they found an island where the plants and animals were completely unaffected by previous human settlement. This soon changed when the rabbits the Portuguese brought with them escaped. In the absence of any natural predators their numbers rose rapidly. Within a few years the land was devastated – large areas had lost all their plants and the soil was eroding very quickly. The situation deteriorated and the settlement had to be abandoned and the settlers were forced to move to Madeira. The island was eventually re-colonised, with great difficulty, thirty years later.

Accidental introduction of mice and rats (who were on nearly every ship) proved to be just as damaging as rabbits. Both the early settlements at Jamestown in Virginia in 1609 and Sydney in Australia in 1790 were nearly wiped out because rats from the ships ate most of the precious stores of grain. During the 1570s, about forty years after the Spanish conquest, Peru was overrun by a plague of rats that had been brought on ships from Europe. The mice that escaped from the ships in Australia flourished in the absence of natural predators. The scale of the problem can be judged from the fact that in just one area of South Australia 32 million mice were killed in four months in 1917. European starlings were first brought to North America in 1891 when Eugene Schieffelin introduced eighty pairs into Central Park New York, supposedly as ornamental birds. They bred rapidly and within thirty years could be found in most states along the eastern seaboard. The starlings were aggressive and took over the ecological niches of a number of native birds and decimated the population of bluebirds and flickers. By the 1950s they had spread across the continental United States and had reached California and Alaska. Some birds travelled in the other direction. Parakeets were first brought to Britain in 1840 as colourful pets in cages. Some escaped almost immediately but they did not flourish until the late twentieth century when average temperatures started to rise. There are now huge colonies – one in south-west London is 7,000 strong and growing at about 30 per cent a year. By 2010 there are likely to be about 100,000 parakeets in the area.

Indigenous plants were also affected by the introduction of European plants and animals, often in unexpected ways. Goats were introduced onto the remote island of St Helena in 1810. In the next two centuries twenty-two of the thirty-three native plants died out because they were unable to withstand the intensive grazing of the goats. In the 1830s when Charles Darwin visited the plains of Uruguay he was astonished to find that hundreds of square kilometres of land were impenetrable because of the growth of the prickly cardoon. It had spread so rapidly because the huge herds of wild horses and cattle had eaten all the edible plants. European plants such as ferns, plaintain, thistles, nettles and sedge flourished in North and South America and Australasia where they were taken accidentally. These plants could spread rapidly. Christmas Island in the Pacific was only settled in 1888 yet, by 1904, thirty species of European weed were well established. In California in 1769 there were only three non-native plant species but within a century this had risen to ninety-one and European species made up half of all the vegetation. On the pampas of South America in the eighteenth century the artichoke and giant Mediterranean thistle went wild and created large impenetrable areas. By 1877 there were 153 different types of European plant growing around Buenos Aires and within fifty years only a quarter of the plants on the pampas were of native origin. In New Zealand, especially once the European bee had been introduced, new plants flourished in a climate very similar to that of Britain. Half the plants now found in the country are of European origin. Even in Australia, where the harsh conditions in the interior limited the area where European plants could grow, the pattern was the same. Within a century of the first settlement there were 139 alien plant species in the south-east of the country – there are now over 800. Some of these plants caused major problems. The prickly pear was introduced in 1839 in order to provide hedges but it soon went wild in both New South Wales and Queensland, creating barriers over two metres high. By 1925 over twenty-four million hectares of land were affected and in about half of this no other plants could survive. It was eventually brought under control by importing South American caterpillars which feed on the plant.

Increased communications between the different parts of the world have also spread pests and diseases. They often proved to be highly damaging in a new environment where there was often little natural resistance. In 1889 the Italian army trying to conquer Somalia imported cattle which were infected with the rinderpest virus which had long been endemic in Europe. It was unknown in Africa and as it spread southwards it had a devastating effect. Millions of cattle and wild buffalo, antelope and giraffe died – probably south of the Zambezi about 90 per cent of the grazing animals died of the disease. The death of cattle devastated the pastoral economies dependent on the animals – probably about two-thirds of the Masai people died as a result. In 1900 the United States Department of Agriculture brought the Asian chestnut to New York. These trees were infected by a fungus to which the Asian variety had developed resistance. It devastated the American trees which were almost eradicated by the end of the twentieth century. The introduction of the potato in Colorado altered the habits of the Colorado beetle, which until then had fed on the wild sand-bur. The beetle rapidly became a major pest destroying potato crops across the United States – it reached the east coast by 1874. Despite rigorous controls it reached France in 1920, spread to the rest of western Europe and arrived in the Soviet Union in 1955. In the 1850s the American vine aphid (phylloxera), which normally lived on wild vines east of the Rocky Mountains, was brought on board ships to Europe. It spread rapidly through the vineyards of Europe causing extensive destruction and threatened the future of the wine industry. The outbreak was only brought under control by grafting European vines on to American rootstock which was resistant to the aphid.

The effects of the expansion of Europe – its people, plants and animals – were far reaching. The wildlife of the world was never the same again. Many species were driven to extinction and others were so reduced in numbers that they could survive in only a few limited areas. In addition many European plants and animals were spread around the world and adapted quickly and easily to the new conditions. The last five hundred years has therefore seen a reduction in the variety of species around the world and an increasing homogenisation of ecosystems.

'The problem of the commons'

The extinction of the passenger pigeon (and others such as the great auk and the dodo) and the near extinction of the American bison pose a series of quite difficult questions. Why did the slaughter continue so indiscriminately? Why was it not seen as ultimately counterproductive in that it would destroy the foundations of the industry that was exploiting these resources? Why did the killing not stop when it was clear what would happen and be followed by a shift to sustainable exploitation? These are examples of what the American ecologist William Ophuls called 'the problem of the commons'. Strictly speaking these herds of animals were not 'commons' in that they were not owned on a communal basis. It is therefore better to speak of 'the problem of open access regimes'. No one 'owned' these animals and therefore no one had an interest in controlling the rate of killing and ensuring that there was sustainable exploitation. Because there was no ownership and the cost of exploitation was low (a horse and a rifle for the bison and a net for the passenger pigeon) a large number of hunters were tempted into the trade. In a highly competitive situation the most rational action for any individual hunter was to maximise their immediate kill before a rival did the same. The faster the population fell the greater the pressure to kill as many as possible as quickly as possible. All the pressures of an 'open access regime' encourage people to exploit resources as quickly and on as big a scale as possible. Anybody who took the opposite view and limited the numbers they killed would merely reduce their own income, increase opportunities for their rivals and not affect the overall rate of killing.

This pattern of maximising short-term gains at the expense of longer-term considerations, even if it meant the ultimate demise of the resource, is a central feature of the way in which humans have hunted animals. Little or no effort has been made to work out mechanisms that would ensure that species would not be over-exploited and driven to the point of extinction. The history of four major areas of exploitation – fishing, the fur trade, sealing and whaling – all illustrate the same dismal truth.

Fishing

As only a small number of domesticated animals could be supported in medieval Europe, fish formed an important part of the diet. (The situation was the same in Japan where the amount of agricultural land was severely limited.) Until the sixteenth century fishing was largely confined to the coastal waters of Europe but even here tell-tale signs of over-exploitation were beginning to appear. By 1500 fishing for herring had almost ceased in the Baltic and a few decades later the same problem was affecting cod fishing off the coast of western Europe. This was one of the pressures that drove fishermen further westwards and led to the exploitation of the rich resources of the Grand Banks off Newfoundland – originally the fish were so plentiful that they could be scooped out of the sea in buckets.

Until the twentieth century the stocks of fish in the vast oceans of the world seemed to be inexhaustible. (In practice fish are only found in substantial quantities in limited areas of the oceans – mainly along the continental shelf.) It was after the development of mechanisation and factory fishing by fleets from the major fishing countries that the first serious problems began to emerge. There was no attempt to limit catches and all the effort was put into maximising exploitation. Ships naturally started in the most productive areas but as these were over-fished and catches started to fall, ships caught younger and younger fish (causing numbers to fall even more rapidly) and then moved off to other less productive areas. The collapse of fish stocks can be traced in area after area. It was first noticeable for plaice in the North Sea in 1890, for haddock in 1905 and cod in 1920. By the 1920s stocks of hake off Ireland and south-west England, cod and haddock east

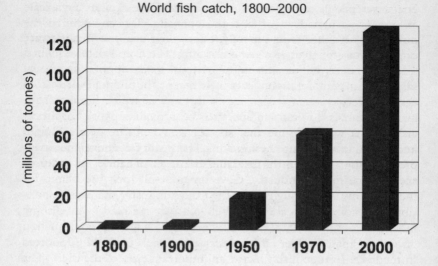

World fish catch, 1800–2000

of Iceland, plaice to the west of Iceland, haddock off the North Cape and plaice off the Kola peninsula had all fallen drastically because of over-fishing. In California the boom and bust of the sardine fishery (immortalised in John Steinbeck's *Cannery Row*) was very concentrated. The industry started around 1900 but took off after 1915 with the development of an export trade in cheap canned fish. By the mid-1930s about 600,000 tonnes of sardines were being caught every year. Within ten years the industry had collapsed through over-fishing and many boats moved south to exploit the stocks of the west coast of South America.

Overall the world fish catch was still relatively small until after the Second World War. Catches rose from about one million to two million tonnes in the nineteenth century and were about twenty million tonnes by 1950. In the next fifty years it rose more than six-fold to 130 million tonnes by 2000. This was only achieved by the intensive short-term exploitation of a number of different fisheries and the effects of this over-fishing became apparent very rapidly. The fish catch from the North Atlantic is now at half the level of 1950 and stocks of cod (the prime fish of the area) are at no more than a sixth of their level a century ago. The haddock catch fell from 250,000 tonnes in 1965 to less than 20,000 tonnes in a decade. The North

Sea industry, the scene of the longest exploitation, was very badly affected. The industry was unable to devise effective controls and many fishing fleets went out of business because it was no longer economic to go fishing – the value of the catch no longer equalled the costs. Those that did stay in the business were adopting high technology in an attempt to find the remaining fish. Ships now use nets almost two kilometres wide or trail long lines stretching for about 130 kilometres. Even so a third of the catch is usually discarded as commercially useless. Governments, especially the European Union, introduced quotas but these were usually set far above the levels proposed by the scientists studying the level of stocks in order to appease the fishing industry. Governments still took the blame for the problems even though it was the fishing industry that was responsible. Subsidies worth about £4 billion a year are paid to the fishing industry around the world but this merely encourages the continuation of over-fishing. (The UK gives a subsidy of over £600 million a year to the net fishing industry, a 100-fold increase since the early 1980s, even though it employs fewer people than those in lawnmower manufacturing.) As stocks and catches declined the price of fish (which once was a cheap alternative to meat) rose dramatically – six-fold in real terms in Britain between 1950 and 2000.

As stocks in European waters and the North Atlantic collapsed, fishing fleets moved elsewhere. The fishing grounds off the west African coast, once the preserve of small local boats, have now been over-fished by large factory ships from the European Union whose catches in this region have risen twenty-fold since the 1950s. When the EU buys fishing rights in west Africa (at the expense of the 200 million locals who depend on fish as their primary source of protein) the governments in the area have to use the money to make their fishing industries more environmentally friendly by disposing of the drift nets that were sold to them by the EU when they were banned in Spain and Portugal in 2001. Fleets have also moved even further afield to the South Atlantic and Antarctic to exploit the large stocks of krill in the area (numbers rose dramatically with the demise of the whale population). In the early 1960s just four tonnes a year were caught. By 1982 this had risen to 520,000 tonnes a year but this was not sustainable. Within a few years the catch had dropped to about 100,000 tonnes a year. In the northwest Pacific the Japanese catch of chub mackerel fell from 1.5 million tonnes in 1978 to half that figure within four years. By the early twenty-first century the world's oceanic fisheries were in crisis and their long-term sustainability seemed doubtful. A report in *Science* in

November 2006 reported that a third of all open sea fisheries were in a state of collapse (defined as producing only 10 per cent of their original yield). Global fish stocks had fallen by a third since 1950 and the rate of decline was increasing. It was estimated that within forty years all the world's major fish stocks would be exhausted.

The same pattern of over-fishing and collapse can be seen in major river and lake fisheries around the world. In the Caspian Sea the sturgeon catch (for caviare) in the 1980s was at a quarter of its level in the early twentieth century. Stocks continued to fall and in 2005 the United States took the drastic step of banning caviare imports because of the state of stocks in the Caspian. There was a similar pattern in the Great Lakes. In the eighteenth century sturgeon were so common in Lake Erie that they could be killed by hitting them with an axe handle. At the end of the nineteenth century the annual catch was still about 450,000 kilograms. By the 1960s it had fallen to less than 2,000 kilograms. The collapse of the cisco fishery was even more spectacular. In the 1930s the annual catch was about 7 million kilograms. By the mid-1960s it was less than 4,000 kilograms. The development of salmon fishing as a sport in the nineteenth century had a drastic effect on some rivers, particularly in Scotland. The catch on the river Tweed at Berwick fell from 149,000 fish in 1842 to just 40,000 a year within thirty years.

As oceanic and lake and river fisheries were put under increasing pressure the second half of the twentieth century saw the development of aquaculture – the farming of fish in cages. By the early twenty-first century almost a third of the world's fish catch was 'farmed' (two-thirds of it came from China, long a centre for fish farming, especially carp). Only a limited number of species can be farmed in this way – carp, shrimp, prawn and salmon – and they have to be fed on fish meal which equals the weight of fish actually produced. (The salmon also have to be fed a dye to make their flesh turn pink.) In the next few decades it is possible that 'farmed' fish will account for about half the world's fish catch.

The fur trade

Trapping and killing animals for fur sustained one of the major trading activities of Europe until the nineteenth century. The trade began by exploiting the animals in Europe but once these had been exhausted and driven to the point of extinction the trade became one of the driving forces behind the expansion of Europe. It was central to the eastward drive of Russia through Siberia to the Pacific and across to Alaska and to the westward drive of European settlement across North America. The origins of the trade can be traced back to the Roman empire when merchants obtained their furs from the nomadic tribes in the area of modern Russia. The real growth, however, took place in medieval Europe. Furs were needed to keep warm in a relatively harsh climate when means of heating were limited but they were also an important status symbol for the rich and powerful. In England numerous regulations were issued to restrict the wearing of furs to people of high status. In 1337 the English Parliament reserved the wearing of furs to royalty and members of the nobility with incomes over £100 a year. In 1363 another act was passed limiting fur-wearing to the nobility and clergy. Other regulations followed at regular intervals which suggests that the rules were not being obeyed.

To obtain one pelt meant killing one animal (by traps so as to preserve the skin). Most of the animals which were hunted were small – squirrels, martens, ermine, sable and foxes – and their value depended on their rarity and changes in fashion. In the thirteenth and fourteenth centuries grey squirrel fur (not the red of the then common European animal) was highly fashionable. When it became more widely available in the fifteenth century the wealthy switched to wearing rarer furs such as sable, fox and marten. Large numbers of pelts were needed to make one garment (especially if squirrel pelts were used) and so large numbers of animals were killed. Several hundred squirrel pelts were needed to make the lining for a cloak and 1,400 for the average counterpane. When Henry VIII had a gown made from sable fur it took 350 pelts. A few surviving documents give an idea of how many furs were bought for the English royal household. In the late 1280s Edward I bought 120,000 squirrel pelts a year (plus an unknown number of other pelts). In the early 1390s Richard II was buying about 109,000 pelts a year.

At first much of the trade was local as each country had its own wild areas where trapping could take place. Soon a number of key areas emerged. In the fourteenth century English merchants obtained many of their furs from Scotland and Ireland. Inverness became an international centre for the collection of marten and beaver skins and attracted traders from as far away as Germany. The local trade in skins continued for centuries – in the nineteenth century Dumfries market in south-west Scotland was still dealing in 70,000 hare and 200,000 rabbit skins a year. Increasingly, however, the trade was shifting to the great unpopulated forests of northern and eastern

Europe as the wild areas of western Europe were reduced in size and suffered from over-exploitation.

From the ninth century the Viking traders at Kiev developed an extensive trapping network for sable, ermine, black fox, beaver and squirrel using the local tribes to do the trapping (just as the Europeans were to do in North America several centuries later). Most of the furs went south to the Byzantine empire but by the twelfth century, as western Europe became wealthier, the Baltic was becoming an important trading area under the merchants of the German Hanse. (About three-quarters of the trade of this group of towns was in furs.) There were three main centres for the trade in the primitive states of western Russia. Novgorod concentrated on the more downmarket but flourishing squirrel trade which became the economic basis of the state. The value of land was calculated in pelts and rents were paid in furs. Moscow and Kazan specialised in sable, fox and marten for the luxury market. The size of the Russian fur trade (and therefore the slaughter of animals) was huge as a few of the surviving documents show. In 1393 one ship left Novgorod for Flanders carrying 225,000 furs. At this time London was importing about 300,000 squirrel pelts a year from Russia and Venice was buying about the same amount from the Hanseatic merchants. At the height of the squirrel trade Novgorod was probably exporting about 500,000 pelts a year. There are no reliable statistics for the Moscow and Kazan trade but its scale can be judged from the fact that in the early sixteenth century 40,000 sables a year were being sold just to the traders from the Ottoman empire.

Hundreds of millions of animals must have been killed at an unsustainable rate in Russia and western Europe. As early as 1240 the Dnepr basin around Kiev, the original centre of the trade, had been exhausted – there were no fur-bearing animals left. At this time the networks established by the Novgorod merchants already extended for over 1,500 kilometres and beyond the Urals in an attempt to find skins. From the early fifteenth century imports into London were in decline and prices of Russian pelts were rising as the number of animals fell drastically. By the 1460s London merchants were complaining about inadequate supplies and the volume of exports from Novgorod had fallen by a half to about 200,000 pelts a year. Other areas of Europe were also nearly exhausted. In 1424 the Scottish kings were driven to ban the export of marten skins. By the sixteenth century the beaver trade from southern Europe had collapsed and only low quality skins such as rabbit were available. Sables could no longer be found in western Russia.

By the sixteenth century the only remaining untrapped area was Siberia and it was the continuing demand for furs from western Europe that drove Russian merchants, using native and Russian trappers, into this largely unexplored area. Furs rapidly became both the main trade of the area and its currency. The trade also provided a third of the income of the Russian state. The early trappers could not believe what they found. They describe vast numbers of animals and ermine so tame that they would come up to houses and be caught by hand. The huge numbers encouraged large-scale killing and the trappers moved further and further east as area after area became exhausted. By the end of the eighteenth century even the vast area of Siberia had been denuded of fur-bearing animals and the traders and trappers turned their attention to the sea otters of the North Pacific islands. Between 1750 and 1790 about 250,000 otters were killed and the trade then collapsed because of the over-hunting. By the nineteenth century the heyday of the Russian fur trade was over – only a few furs could be found. The white fox was almost extinct, and only 2,000 blue fox furs a year could be obtained, although about 20,000 each of sables, ermines and red fox were still being killed in Siberia.

From the start of European settlement in North America the search for furs was one of the driving forces behind trade and expansion across the continent. At the first contact between the French and the natives in 1534 the former traded their goods for beaver skins. A flourishing trade developed but for a long period the Europeans did not trap the animals themselves – the locals did that and exchanged furs for European goods. Most trapping was concentrated on the beaver because the animals were sedentary and settled in large colonies so that the furs could be obtained with little effort. However, their low birth rate meant that animal numbers could not recover from over-hunting. The usual pattern predominated – hunters would concentrate on one area until it was no longer economic to continue and then move on. By 1600 the region around the St Lawrence river was exhausted as was upstate New York shortly afterwards – beaver were common in the Hudson river in 1610 but extinct by 1640.

By the mid-seventeenth century the trade was well organised in the interior of North America, mainly along the St Lawrence river, and controlled through a series of fortified trading posts. The rivalry between the French merchants and the British Hudson's Bay Company was intense and ensured a high rate of exploitation. Europeans were also becoming trappers as well as traders as they learnt the necessary skills from the natives. The consequences for wildlife in the areas where the trappers spent the winter could be dire. In the winter of 1709–10 at Port Nelson eighty men consumed 90,000 partridges and 25,000 hares. The scale of the fur trade was even more damaging. In 1742 Fort York traded 130,000 beavers and 9,000 martens and from just one trading post in Canada in the 1760s the Hudson's Bay Company was taking over 100,000 beaver skins a year. In one year (1743) La Rochelle, one of the centres of trade with Canada, imported skins from 127,000 beavers, 30,000 martens, 12,000 otters, 110,000 racoons and 16,000 bears. Similar figures were common at the other French and British ports trading with North America. It is not surprising therefore that this scale of exploitation brought the North American fur trade to the point of collapse by the late eighteenth century. Beaver skin exports from Canada fell by a half between 1793 and 1805.

The American trade was sustained through a last burst following the opening of routes to the west and the Pacific coast. In 1805 the first American explorers (Lewis and Clark) travelled through the area west of the Mississippi into the Rocky Mountains and on to the Pacific. They reported that the area was 'richer in beaver and otter than any country on earth'. Within less than forty years the area would be almost cleared of both these animals and the American fur trade would have nowhere else to go. In 1840, a traveller, Frederick Ruxton, noted the achievements of the trappers:

> Not a hole or corner but has been ransacked by these hardy men. From the Mississippi to the mouth of the Colorado of the west, from the frozen regions of the north to ... Mexico, the beaver hunter has set his traps in every creek and stream.

Trading was organised in the usual way with a mix of native and European trappers. The natives traded their furs for European goods and the European trappers worked either independently or for the main firms involved – the British Hudson's Bay Company or the American Jacob Astor. Unrestrained competition drove the beaver to the point of extinction. By the early 1830s the number of animals killed was already in decline as they became more and more difficult to find. By 1831 the beaver was extinct on the northern Great Plains and two years later the Hudson's Bay Company issued instructions not to waste time hunting in certain areas because the beaver was so rare. 1833 saw the almost complete collapse of the fur trade in the far west of North America. By the late 1830s the situation was so bad that only 2,000 beaver skins a year could be obtained from the whole of the Rocky Mountains. Beavers were saved from extinction only by a change in fashion. Their skins were mainly used to make hats but as supplies dried up prices rose and people switched to the newly fashionable silk hats. The last efforts of the trappers were concentrated on other furs – about 500,000 muskrat skins and 137,000 martens were sent to England in the 1840s and 1850s – but these trades collapsed as supplies ran out.

By the second half of the nineteenth century the world's stocks of fur-bearing animals were so reduced that it was hardly worth bothering to hunt them any more. Attention turned to a few exotic species – the chinchilla is now almost extinct in South America because of over-hunting – and the new land of Australia. The duck-billed platypus, the opossum and different types of wallabies were hunted through the nineteenth century for their fur. At this time the state of Victoria was exporting 250,000 skins a year. In 1919–21 Australia sold five and a half million opossum furs and 200,000 koala skins. Once these supplies had dried up there was no alternative but to turn to the 'farming' of fur-bearing animals, in particular the mink. Eighty per cent of the world's trade in furs now comes from such farms.

Sealing

The demand for fur and specialist skins was the main driving force behind seal hunting. As early as 1610 the Dutch were killing seals along the African coast for their skins but the sealing industry did not develop on a major scale until the eighteenth century when the trade in furs from land animals was in rapid decline. The animals were usually clubbed to death when they were onshore to breed and were defenceless. Once again the characteristics of 'open access' regimes were demonstrated. An area would be exploited rapidly and intensively until the seals were either extinct or so reduced in number that it was no longer economic to hunt them. Then the sealers would move on to another area. Between the 1780s and the 1820s the trade concentrated on killing the southern fur seal which was found in large numbers across the southern hemisphere. One of the first areas to be exploited (and exhausted) was the islands of the South Atlantic. In 1790–91 one American ship took 5,000 skins from the island of Tristan da Cunha, and the Falkland Islands and the mainland of Tierra del Fuego were exhausted at about this time. In the first quarter of the nineteenth century South Georgia was a major centre of the trade and, in total, well over a million seals were killed there. The South Shetland Islands were exhausted in just two years – one ship took 9,000 seal skins in three weeks and two ships took 45,000 between them in one season. Hunting then moved on to the southern Indian Ocean with the trade centred around Kerguelen Island. By the mid-1820s the seals had been hunted to extinction. In the Pacific, sealing was centred on the islands off the coast of Chile, in particular Mas Afuera in the Juan Fernandez Islands. One account describes how a single ship killed 100,000 seals in one visit and at times there were fourteen ships operating around the island. Between 1797 and 1803 over three million seals were killed on this one island and the herd was on the point of extinction. When the first Europeans arrived in Australia in the late eighteenth century they found huge seal colonies – they wiped them out within a couple of decades. In a single season (1805) 100,000 seals were killed along the Bass Strait and it was no longer economic to hunt them. Macquarie Island was only discovered in 1810 but 180,000 seals were killed in three years and by the early 1820s the herd on the island was extinct. By the 1820s the southern fur seal was almost wiped out after only three decades of hunting. It was no longer economic to hunt it anywhere in the Atlantic, Indian or Pacific Oceans. Probably about six million seals were killed in the first two decades of the nineteenth century.

In the North Atlantic sealing concentrated on the harp seal. In the autumn and winter the animals migrate south from the Davis Straits to Labrador, the St Lawrence estuary and Newfoundland, where the young are born on the pack ice in February. For ten days the newly born seals have their coveted white fur and they were the focus of the slaughter. Adults were also killed for their coarser fur and their oil. The Newfoundland sealing industry began in the early nineteenth century and by the 1830s about 80,000 seals a year were being killed. The peak of the trade came in the 1850s when 600,000 seals a year were being killed. The introduction of larger steam-powered ships, capable of processing the seals onboard, meant that a single vessel could deal with 20,000 seals a day. The huge herds could not withstand killing on this scale for long and the industry was soon in decline. Between 1800 and 1915 about forty million seals were killed in the area and the herds were reduced to about about a fifth of their original size. In the far north the exploitation of the harp seal was concentrated on Jan Mayen Island inside the Arctic Circle. It was a brief affair. In the 1840s about 400,000 animals a year were being killed. A decade later the industry collapsed as the animals neared extinction.

Sealing in the North Pacific exploited the northern fur seal. The animals mainly stayed at sea (migrating from the Bering Sea to the coast of central California) but came ashore to breed in the summer – four-fifths of the seals bred on the Pribilof Islands in the Bering Strait. The first hunters to reach the islands were the Russians. They originally concentrated on the more valuable sea otter but once these had been killed they switched to the fur seal. The killing was on a vast scale – 127,000 animals were killed in 1791 alone and soon stocks were far more than could be sold. In 1803 there was a pile of over 800,000 skins on the islands but 700,000 of them were rotten and had to be destroyed. By the 1820s the number of seals was so low that only 7,000 a year could be killed – by then about 2.5 million seals had been killed on the Pribilof Islands. The Russians moved across to the mainland and Alaska. By the time the territory was sold to the Americans in 1867 about four million seals had been killed. Meanwhile seal numbers on the islands had recovered and in the first year of American control they killed 250,000 seals on just one of the Pribilof Islands. The Americans were able to kill about 100,000 a year until the 1890s when the trade collapsed through a shortage of animals. By 1910 the herds on the Pribilofs, which had once numbered several millions of animals, had been reduced to a remnant of just over 100,000. From the 1890s the herds were so diminished that the sealers had to turn to the more difficult task of killing the animals at sea – they killed about four million in this way.

The elephant seal, the largest of the seals, was hunted not for its fur or skin but for its oil. It was found in large herds, but only on a few islands around the Antarctic and off the west coast of North America. It became the prey of whalers anxious to supplement their catch as whale numbers fell. About a million of these seals were killed in the nineteenth century in the South Atlantic and they were only saved from extinction when both Kerguelen and Macquarie Islands, where the last few animals survived, were turned into nature reserves. About 250,000 elephant seals were killed along the coast of California and large numbers were still found in the mid-1840s. A scientific expedition sent out in 1884 to report on the state of the herds failed to find any. In fact a small colony of about twenty animals had survived and since then elephant seals have been protected and numbers have risen. Sealing still continues around the world but at a low level because of much diminished numbers of seals and an increasing consumer boycott of furs and skins. This was, in part, prompted by the sight of young seals being clubbed to death on the ice in Canada. Nevertheless in the early twenty-first century Canada was still authorising the killing of 330,000 seals a year. It is difficult to estimate the total number of seals killed across the world during the height of the trade in the hundred or so years after the late eighteenth century. The best estimate would suggest about sixty million.

The killing of walruses for their oil, skin and the ivory from their tusks began much earlier than sealing but collapsed much earlier, the industry lasting little more than three centuries. In 1456 walruses could still be found in the Thames and even in the mid-nineteenth century they were common in both the Hebrides and the Orkneys. Now they number no more than about 25,000 in the whole of the North Atlantic. Exploitation followed the usual pattern. In the seventeenth century the English Muscovy Company destroyed a herd of 20,000 walruses on Bear Island, south of Spitzbergen, inside ten years. At the same time the herds that lived around the mouth of the St Lawrence were also exterminated – about 250,000 animals. Hunting then moved on to Labrador and the Arctic but by the 1860s nearly every walrus here had also been killed. The last major herd in the North Atlantic survived on Spitzbergen until the 1920s when they were also hunted to extinction. With the North Atlantic in severe decline walrus hunting in the late nineteenth century shifted to the herds of the North Pacific. In the years around 1870 about 85,000 animals a year were being killed but this could not be sustained for long. By 1891 the walrus was extinct on the Pribilof Islands. Again the exact number of walruses killed before the industry collapsed at the end of the nineteenth century is unknown – it was probably at least four million.

Whaling

One of the longest-lived human attacks on animals has been against the whale. Whales are some of the largest animals to have lived – a sperm whale can weigh about 100 tonnes and measure about thirty metres long. They have few predators and numbers were high despite their position at the top of the food chain. Whales are long-lived but their reproduction rate is low and their numbers take a very long time to recover from any attack. Sustained killing, particularly if concentrated around breeding grounds, could easily drive a local population to extinction. Whales fall into three main types and their characteristics have determined the development of whaling. The right whale is slow and easy to catch – they were the 'right' whale to catch and when dead they floated and so could be towed ashore to be cut up and processed. The fin whales (or rorqual from their Norwegian name) include the blue, fin, humpback, sei and minke whales and are difficult to catch because they are fast swimmers. The third type is the toothed sperm whale. Whales were hunted less for their meat (except in Japan where it was important because of the shortage of domesticated animals) than for their oil, normally boiled down from their blubber. Before the major exploitation of fossil fuels whale oil provided one of the major sources of lighting – the sperm whale produced the highest quality candles from the spermaceti found in its head. Millions of candles were manufactured from all kinds of whale oil and London even had 5,000 street lamps lit by whale oil in the 1740s. As well as providing almost the only source of lubrication for industrial machinery whale oil was also used to clean coarse woollen cloth. An important supplementary part of the industry was the use of whale 'bone' in corsets, umbrellas, whips, fishing rods and cutlery handles.

Until the eighteenth century whalers concentrated on the right whale because their slow swimming speed meant they could be caught, especially around their breeding grounds, with the primitive technology then available. They were chased in rowing boats and harpooned with barbed spears and the whalers could then attach more ropes and hang on, for days if necessary, until the animal died from exhaustion. Right whales had a thirty to fifty centimetre thick layer of blubber (ten times that of a sperm whale) which produced a large quantity of oil. They also produced about a tonne of whalebone per animal. Whales were hunted and exterminated in the Mediterranean by the first centuries CE. From about 900 CE hunting was concentrated in the Bay of Biscay and was dominated by ports along the northern Spanish coast. Even this small-scale industry so depleted stocks that by the fifteenth century the whaling ships were exploiting the area around Newfoundland as the Biscay whales neared extinction. In the sixteenth century a new large-scale whaling industry developed. It was dominated by the Dutch but the English and Germans were also important. Hunting concentrated around Spitzbergen where the right whales had their breeding grounds. From 1600 they were slaughtered in large numbers, including the young and the pregnant females. The beaches of Spitzbergen were an abattoir where the whales were dismembered, their blubber boiled down and the remains left to rot after the whalers left. Within twenty-five years the whales had almost been exterminated and the hunters had to move further afield to Greenland and the migration routes of the whales. Arctic whaling was more difficult because the whales had to be processed while they were tied up alongside the ship rather than onshore as at Spitzbergen Island. Nevertheless the industry flourished

– by the end of the eighteenth century it employed about 10,000 men, involved several hundred vessels and killed about 2–3,000 right whales every season.

The American whaling industry began about 1650 and initially concentrated its efforts along the eastern coast of North America. By 1700 there were no whales left in the area and the ships moved further north to Labrador and the Davis Straits. The industry expanded rapidly – the major whaling port of Nantucket had twenty-five ships in 1730 but over 130 by the end of the century. By 1830 Arctic whaling was on the point of collapse as even immature juveniles were being killed, thus ensuring that numbers fell even more quickly. In desperation the whalers moved to the difficult area around Baffin Bay but here too stocks were rapidly depleted.

The main development in the industry in the eighteenth century was the increasing concentration on killing sperm whales. They produced less oil than right whales but the spermaceti could be stored without processing and this, together with the high prices it fetched, made long voyages economically feasible. Sperm whales were killed off the east coast of North America from 1712 but stocks were soon exhausted and the hunt moved into the South Atlantic, first along the African coast (after 1763) and then off Brazil (from 1774). Once again stocks were quickly exhausted and the whalers moved to the area around Australia. Many of the first settlements, particularly Hobart on Tasmania, were dependent on whaling. The first half of the nineteenth century was the great age of Pacific whaling (the setting for Herman Melville's *Moby Dick*). At the peak of the trade around 1840 the Americans had over 700 vessels hunting across the Pacific and every whaling ground from 88°N to 55°S was fully occupied. The sperm whale hunting area stretched from California to Japan, from Peru to the Gilbert Islands and from Chile to Australia. On average each vessel killed about 100 whales a season, which means that the Americans alone must have been killing about 70,000 whales a season. Soon the Pacific industry was in decline. In the 1850s the north-west coast of North America could still support about 600 whaling ships every year but a decade later hunting ceased – there were not enough whales left to make hunting worthwhile. By 1880 all that was left of the Pacific industry were a few small-scale grounds off the coasts of Peru and Australia.

The whaling industry might have collapsed entirely at this point but instead it was transformed by a series of technological changes that ensured that the first half of the twentieth century saw an even greater slaughter of whales. The introduction of steamships and then factory ships capable of processing the whales at sea and storing the oil on board meant that long voyages to distant whaling grounds became economic. The development of the explosive harpoon – essentially a hand grenade that exploded inside the whale – made it possible to hunt the fast-swimming and deep-diving rorqual whales which had been almost immune to attack until then. The market for whale products was also transformed. The traditional uses of whale oil – lighting and lubrication – were in decline as petroleum-based products took over and the market for whalebone in corsets was falling as fashions changed. However, two new, large markets developed for whale oil – it was processed into soap and margarine. In addition whale blubber produced glycerine, a key constituent of the new high explosive nitroglycerine.

The attack on the rorquals showed the same pattern as in the past: concentration on the largest whales (blue and humpback) in the most accessible grounds and then a move into more distant waters after the smaller whales (sei and minke) that had to be killed in large numbers to make a voyage profitable. The technological innovations began in the Norwegian fleet in the 1870s and within a decade they had eliminated the rorquals along their coastline before moving further afield to the Faroes and Iceland. By around 1900 the European whaling grounds, long denuded of right and sperm whales, were also cleared of rorquals and the Norwegians and the British in particular moved into the last great whaling ground – the South Atlantic. The initial catches were, of course, high but over-exploitation meant that within a decade some of the grounds were exhausted. The first base was South Georgia where in 1906–11 the Norwegians alone were killing 6,000 humpbacks a season. By 1913 the total catch for all nations was down to 500 a year and by 1917 the humpbacks were exhausted and the attack turned to other species. Within less than a decade blue whales could hardly be found around South Georgia. The whaling fleets moved on to the South Shetland Islands, the Ross Sea, the Antarctic and then the breeding grounds off southern Africa, Madagascar and Peru. In the 1920s fleets from Germany and Japan entered the hunt and increased the the scale of competition and the slaughter. By the 1930s there were about 200 whaling ships operating in the Antarctic, killing about 40,000 whales a year. Much of the killing concentrated on the blue whale – in the 1930–31 season over 19,000 were killed, a sixty-fold increase on the number killed just twenty years earlier. However, a crisis was rapidly developing as whale numbers declined drastically. The number of factory ships operating in the 1930s rose by almost 60 per cent but the amount of oil obtained rose by only 11 per cent.

It was clearly in the interests of the whaling industry to devise a scheme to control catches so that whaling could continue at a sustainable, albeit reduced, level. Some attempts were made in the 1930s but they failed and the most that was achieved was that some countries banned whalers from other countries from their territorial waters but this was merely an attempt to keep the whales for their own nationals. The industry was largely moribund during the Second World War and there was therefore a slight recovery in whale numbers. In 1946 a body was created to regulate the whaling industry – the International Whaling Commission. It was a grouping of whaling nations and its powers were very restricted. Quotas were set, supposedly on the advice of a scientific committee that decided what level of whaling was sustainable. Its recommendations were usually ignored and higher quotas agreed. Indeed the quotas were usually so high that most countries could not catch enough whales to meet them. A three-quarters majority was needed before a new policy could be adopted but members were not obliged to follow a majority decision – they could simply ignore it (as the Dutch did in the 1950s when they thought their quota was inadequate). Even from the narrow perspective of the whaling industry the IWC failed in its first twenty years. Stocks fell rapidly after 1946 – of the total number of whales killed in the Antarctic between 1900 and 1970 more than half were killed during the first twenty years of the IWC's existence. As the large whales were driven to the point of extinction (blue whales in particular), the whalers turned to the smaller minke and sei whales but although the total number of whales caught in the early 1960s was twice that of the early 1930s the amount of oil recovered fell by almost a half. The whaling industry effectively destroyed itself in the first twenty years of the IWC. By the 1950s many whaling companies were going bankrupt – the cost of sending out a ship was no longer covered by the returns. By the early 1960s many countries such as Britain stopped whaling – this was not a policy decision, simply the result of basic economic forces.

Under the influence of various environmental and wildlife groups public opinion about whaling was changing, especially because its products were now marginal (it was easier and cheaper to make margarine from palm oil and detergents were replacing soap). Governments that no longer had a whaling industry to support found that they could improve their environmental credentials by supporting anti-whaling moves. Gradually from the early 1960s the IWC set quotas for individual species rather than the total catch and stopped hunting of blue whales (they could not be found anyway). Once again the quotas that were set were too high to provide effective protection and they were ignored by the countries that continued whaling – Japan, Iceland, Norway and the Soviet Union. Finally, under continuing external pressure, the IWC agreed in 1982 to a four-year moratorium on commercial whaling to start in the 1985–86 season. However, the ban on whaling was not total – the IWC allowed 'scientific' whaling to continue (it had started in 1962 to help the IWC scientific committee try to determine whale population size and breeding rates). The three countries still whaling in the 1980s (Japan, Iceland and Norway) seized on this loophole as a way of keeping their industries alive. The Japanese Joint Whaling Company was miraculously transformed into the Cetacean Research Institute and its whaling ships became research vessels. In addition aboriginal groups dependent on whaling for their existence – in Alaska and Siberia (and remarkably St Vincent and the Grenadines) – were allowed a limited catch of about 200 whales a year.

The ban on whaling (technically a zero quota because the IWC does not have the power to ban whaling) has now been in force for about twenty years. In 1994 a 'no whaling zone' – the Southern Ocean Whaling Sanctuary – was established. In the last two decades Japan, Norway and Iceland have killed nearly 30,000 whales. Most of these have been the responsibility of the Japanese 'scientific' programme – under IWC rules the meat and other products of whales killed for scientific purposes can be sold commercially. Whale meat remains a delicacy in some Japanese restaurants but there is a surplus of meat in most years and this is processed into school dinners. The Iceland whaling industry continues a similar 'research' programme but it is small-scale. In 1993 Norway decided to restart whaling and openly defied the IWC – it awarded itself a quota of about 1,000 whales a year. Iceland did the same in 2006. The number of whales killed has risen steadily every year since the ban was introduced – less than 600 were killed in 1992 but by 2005 the figure was over 2,000. Whale numbers have recovered in the last two decades but for many species they are still at very low levels that might not be enough to ensure long-term viability. There are probably no more than 500 blue whales left in the world (350,000 were killed between 1910 and the early 1960s) and the fin whale population is no more than 20,000. Within the IWC there is growing pressure for a resumption of whaling (especially from Caribbean countries that would be able to sell their quotas) now that stocks have recovered somewhat. In 2006 the annual IWC meeting voted to end the moratorium on whaling. However, the majority was not sufficient to meet the three-quarters needed to overturn the ban. Most observers expect commercial whaling to resume in the next few years.

Conservation and extinction

The same pattern has been repeated in industry after industry – the bison, the passenger pigeon, fishing, the fur trade, sealing and whaling. For centuries humans have acted as though the supplies of animals were infinite or if they were not it did not matter. It has been a monument to human short-sightedness. Not only have the industries involved declined or collapsed altogether but there has been a catastrophic loss of wildlife around the world. Across large areas of the world animals have been driven to extinction and overall numbers have fallen dramatically – total extinction was only avoided in some cases because it became uneconomic to hunt the last remnants of a once thriving species. The result has been the creation of an impoverished world.

A reaction to much of this slaughter and the continuing destruction of forests and other wild areas developed in the late nineteenth century with the rise of movements demanding greater efforts at conservation. This led to the creation of special areas to preserve natural ecosystems and their wildlife. One of the earliest was in the United States and led to the creation of National Parks such as Yellowstone where development and exploitation were prohibited. Other countries reacted more slowly – National Parks were first set up in Britain in the 1940s but it was not until the 1980s that a few hundred small sites were designated as being of special scientific interest. Other outstanding areas such as the Great Barrier Reef, the Galapagos Islands and the Serengeti park in Tanzania have been designated as being of worldwide importance. In the 1990s Antarctica was declared a conservation area that would not be open to development. However, most of the world's ecosystems remain unprotected. Even those that are protected still face problems. In Britain major developments inside the National Parks are not prohibited and sites of special scientific interest can be, and have been, destroyed, often by the government itself. In many developing countries a shortage of resources means that effective management is impossible and destruction and the killing of animals by poachers has continued. In many instances the parks are little more than lines drawn on a map. The late twentieth and early twenty-first century has seen a growing reaction against conservation efforts, particularly in the United States led by the 'new right' of the Republican party. They argue that continued economic growth and giving corporations access to the raw materials that may be in conservation areas is the most important objective. In 2003 the Assistant Secretary of the Interior responsible for wildlife and National Parks in the Bush administration, Craig Manson, said: 'The interests of developers should prevail over endangered species.'

Part of the growing movement to conserve wildlife and the natural world has been the increasing importance of citizen's groups and campaigning organisations. In the United States bodies such as the Audubon Society and the Sierra Club and in Britain the Royal Society for the Protection of Birds and nature conservation trusts have attempted to preserve individual species or have bought small areas to protect important habitats. Since the 1960s global organisations such as the Worldwide Fund for Nature, Friends of the Earth and Greenpeace have raised large amounts of money and they campaign not just to preserve species and habitats but on a much wider range of 'green' issues. In the last few decades there has been a series of international treaties and conventions designed to safeguard important habitats and preserve wildlife – many have been negotiated by the UN Environment Programme. These agreements have included conventions on biological diversity, the international trade in endangered species, migratory species, wetlands, African–Eurasian migratory waterbirds, and bats in Europe. The problem with these international agreements is that although countries have been willing to sign them, they have been less willing or able to enforce them. Many have found it easier, especially where there are high levels of corruption, to turn a blind eye to the destruction of nature reserves and the trade in items such as crocodile skins, rare furs, ivory and rhinoceros horn.

The conservation and protection movement has raised public awareness and achieved a number of small-scale victories. It has, however, been overwhelmed by the tidal wave of destruction that continues to sweep across the world. There is no doubt that the world is now facing its sixth great extinction of animals and plants (the last one was 65 million years ago when the dinosaurs became extinct). It is caused by human actions – habitat destruction (especially the clearing of the tropical forests), hunting and, increasingly, climate change. Species become extinct naturally for a variety of reasons and this 'background' rate is about one to three species a year. Between 1600 and 1900 human actions increased this rate slightly but not to a catastrophic extent. The twentieth century has seen this rate rise to unprecedented levels. Now the extinction rate is about 1,000 times greater than the natural rate and the human impact on the natural world is even greater than these bare statistics would suggest because other species are being eliminated in many areas of the world and reduced to living in a few remote, isolated refuges. In the last 400 years 83 mammals, 113 birds, 288 other animals and 650 plants have become extinct. But nearly all of these have occurred in the last century – of the twenty-one marine species known to have become extinct since 1700, sixteen have occurred since 1972. The extinction rate for mammals in the twentieth century was forty times the 'background' rate and for birds it was about 1,000 times higher. It is clear that this rate will rise even further in the twenty-first century. The best estimates (as published in *Nature* in January 2004) are that as the world's climate changes rapidly and the remaining tropical forests are cleared, about half of all the world's existing species will be extinct by 2100.

Many would argue that this mass extinction either does not matter or is not happening. As Craig Manson said about extinction, 'It's presumptuous to suggest we know for sure it's a fact.' The scientists who study in this field would beg to differ. Animals and plants will survive and some will move into the ecological niches vacated by those that have become extinct. But it will be an impoverished world and one where the diversity of life is significantly diminished – animals and plants will look increasingly the same throughout the world. There is also a utilitarian argument that we do not know anything about many of these species (the majority of species that live in tropical forests have not even been catalogued) and that many might contain chemicals useful as drugs or for other purposes. However, the future of these species will not be decided on the basis of a balance of arguments. The economic forces promoting habitat destruction and climate change will be the driving force behind species extinction.

9
THE FOUNDATIONS OF INEQUALITY

The rise of western Europe after 1500, from being a backward area of the world to one which dominated the rest of the globe, not only drastically affected a range of ecosystems but also reshaped the relationship between the different regions of the world. The states of Eurasia were increasingly linked together in the years after about 200 BCE (following the rise of the Han and Roman empires) and over the next thousand or so years there was an increasing interchange of technologies, scientific knowledge, crops and religions. However, the Americas, Australia, most of the Pacific and much of Africa were isolated until the expansion of Europe brought about a gradual integration of the different parts of the world into a single system and created a world economy. This process took time. Until about 1750 European influence was largely confined to the Americas and a few coastal trading posts in Africa. The European impact on the long-established trading systems and wealthy states of Asia was very limited. Only after 1750 did the Europeans settle in Australasia and exert an increasing influence over Asia and then Africa. Overall the impact of this process was drastic. The world economy that emerged over the four hundred years after 1500 was dominated by the states of western Europe and the areas where extensive European settlement took place – North America, Australia, New Zealand and South Africa. The tropical colonies in Africa and Asia and large parts of Latin America were forced into a subordinate position. Japan was one of the few non-European states to avoid this fate mainly because it did not come under external political and economic control. (China was only partially successful.)

In the earliest phases of European expansion, from the sixteenth to the mid-nineteenth century, western Europe itself was still characterised by predominantly agricultural economies though with increasingly important commercial sectors. The newly established colonies in the Americas provided an opportunity to grow crops (mainly for the luxury market) that could not be grown at home either because the climate was unsuitable or because the necessary cheap labour was not available. In addition the colonies provided some raw materials, particularly gold and silver from Mexico and Peru, together with timber. Increasing political control and the expansion of settlement combined with the industrialisation of western Europe in the nineteenth century intensified this process. The economies of the colonies were moulded and re-oriented to suit the demands of the home economy. Agricultural production for the European market was expanded and new crops introduced to meet changing demands and new industrial processes. In parallel Europe's demand for raw materials increased significantly and the colonies and dependent economies around the world provided an ideal source of supply. These countries and colonies also became ideal dumping grounds for European industrial production not required in the home market. By the mid-twentieth century the economies of many of the colonies had been developed as the colonial power demanded. The result was that when the colonies became independent they found it very difficult to remould their economies and escape from the constraints of a global economy constructed by the wealthy and industrialised states.

The first stages of European expansion

The creation of inequality in the world through the creation of dependent and lop-sided economies was a complex process that took many centuries. Some of the most important features can, however, be identified in the very first decades of European expansion in the fifteenth century, long before the Portuguese sailed into the Indian Ocean and the Spanish conquered Mexico and Peru. In the fifteenth century the Portuguese and Spanish discovered, conquered and settled in the Atlantic islands – the Azores, the Madeiras, the Canaries and the Cape Verde Islands. The Madeiras were unoccupied before Portuguese settlers arrived in the 1420s. The landscape of Madeira was transformed by the fires the first settlers started deliberately to clear land for agriculture. The original forest cover was almost totally destroyed. The settlers also introduced pigs and cattle which caused irreparable damage to the ecosystems of the island. In the 1450s the Portuguese began sugar cane cultivation in order to compete with the produce from Cyprus. This required a labour force to build terraces and artificial watercourses, cultivate the crop and undertake the back-breaking labour over many hours needed to process it and boil down the juices. Once the sugar had been processed it had a very high value for its bulk and could be traded over long distances by ship at a considerable profit. The Europeans (mainly Venetians) had already found on Cyprus that neither they nor the local population were prepared to undertake the heavy labour involved in sugar production. Instead they moved in slaves (mainly from around the Black Sea) to do the work. At this time the demand for sugar was still low and the slave trade probably did not involve more than about 1,000 people a year. This was to change in the Atlantic islands as sugar production (and later cotton production in the Cape Verde Islands) took off. By 1500 Madeira had changed from being a small colony of largely self-sufficient farmers into one of plantation agriculture with 20,000 settlers, thousands of whom were slaves labouring on the sugar plan-tations. Its sugar production (by then larger than that of Cyprus) was increasingly sold to north-west Europe where Antwerp dominated the trade.

The Spanish conquest of the Canaries was a more difficult oper-ation, lasting from 1402 and the capture of Lanzarote, to 1496 and the final subjugation of Gran Canaria. The seven islands of the group had been inhabited for more than a thousand years by the Guanches, who came originally from North Africa and numbered about 80,000 when the Spanish arrived. The Guanches hold the dubious distinction of being the first people to be driven to extinction by the Europeans. They were enslaved and worked to death on the sugar plantations that the Spanish established – the last of them died in the 1540s. The islands were rapidly cleared of forests to provide fuel for the sugar boilers and a thriving population of rabbits, another Spanish import, prevented any natural regeneration.

The bringing together of two distinguishing characteristics of the Atlantic economy that the Europeans developed after 1450 – plantation agriculture and large-scale slavery – happened for the first time in the mid-fifteenth century when the Portuguese brought African slaves to work on the sugar plantations of Madeira. Between the 1440s and the end of the fifteenth century the Portuguese took about 140,000 slaves from Africa to their Atlantic islands. Other important aspects of the world the Europeans created were already present – the conquest and elimination of the native peoples, the growing of crops for export on the best land and the displacement of indigenous, subsistence agri-culture to the poorest land. In most areas, apart from North America and Australasia, European settlers were a small percentage of the population. Over the next few centuries this system was to be replicated on a gigantic scale.

Slavery and indentured labour

Western Europe, originally a source of slaves for the Islamic world, began to use slaves on the sugar plantations on Cyprus and Sicily and then on the Atlantic islands. The original intention of the Spanish after the conquest of Mexico and Peru was to use the natives as slave labour. The original method of organising the newly conquered lands was to grant *encomiendas*, which entitled the colonists to the labour of a defined number of natives (but in theory not their land) – Hernando Cortés for example was granted the right to the labour of 115,000 people in Mexico. This system broke down because the level of exploitation was so high and because the diseases the Europeans brought with them caused a catastrophic fall in the population. Some countries, in particular England, used a system of indentured labour – people from England went to the colonies on a passage paid by their employer in return for a period of unpaid labour in the colony (usually seven years). When this period was over they were free (if they survived and most did not) and entitled to a grant of land. The majority of whites who went to North America before the 1780s were indentured servants. However, labour for the sugar plantations could not be found in this way – people were simply unwilling to undertake the back-breaking labour involved and the few who did go usually died very quickly from tropical diseases.

In these circumstances the Europeans turned to the increasing import of slaves from Africa. At first the Portuguese dominated the trade which was mainly from the region around Angola to Brazil. They were followed by the Dutch and then the British and French to the increasingly important islands of the West Indies and then to North America as the cultivation of sugar, tobacco and cotton took off. The slave trade was on a relatively small scale between 1500 and 1600 – about 370,000 people were taken forcibly from Africa to the Americas (mainly Brazil). In the next hundred years to 1700, as sugar production in the West Indies on islands such as Barbados and Saint Domingue expanded rapidly, the number of slaves taken to the Americas rose to 1,870,000. The eighteenth century witnessed the slave trade at its height – it involved the transport of 6,130,000 people at a rate of about 80,000 a year by the last two decades. In total about twelve million people were bought by the Europeans and taken to the Americas as slaves. About a fifth of these people died in the terrible conditions on board the slave ships as they crossed the Atlantic – the longer voyages to the West Indies produced the highest mortality. However, far more people were enslaved in Africa than began the voyage to the Americas because of the conditions on the journey to the slave ports – in total about twenty million people were enslaved to provide the labour the Europeans demanded.

In the period up until 1820 (when the large-scale trade in slaves came to an end) three and a half times as many Africans were taken to the Americas as the number of European immigrants – 8.4 million compared with 2.4 million. The brutal conditions under which the slaves lived can be judged from the fact that in the 1820s the population of the Americas was half black and half white. The average life expectancy of a slave in the Caribbean, even a young, healthy male, was about seven years. In the 1690s such a slave could be bought for about £20, which was the value of the sugar each slave produced every year. Even allowing for other expenses this left a huge profit for the slave owner and a high death rate did not matter – fresh slaves could always be bought. Plantations and the slaves who worked on them became central to the growing wealth of western Europe as a few examples will demonstrate. In 1638 no sugar was grown on Barbados yet within seven years plantations covered half the island and they were soon producing two-thirds of the sugar consumed in England. By the 1680s there were over 50,000 slaves and they made up three-quarters of the population. Sugar production spread across the West Indies and sugar consumption in Britain rose five-fold between 1710 and 1770. By the middle of the eighteenth century sugar and tobacco (grown in North America) made up over half of Britain's overseas trade and tobacco, indigo and rice accounted for three-quarters of the exports of all the British colonies on the North American mainland. In the first half of the nineteenth century it was the export of cotton to the growing textile industry of Britain that provided the impetus for a massive expansion of slavery in the United States. The number of slaves rose from 700,000 in 1790 to over four million by 1860. Slavery spread into nine new states and halfway across the continent to the western border of Texas. After 1830 there were more slaves in the United States than in all the rest of the Americas combined. Slavery was finally abolished in the British territories in 1833, in the United States in 1863–65, in Cuba in 1886 and finally in Brazil in 1888.

When slavery was abolished the Europeans looked elsewhere to secure a cheap labour force that could be kept under tight discipline and provide the crops, resources and profits the Europeans demanded. The main sources for this revitalised system of indentured labour were India, China and the Pacific islands. Indian labour was particularly important on the sugar plantations of the West Indies, Mauritius, Natal and Fiji and later for other types of plantations in Malaya, East Africa, Sri Lanka and Burma. In the century after 1834, thirty million people left India to work as indentured labour across the world. By 1861 Indians made up two-thirds of the population of Mauritius and they now make up about half the population of Guyana, Trinidad and Fiji. Most of the thirty million Chinese labourers recruited for work abroad went to south-east Asia but many were also taken across the Pacific. Between 1849 and 1874, 90,000 went to Peru to replace the Hawaiians who had died there digging out the guano beds to provide fertiliser for Europe. The sugar and pineapple plantations on Hawaii sucked in 46,000 Chinese, 180,000 Japanese and 160,000 Filipinos as cheap workers. The sugar plantations of Queensland on the other hand used islanders from the Pacific. In the last four decades of the nineteenth century 60,000 islanders worked on the plantations (about 10,000 at any one time) and a quarter of them died before they could return home.

The European impact on traditional agriculture

When the Europeans took control of other parts of the world they inherited well-adapted traditional agricultural systems. Although all agriculture involves major disturbance to natural ecosystems, most of the traditional methods had, by a series of techniques, evolved over a long period of time, limited the damage to the environment and produced an agriculture that was stable, resilient and diverse, and capable of maintaining output over the long term. The exact techniques involved and the crops grown varied from area to area depending on climate and soils but in general external inputs were kept to a minimum and a wide variety of crops were grown to meet the needs of the farmer and the local community. Hand cultivation reduced soil disturbance and erosion (as did careful terracing); crop diversity avoided soil exhaustion and reduced the impact of pests and diseases and recycling and composting of wastes maintained soil fertility. These systems usually involved a high degree of local self-sufficiency with only a minimal contact with external markets. These systems did not meet European requirements. They wanted the colonies to produce crops for the domestic market in Europe or ones which could be sold on the world market to provide an income that would help pay for the colony. These traditional agricultural systems were therefore deliberately disrupted and agriculture in the colonial era became more specialised – a limited range of crops (often only one) were grown for export. This led to environmental problems by reducing soil fertility through the continued growing of a single crop, greater susceptibility to pests and diseases and profound changes to subsistence agriculture as it came under pressure from European demands.

The way in which the agricultural economies of the colonial world were transformed can best be studied from three perspectives. First, British policy in Kenya illustrates both the process and the speed with which fundamental changes could be introduced. Second, the development of plantation agriculture and the balance between this sector and the peasant smallholders illustrates both the environmental and social consequences of European policies. Finally, it is possible to trace the cultivation and export of the key crops the Europeans wanted – sugar, tobacco, cotton, rice, tea, coffee, bananas, rubber, cocoa and palm oil – around the world.

The reshaping of the Kenyan economy and society between the establishment of formal British control in 1895 and the 1920s was radical and highly concentrated. The interests of the African population did not determine the direction of development. The key factor was the British requirement that the colony should contribute to the overall development of the empire and produce commodities that Britain required. The government commission on the development of east Africa wrote in 1925:

> Britain possesses a rich potential heritage in tropical Africa. From it, with wise capital expenditure, she can expect to receive in ever-increasing quantities supplies of those raw materials and foodstuffs for which she is at present so dependent on foreign countries.

Although the number of white settlers was small in the early years of the colony (less than 2,000 by 1906), the British allocated them the best land (where most of the Africans lived) on long leases. By 1910 about 250,000 hectares a year were being granted to the whites. Both the settled Africans (the Kikuyu) and the pastoral tribes (Nandi and

Masai) were removed from this land even though as late as 1930 two-thirds of the white-owned land was still not in use for agriculture.

From the start of the colonial period the desired pattern of development was the establishment of large plantations run by Europeans and using cheap local labour. The first crops selected were coffee, sisal and maize. Until the late nineteenth century the main centre for coffee growing in the British empire had been Sri Lanka. However, the spread of the coffee leaf disease reduced crops to a fifth of previous levels and Britain became dependent on Brazil. In order to reduce this dependence, coffee cultivation was strongly encouraged in Kenya after 1907 and within fifteen years over 700 estates had been established. Rubber plantations were tried but those in Malaya did better and Kenya turned to sisal (to break an American near-monopoly) and to maize which could be sold in Africa. The main problem was finding a cheap labour force. The white settlers rejected any import of cheap Indian labour on racial grounds and looked to the colonial government to ensure that native labour was available on the right terms. The first step was to restrict the amount of land available to the Africans because it would, they argued, undermine 'the foundation on which the whole of this enterprise and hope is based, namely, cheap labour'.

A variety of measures were introduced to ensure that the Africans needed to work to earn money and did not remain as subsistence farmers. Both a hut and a poll tax, both of which had to be paid in cash, were introduced. When this did not produce enough labour, tax rates were raised until it did. The size of the 'native reserves' was further reduced. Import duties were imposed to raise the cost of goods to the Africans and only agricultural tools and machinery intended for European farmers were exempted. Taxes rose sharply after 1920 and all Africans were required to carry a pass but these could only be obtained if they had a job. Forced labour schemes for the government were introduced and little or no help was given to native agriculture which was confined to the less productive land on over-crowded reserves. In these conditions it is hardly surprising that the African population of Kenya fell from four million in 1902 to two and a half million in 1921. By 1930 the transformation of a traditional African economy into one controlled by the whites and integrated into the international economy was largely complete. The average income per head of the whites was two hundred times that of the Africans. Agricultural products from white-run plantations rose from a mere 5 per cent of Kenyan exports in 1913 to three-quarters of the total in 1932 when coffee alone made up 40 per cent of all exports.

The economies of nearly every European colony and many nominally independent countries in Latin America were drastically altered to provide the products that western Europe (and increasingly the United States) wanted. The pace and nature of change varied from area to area depending on the type and timing of European control and the characteristics of the crops that were grown. Until the nineteenth century the development of large plantations growing a single crop was mainly confined to the Americas where land was plentiful, the climate (apart from some tropical areas) was suitable for Europeans and large numbers of slaves could be imported from Africa. Most of the territories in south-east Asia remained under local rule until well into the nineteenth century although they produced and traded in many goods for the European market. Africa, apart from a few trading posts and forts along the coast and the settlement at the Cape, was not subject to European control until the late nineteenth century – the climate was unsuitable for Europeans until medical advances were made and the resources they wanted could be obtained through trade. In all three continents methods of production were largely determined by who had access to land and capital, the type of crop and how it was processed. Plantations were European and American owned, financed and managed and increasingly run by commercial companies and large corporations. Smallholders were usually in a minority, although they were important for the production of rubber in south-east Asia and some crops in west Africa. The main tropical tree and bush products – rubber, coconut, oil-palm, sisal, cacao and coffee – all had a considerable time lag between planting and full-scale production. They needed substantial investment and risks were high because supply and demand, and therefore prices, could fluctuate dramatically. Annual crops such as sugar, cotton, jute and tobacco could adjust more easily to changes in demand but substantial investment was still required – sugar had to be processed on the spot because it was too bulky to move over any distance. These crops required intensive labour over a short period, unlike tea and rubber where the demand for labour was high but fairly evenly spread through the year.

The crops of the Americas

As we have seen, sugar was the key crop in the development of the Atlantic economy from the seventeenth century. The centre of production moved from Brazil to the islands of the West Indies where the French colony of Saint Domingue was the largest producer until the great slave rebellion of the 1790s that led to the first independent state in the area, Haiti, a decade later. In the nineteenth century the Spanish islands of Puerto Rico and Cuba where slavery still flourished and the soil was not exhausted became the centres of production. The second export crop to become important was tobacco as the habits of smoking and snuff-taking spread in Europe. Originally a Spanish monopoly, it was introduced into the failing English colony of Virginia where it revitalised the economy in the decades after the first settlement. Although it exhausted the soil very quickly, it was easy to grow and did not need the large investment in processing and storage required for sugar cultivation. At first it was grown on small-holdings but later large plantations worked by gangs of slaves dominated the industry. During the seventeenth century tobacco became the foundation of the economy of Virginia and later the neighbouring colony of Maryland and it was used as the local currency. Imports from the American colonies into London rose from 9,000 kilograms in 1619 to ten million kilograms by 1700. As the soil was exhausted the frontier of cultivation moved steadily westwards across Virginia, Maryland and the other tobacco-growing colonies.

By the end of the eighteenth century slavery in the newly independent United States appeared to be in decline – tobacco was no longer such an important crop in the economy and other plantation crops such as indigo and rice (along the seaboard of the southern states) remained relatively small-scale. Slavery was not only revitalised but vastly expanded by the growth in cotton production for the British market – by the mid-1830s raw cotton made up a fifth of all Britain's imports. In 1807 the United States supplied 60 per cent of Britain's cotton and by 1820 the US was the largest producer in the world. US exports of cotton rose from 3,000 bales in 1790 to four and a half million bales in 1860. Most of the cotton was grown on plantations dependent on slave labour. Like sugar and tobacco, continuous cotton production rapidly exhausts the soil. As a consequence the frontier of production moved steadily westwards in the first half of the nineteenth century from the eastern seaboard states of Virginia, the Carolinas and Georgia into the new states of Alabama, Mississippi, Louisiana and Texas. The older states switched to breeding slaves for the plantations in the west – nearly a million slaves were forcibly moved westwards in this period and this normally involved the break-up of the families in which they lived. Even after the abolition of slavery the ex-slaves normally had little alternative but to continue working on the plantations or become impoverished sharecroppers.

The area under cotton continued to expand from one and a half million hectares in 1830 to twelve million hectares in 1900. Cotton production then fell as lower demand reduced prices, soil exhaustion became a major problem and a devastating pest, the boll weevil, spread from Texas after 1894.

The last of the great plantation crops in the Americas was the banana. It was first brought to the Americas from the Canary Islands by the Spanish in 1516 but it remained a purely local food crop. The development of refrigerated ships in the late nineteenth century made it possible to transport the fruit to North America and western Europe. By the 1890s bananas were grown extensively along the Atlantic coasts of Central America, often using indentured labour imported from India. In the West Indies the British relied on relatively small-scale production on a number of islands, in particular Jamaica. On the mainland the need for capital investment in refrigeration and a large labour force to crop the plantations throughout the year meant that a few large companies came to control the industry. One in particular, the US-owned United Fruit Company (UFC), which set up its first plantation in 1889, soon became by far the most important. Contractors were brought in to supply the labourers who were housed in company barracks, paid a small wage and given notes to exchange for goods (at inflated prices) in UFC-owned plantation stores. Such treatment of workers would have been illegal in North America and western Europe. In the 1930s soil exhaustion and an outbreak of both Panama and sigatoka diseases badly damaged the plantations and new ones were established along the Pacific coasts of the Central American republics. The economies of these states ('the banana republics') came to rely on this one export and the UFC (backed by the American government) came to dominate the politics of many of these countries. After 1945 Ecuador became one of the major producers in the world, relying on smallholdings rather than plantations. However, a few large companies such as UFC and Fyffes dominated the sale and marketing of bananas.

Plantations and cash crops in south-east Asia

Three major crops characterised the plantation and cash crop economy of south-east Asia from the nineteenth century – tea, rice and rubber. Until the early nineteenth century the cultivation of tea was restricted to China and Japan. As tea drinking rose in popularity in eighteenth-century England, the value of annual imports rose two hundred-fold. Once the East India Company's monopoly of the tea trade was abolished in 1833 cultivation spread rapidly into Assam and later into Sri Lanka and southern India. In Assam tea plantations were established by clearing the forests and by 1900 there were 764 estates covering about 150,000 hectares. Tea cultivation moved into Sri Lanka and southern India to replace the coffee bushes that had been devastated by coffee blight. In the last quarter of the nineteenth century tea plantations on Sri Lanka expanded from about 400 hectares to about 150,000 hectares. Harvesting the crop is a highly labour-intensive operation – it needs about a hundred people per hectare per day. The tea companies which owned the plantations brought in large numbers of labourers – by 1900 in Assam over 400,000 people had been brought in to work on the tea plantations. On Sri Lanka Tamils were brought from southern India; they now form the main minority population and the tensions between the two communities led to a long-running civil war by the 1980s. The conditions under which these tea plantation labourers lived were wretched – low-quality company barracks and very poor wages. Pay was low because labour costs made up about two-thirds of the cost of production and the tea companies were determined to keep prices as low as possible to ensure tea remained the most popular drink in Britain.

Rice had been the staple food of south-east Asia for thousands of years before the Europeans arrived. It was grown by peasants mainly for their own use and with a small surplus sold in local markets. The first country to be transformed by rice production for export was Burma after the establishment of British control over the southern part of the country in 1852. Britain was cut off from its usual supplies from South Carolina by the outbreak of the American Civil War in 1861, and the opening of the Suez Canal in 1869 made it easier, quicker and cheaper to ship goods to Europe from south-east Asia. The area under rice production in Burma rose twenty-fold between 1855 and 1920 and by the latter date half of Burma's crop was being exported every year. These changes had devastating social consequences for Burmese peasants because the new trade was controlled by wealthy absentee landlords or the owners of rice mills. These groups had access to finance and could afford to buy up peasant land to establish large estates on which the former peasants were employed as cheap labour. The peasants who tried to compete in this process usually went into debt to moneylenders who finished up owning a third of the land in Burma. The Burmese peasants were turned into either badly paid landless labourers or poverty-stricken sharecroppers permanently in debt.

The French imperial authorities brought about remarkably similar changes in southern Vietnam after they seized control in 1861. The area devoted to rice cultivation increased five-fold between 1880 and 1940 while exports rose six-fold. Production was concentrated in the Mekong delta, where the French divided up the land and sold it in large blocks. Only the rich could buy the land and they set up large estates usually worked by sharecropping tenants, kept in a position of quasi-serfs and tied to the land in a state of permanent debt. The experiences of Burma and Vietnam can be compared with Thailand which remained nominally independent in this period. Here too the increasing demand for rice saw a six-fold increase in the area under cultivation in the century after 1850. Exports also rose from about 50,000 tonnes to 1.5 million tonnes, causing problems in feeding the local population. However, there was no colonial power to back the creation of large estates and so small peasant cultivators remained the backbone of rice production.

Rubber production in the nineteenth century was transformed by the accidental discovery in 1840 of the vulcanisation process. This made rubber lighter, more flexible and less affected by heat and cold. It greatly stimulated demand as did its use in new products, in particular bicycle and car tyres. British imports of rubber rose sixty-fold between 1840 and 1900. For most of the nineteenth century rubber was gathered in Amazonia from wild trees. The economy of this part of Brazil was transformed by a rise in rubber exports from thirty-one tonnes in 1827 to 27,000 tonnes by the end of the century. In the early twentieth century the town at the centre of the trade, Manaus (over 3,000 kilometres up the Amazon), had eight daily newspapers, an opera house, telephones and it was one of the first cities in the world to operate electric trams. However, the British were determined to set up their own source of supply within the empire. Seeds of the rubber tree were stolen from Brazil in 1876 by Henry Wickham – he smuggled out 70,000 seeds wrapped in banana leaves and was knighted for his efforts. The seeds were germinated at the Royal Botanic Gardens at Kew and then used to start the first experimental plantations in Malaya. By the 1890s it was clear that rubber could be grown in south-east Asia and the Dutch set up their own plantations on Sumatra after 1906. South-east Asia soon became the centre for world rubber production. At the end of the nineteenth century Malaya had only 120 hectares of rubber plantations – by 1940 the area had expanded to over 1.3 million hectares and as early as 1919 half the world's rubber came from the colony. In the first two decades of the twentieth century production was concentrated on large estates and when the local peasants showed little interest in working on them as labourers the British brought in Tamils from India to do the work. From the 1920s some locals did take up growing rubber on small landholdings but they had little independence – they were dependent on the large traders and firms to buy their rubber.

The growth of rubber production in south-east Asia dealt a fatal blow to the Brazilian trade. In 1910 there were still over 150,000 tappers collecting it from the trees growing wild in the forests but this was far less efficient than gathering rubber from the neat rows of the Malayan estates. Demand for Brazilian rubber fell steadily and by 1930 output was at about a third of the level in 1900. Brazil (and other South American countries) tried to copy Malaya and set up plantations but environmental factors intervened. Rubber trees flourished in the wild when they were scattered through the forest but when planted close together leaf blight disease spread rapidly and destroyed the trees. The disease did not spread to south-east Asia. Brazil now imports more rubber than it produces. The Americans, who backed the Brazilian attempt to set up plantations (the Ford car company provided the initial investment), were determined to break the British and Dutch stranglehold on the international rubber trade. In the 1920s the US tyre company Firestone, backed by the US government, turned to the American semi-colony of Liberia in west Africa. The Liberian government granted a concession over 400,000 hectares of land at a price of less than one cent per hectare and by the late 1920s Firestone already had over 30,000 hectares of rubber plantations. The workers on the plantations lived in terrible conditions and were paid a pittance. The domination of the Liberian economy by US companies such as Firestone was such that in 1943 the US dollar became the official currency of the country.

Plantations in Africa

Plantations and the growing of cash crops for export were a comparatively late development in Africa, reflecting the fact that European political control over the continent was not established until very late in the nineteenth century. Although coffee is indigenous to Africa it was not grown there in any quantity and the Dutch began extensive cultivation in south-east Asia – in Sri Lanka in the late seventeenth century and on Java after 1712. It rapidly became the main export of the Dutch East Indies, which had over 300 million coffee bushes in the mid-nineteenth century. Production across the region collapsed with the spread of coffee blight disease in the 1870s and Brazil became the major producer in the world. Coffee had been introduced into Brazil by the Portuguese in the 1770s and cultivation moved steadily westwards as the soil was exhausted and railways opened up new areas. Coffee was grown on large plantations dependent on slave labour to keep costs as low as possible. After slavery was finally abolished in 1888 the industry relied on a flood of immigrants from Europe who worked in very poor conditions and were paid very little. In the late nineteenth century Brazil dominated the world coffee market (its output was three-quarters of the world total). The British were determined to establish their own sources of supply by setting up plantations in their newly conquered territories in east Africa. Coffee cultivation began in Malawi in 1878, Kenya in 1895 and Uganda in 1900. Coffee was grown on estates owned and operated by the British using African labour – often provided free as forced labour by the British authorities. Only after 1950 did the number of smallholders growing coffee expand but they remained dependent on the large marketing companies to buy their output and suffered badly from any fluctuation in the world price.

Cocoa, which was grown in Central America before the Europeans arrived, was first cultivated to meet European demand by the Portuguese on their Atlantic islands and then in their colony of Brazil using slave labour. Brazil remained the main producer in the world until the 1880s. It was introduced into west Africa in the 1870s and soon the British authorities realised that it would provide export revenue for what had until then been relatively unproductive colonies. British chocolate producers, in particular Cadbury, wanted their own secure supplies and also backed these efforts. European-controlled plantations, often owned by the chocolate manufacturers, were set up and by 1911 Ghana was the world's largest producer. The French colony of Ivory Coast began cocoa production as a local initiative but it was soon taken over by French-owned plantations. Across west Africa production of cocoa was soon controlled by European-owned plantations on which the locals were employed as seasonal labourers and occasionally as sharecroppers.

Palm oil in west Africa was originally grown on a limited basis because demand was low – it was only used as a minor lubricant and in soap production. In the late nineteenth century technical advances in the European food industry enabled it to be used as the main constituent in margarine. By 1900 west Africa was exporting 50,000 tonnes of palm oil a year to Britain. Production continued to increase, especially with the development of large European-owned plantations after 1920. Palm oil production also spread into south-east Asia where by the late twentieth century it was one of the major forces behind the clearing of the forests and a major threat to the continued existence of the orang-utan. There are probably only 60,000 left in the world and they are dying at the rate of about 8,000 a year as the forest is cleared and replaced by palm oil plantations.

Cash crops and underdevelopment

By the early twentieth century western Europe, and increasingly the United States, had brought about a major transformation in the agricultural economies of much of the rest of the world. Because agriculture dominated these economies it also had important social effects too. Countries and societies which had been largely self-sufficient in food were increasingly integrated into a world economy dominated by the industrialised countries. Through a powerful mixture of political control, economic pressure, investment and the structure of the world market, 'development' in these countries took the form of growing crops for sale to other countries. The crops were either to provide luxury items in the diet of people living in western Europe and North America – sugar, coffee, tea, cocoa and bananas – or to provide raw materials for manufacturing – cotton, rubber and palm oil. In the industrialised countries 'development' meant something very different – the building of a thriving and varied industrial base and rapidly rising levels of consumption and affluence for the population. In this process the dependent and colonial economies were restructured to specialise in a few commodities or in some cases a single crop. The environmental effects of this transformation were harmful. A diverse agriculture was increasingly displaced over wide areas by a monoculture which often exhausted the soil, reduced biodiversity and was vulnerable to pests and diseases. Socially the consequences were equally harmful. Self-sufficient peasants were turned into landless labourers or sharecroppers tied to the land and often heavily in debt. They became more susceptible to market fluctuations as the production of export crops rose at 3.5 per cent a year in the first half of the twentieth century while food production for home consumption rose at a slower rate than the growth in population. As a result these countries had to import food, often at high prices.

The perverse effect of cash-crop-dominated agriculture on the population at large can be seen in many countries. In the 1950s sugar cultivation took up 60 per cent of all the cropland on Cuba and made up three-quarters of the country's exports. On Fiji by the early 1980s sugar production was even more dominant. It made up four-fifths of the country's exports and employed a fifth of the population. Fluctuations in the sugar price therefore had highly magnified effects on the economy – if prices fell it was difficult to pay for imports and wages of labourers were either cut or they were thrown out of work. In east Africa in the late twentieth century agricultural land was taken over to set up huge farms growing out-of-season flowers for sale in Europe. Not only did the peasants lose their land and become labourers dependent on their wages to buy food, they were also exposed to very high levels of pesticide use, usually without any protective clothing.

The achievement of political independence by the African and Asian colonies in the twenty years after 1945 did not transform their economic situation. The experience of the Latin American countries, which had been independent since the 1820s, had already demonstrated how difficult it could be to achieve such a structural transformation. The newly independent countries found that agricultural, trading and landowning patterns were well established and that there were strong forces, both internal and external, opposing change. Once an economy had been set in a particular mould by the colonial or neo-colonial power and when the industrialised countries and their corporations retained overwhelming financial and commercial control with the terms of trade set in their favour, it was very difficult to change course. In this situation many of the countries had little alternative but to try to increase export earnings by producing more of the cash crops introduced by the colonial powers. Just before independence the Ivory Coast produced 75,000 tonnes of cocoa and 147,000 tonnes of coffee a year. By the mid-1980s production had risen to 228,000 tonnes and 305,000 tonnes respectively. The result was an economy even more dependent on these two crops. Many other countries were even more dependent on a single crop. In the 1980s coffee made up 93 per cent of Burundi's exports. Attempts at co-operation by producing countries in an attempt to stabilise prices failed and fluctuating commodity prices (which have steadily fallen in real terms) continue to undermine vulnerable countries. Even when the major corporations that owned large plantations have been nationalised and their land redistributed the poorest countries still exercise little control over the trade in the commodities they produce because they remain dependent on these corporations for processing, manufacturing and marketing. One of the major tea companies, Brooke Bond, now only owns 1 per cent of Sri Lanka's tea plantations but it controls a third of the country's tea exports. In many areas the large supermarkets who control distribution have forced lower and lower prices on the producers. By 2004 banana producers in Costa Rica could not get the legal minimum price for a box of bananas (the supermarkets would not pay it) and therefore they could not pay their workers the legal minimum wage. Nearly 90 per cent of the retail price of bananas now goes to the trading companies, distributors and retailers even though they are not involved in any form of processing. Just 2 per cent of the retail price is paid to the workers on the banana plantations.

Despite major problems of hunger and malnutrition the developing world and the poorest countries continue to be net exporters of food. twenty per cent of the world's food trade flows from these countries to the industrialised world and only 12 per cent in the opposite direction. This imbalance does not just apply to the tropical crops that cannot be grown in western Europe, North America and Japan. Within a year of the opening of the Suez canal, India became a wheat exporter to help feed the population of Britain. Even in the acute famine of 1876–77 India continued to export wheat to Britain and in the 1880s India was providing 10 per cent of the world's grain exports. In the second half of the twentieth century Latin America increasingly provided large quantities of beef for the US market at the expense of home consumption. Between 1960 and 1972 Guatemalan beef production doubled but home consumption fell by a fifth. The situation in Costa Rica was even worse – in the same period beef exports quadrupled while domestic consumption fell by 40 per cent. The average American cat now eats more beef than an inhabitant of Costa Rica.

Timber

The Europeans saw the rest of the world not just as potential suppliers of luxury foods and industrial crops but also as a source of timber, minerals and other raw materials. The impact of these trades also produced dependent economies characterised by underdevelopment and poverty. Timber was one of the most important products sent to Europe from the early colonies. Indeed the colony of British Honduras (later Belize) only existed as a result of a settlement by traders looking for mahogany for the European market. The scale of these operations can be judged by British activities in India and Burma. By the early nineteenth century British merchants had almost completely destroyed the teak forests of India's Malabar coast in order to supply European demand and they needed to find another source of supply. The unexploited forests of Burma provided a strong motive for the initial British conquest in 1826. The first area to be opened for exploitation by the new colonial government (the province of Tenasserim) was stripped of teak within twenty years. The annexation of Lower Burma in 1852 was the next stage in the process and it allowed the huge forests of the Irrawaddy delta to be stripped of hardwood to supply the European market. By the end of the nineteenth century about four million hectares of forest had been cleared. In the western Himalayas, after the British gained control of the Gorakhpur district in 1801, over one million trees were felled in twenty years. In other parts of the region local rulers sold felling rights to British merchants and by the 1850s severe destruction of the forests was apparent across the whole area. In the 1850s the demand for railway sleepers rose rapidly as the British built railways to move crops to the ports for export to Europe. Tree felling moved further inland and up into mountainous regions – by the 1870s half a million trees were being cut down every year just to provide railway sleepers.

Even very specialised timber trades could be highly destructive. In the early nineteenth century sandalwood was a highly prized commodity not just in Europe but also in China. It was obtained mainly from the islands of the Pacific but the trade lasted for less than a quarter of a century before all the trees were cut down. There was a similar pattern to the hunting of many animals – European and American traders would systematically exploit an island until there were no suitable trees left and then they would move on to another. The sandalwood trees on Fiji were all destroyed between 1804 and 1809, those on the Marquesas Islands lasted for three years after 1814 and those on the Hawaiian islands for slightly longer, from 1811 until 1825. With no trees left to fell the industry collapsed.

The establishment of American control over the Philippines after the Spanish-American war of 1898 provides a good example of the development of modern logging. Within two years a Bureau of Forestry was set up and commercial logging started in 1904. At that time about 80 per cent of the virgin forests were still intact. Half had been destroyed by the early 1950s and the process continued after the Philippines became independent. By the 1980s less than a third of the forests remained and by the early twenty-first century just 3 per cent of the original forests were still intact. What happened in the Philippines was not unusual – the newly independent countries treated timber as just another crop and one of the few sources of export revenue that they had available. Timber exports from Indonesia rose two hundred-fold between 1960 and 1980 – much of the timber went to Japan, which had one of the strictest forest conservation policies of any of the indus-trialised states in order to protect its own forests. The amount of forest in Indonesia fell from 162 million hectares in 1950 to ninety-eight million hectares in 2000 when the rate of destruction had risen to two million hectares a year. The Ivory Coast exported just 42,000 tonnes of timber in 1913. This had risen ten-fold to 402,000 tonnes in 1958 at the end of the colonial period and then quadrupled to over 1.6 million tonnes by the early 1980s. Rainforests still covered about twelve million hectares of land at the end of the colonial era but had fallen to four million hectares within less than two decades of independence. By the end of the twentieth century less than a fortieth of the Ivory Coast's original forests were still in existence.

Minerals

Mineral exploitation was another factor in the creation of a highly unequal world. It was a driving force behind early Spanish colonial expansion and once the treasures of the Aztec and Inca empires had been melted down the colonial authorities turned to the exploitation of the great silver mines at Potosi about 4,000 metres up in the Andes of Bolivia. The local population was subjected to forced labour – the workers had to stay underground for a week at a time and processed the ore using highly toxic mercury. Not surprisingly the death rate was very high. The population of Potosi was about 160,000 (nearly all of whom were forced labourers), making it one of the largest cities in the world – it was bigger than Paris, Rome, Madrid, Seville and London. In the three centuries after 1500 about 85 per cent of the world's silver production and 70 per cent of the gold output came from the Americas. On average the Spanish officially shipped 330,000 kilograms of silver every year to Europe and about 150,000 kilograms was sent from Acapulco across the Pacific to Manila. The contraband trade probably doubled these figures. The control Europe exercised over this production enabled it to buy its way into the well-established and rich Asian trading system at a time when it made very few products that were wanted in the more sophisticated markets of the East. Gold was also obtained along the west African coast (hence the name Gold Coast – the colonial name for Ghana) at the rate of about 40,000 kilograms a year at the peak of the trade in the early seventeenth century.

The final partition of Africa among the European powers (with the Americans in attendance) at the Berlin Congress of 1885–86 marked the beginning of the large-scale exploitation of the mineral resources of the continent. Some areas were even controlled by mining companies – King Leopold of the Belgians sold Katanga with its rich copper deposits to a mining company in return for the company financing his conquest of the rest of the Congo. It remained his private property until the level of exploitation and inhuman treatment of the natives forced the Belgian government to take over the colony. Minerals were the core of European investment in Africa – it accounted for two-thirds of all investment in the continent until the 1930s and mineral exports rose seven-fold between 1897 and 1935 when they made up half of all exports. Production was dominated by copper from the Belgian Congo and Northern Rhodesia (Zambia) together with gold and diamonds from South Africa. Other investment was largely limited to railways and ports to help bring the raw materials to Europe but this did little to help the local economy – copper from Katanga in the Belgian Congo went via the Benguela railway to the Portuguese colony of Angola for export. Processing plants were noticeable by their absence. The Europeans provided the skilled workers and the Africans the unskilled workforce. The introduction of hut and poll taxes by the colonial authorities, which had to be paid in cash, forced Africans into the labour market to work either on plantations or in the mines. The native mineworkers were housed in squalid barracks with minimal facilities, lived on a debilitating diet of maize mealie meal, were separated from their families and were usually working hundreds of miles from their homes and often in a foreign country. By the 1950s two-thirds of the workers in the South African mines came from outside the country.

In the same way that cash crops for export dominated the economies of some colonies, mineral production dominated others. In both cases it proved almost impossible to alter this structure after independence. Minerals still provide over 90 per cent of the exports of both Zambia and Mauretania. Such countries became increasingly important in world mineral production. Between 1913 and 1970 the share of the world's iron ore mined in the developing world rose from 3 per cent to 39 per cent and the amount of bauxite from less than 0.5 per cent to over 60 per cent. However, the producing countries see few of the benefits from this output because mining remains under the control of a few transnational corporations located in the industrialised world. Governments do not have the power to exercise effective control and are usually dominated by the corporations. These countries only process about 10 per cent of the copper ore they produce and less than a fifth of the iron ore. Transnational corporations usually refuse to build smelters and processing plants even when cheap energy is available from hydroelectric projects as both Guinea and Ghana found to their cost shortly after independence. Alumina is worth six times as much as raw bauxite and the final product (aluminium) is worth twenty-five times more. However, these high-value operations are largely confined to the industrialised world.

The environmental and economic effects of mining can be seen in three cases – Liberia, Mauretania and New Caledonia. After 1945 Liberia awarded concessions to transnational corporations to exploit the large iron ore reserves in the country. Four huge opencast strip mines (the most environmentally damaging form of mining, which strips away huge quantities of topsoil and rock, leaving large pits and canyons) were opened and railways were built to take the ore to the coast. Little local labour was used because the corporations preferred to bring in outsiders. The Liberian economy appeared to grow substantially and exports rose but few benefits filtered down to the local population. The same effects could be seen in Mauretania when the large iron ore deposits were exploited after 1959. The Mauretanian government could obtain no more than a 5 per cent stake in the company set up to run the operation. The company built its own 650-kilometre-long railway to the port of Nouadhibou and ran its own army to protect the mines. After seven years of mining the Mauretanian economy had, on paper, grown two-and-a-half-fold but almost none of these benefits reached the local population. The mining operation was a largely autonomous unit employing little local labour, importing most of its requirements and repatriating its profits out of Mauretania. New Caledonia was annexed by France in 1853 and two decades later large nickel deposits were discovered. Nickel became a key metal in the twentieth century because it was so hard and resistant to corrosion – it was widely used in aircraft, armaments and nuclear power plants. New Caledonia had between a quarter and a third of the world's known nickel reserves and the French set up a company, Société le Nickel (SLN) to exploit the reserves. By the 1920s New Caledonia was the largest producer of nickel in the world using labour imported from Japan, Java and Vietnam. In the twentieth century over 500 million tonnes of rock were moved to find 100 million tonnes of nickel ore which produced 2.5 million tonnes of nickel. The landscape around the nickel mines was devastated by opencast mining, silt and debris clogged up streams, much of the waste was dumped on arable land and offshore coral reefs were ruined. Because of the huge quantities of ore needed to produce nickel, smelting was done on the spot but there were no environmental controls until the 1980s and the resulting acid rain further ruined the landscape. There has been no effective clean-up of the damage wrought by a century of nickel mining and smelting. The local population has borne the cost of the mining while the benefits have accrued elsewhere.

Fertilisers

In the late nineteenth century the use of fertilisers to increase agricultural output rose dramatically. The United States had its own internal sources of supply but western Europe turned to Morocco and Tunisia and the large guano deposits off the Pacific coast of South America. The latter were originally part of Bolivia but Chile's victory in the war of 1881 (fought because of the deposits) gave it control over the coast and the guano islands and turned Bolivia into a landlocked country. The guano was mined in appalling conditions by imported Chinese labourers and Chile was soon exporting over a million tonnes a year – the tax on the exports made up 80 per cent of the government's revenue. The British empire remained dependent on these external sources until the discovery, early in the twentieth century, of the huge phosphate deposits in the Pacific on Nauru and Ocean Island. This opened up the prospect of a cheap source of fertiliser to improve Australian and New Zealand agriculture, which were key sources of imported food for Britain. The story of these two islands illustrates in dramatic form the impact of the industrialised world's demand for resources on the people and the environment of the rest of the world.

Ocean Island was small (about five kilometres long and four kilometres wide), covered in lush, tropical vegetation and inhabited by about 2,000 Banabians following a typically Polynesian way of life. Nauru was slightly bigger with about 1,400 people. Ocean Island was annexed by Britain in 1901 whereas Nauru was a German possession until 1914. The two islands consisted almost entirely of solid phosphate deposits, perhaps the richest in the world. In 1900 the British-owned Pacific Islands Company bought the right to all the minerals on Ocean Island for £50 (made up of over-priced company trade goods). It was a 'treaty' of dubious legality – it was made with the local chief even though he did not have the authority to lease land belonging to others. Within five years 100,000 tonnes of phosphate was being mined every year. The company reached an agreement with the German administration to begin mining on Nauru in 1907. On both islands the company did not employ the islanders but brought in about 1,000 labourers (mainly Chinese). About eighty Europeans supervised the operations and a detachment of police from Fiji kept order. Nauru was occupied by Australian forces early in the First World War. The company was bought out in 1919 and all its rights and assets passed to the British Phosphate Commission (jointly owned by the British, Australian and New Zealand governments). Its purpose was to supply phosphate at cost price and therefore well below the world price. By the early 1920s mining was producing about 600,000 tonnes of phosphate every year.

The price for this subsidy to Australian and New Zealand farmers (and hence British food prices) was paid by the islanders. Mining involved clearing away the vegetation and stripping out the top fifteen metres of land leaving an uninhabitable wasteland of jagged pinnacles on which nothing would grow. Mining was steadily destroying the two islands. The Banabians refused to sell or lease any more land to the Commission. This defiance could not be accepted and in 1927 the British government authorised deep mining over the whole of Nauru. In 1928 the British government gave itself the power to confiscate all the land the Banabians refused to either sell or lease for mining. By the early 1930s mining was producing about one million tonnes of phosphate a year.

On the outbreak of war with Japan in late 1941 the Europeans and most of the Chinese labourers were evacuated but both sets of islanders were left behind. The Japanese occupied the islands and transported the islanders to the Caroline Islands. Before the war the British had wanted to remove the Banabians from Ocean Island so that mining could proceed unhindered. The Japanese actions provided a convenient opportunity to do this and after the war the Banabians were forcibly moved to Rambi Island in Fiji. One thousand five hundred labourers were moved to the island and deep mining across the whole landscape began in 1947. The inhabitants of Nauru were in a better position because the territory was, as a former German colony, a League of Nations and then United Nations mandate only administered by Australia. They had to be allowed to return home but they were treated as second-class citizens. Like the one thousand three hundred Chinese labourers the natives were excluded from all company facilities (shops and recreation) on racial grounds. Throughout the 1950s about one million tonnes of phosphate a year was being mined on the two islands, rising to nearly three million tonnes a year by the mid-1960s. It was clear that at this rate the deposits would soon be exhausted. The last shipment from Ocean Island was made in 1980 and mining on Nauru ceased at the end of the decade. In the eighty years of mining about twenty million tonnes of phosphate was dug out of Ocean Island and about three times that amount from Nauru giving a total of about eighty million tonnes from these two tiny Pacific islands.

The imminent exhaustion of the deposits on the two ravaged islands raised in an acute form the question of how to treat the inhabitants. The Australians wanted to resettle the people of Nauru on the mainland but the locals rejected the plan. They were in a much stronger position than the Banabians because they were protected by the United Nations. In 1965 they, rather than the Australian government, were given the right to decide how the small royalties they received on each tonne of phosphate should be spent. After a long struggle Nauru was granted independence in 1968 and two years later the management of the phosphate operation was transferred to them. The islanders continued to live along a narrow coastal fringe, the only part of the island not devastated by mining. Their traditional way of life was gone and their only means of subsistence came from the royalties and profits from phosphate mining. This provided enough revenue to finance a parody of European-style development. The islanders did not need to work but their material standard of living was high. There was one road on the island, which went nowhere, but the island had one of the highest rates of car ownership anywhere in the world. The population depended on imported food and developed many of the diseases associated with the affluent lifestyles of the industrialised world. However, the revenues were mismanaged and by the early twenty-first century Nauru was bankrupt. There was no alternative but to reach an agreement with an Australian company in November 2005 to re-open phosphate mining in a desperate attempt to raise some revenue.

The treatment of the Banabians, who were not protected by any international organisation, was far worse than that of the Nauru islanders. In 1911 the British government suggested that a trust fund, financed by phosphate royalties, should be set up for the Banabians. The British Phosphate Company proposed a munificent annual payment of £250 at a time when it was making a profit of £20 million a year and paying dividends of 40 to 50 per cent every year to its shareholders. Eventually the government persuaded the company to raise the royalty to six pence a tonne to be placed in a fund for the Banabians when the phosphate deposits were exhausted. The government's actions were far less philanthropic than they seemed. Ocean Island was incorporated into the Gilbert and Ellice Islands colony and most of the phosphate royalties were allocated to pay for the administration of a colony that had previously run at a loss. The Banabians were never told that 85 per cent of their royalties were being spent in this way. Indeed they were not told how much they were earning or how any of it was spent because the government (in the patronising manner typical of European colonialism) thought they were 'feckless'. They were occasionally given very small sums of money. The Banabians were not consulted about their forcible move to Rambi Island after the Second World War yet it was their royalty money that was used to buy the island (the colonial government of Fiji kept the proceeds). They were left on Rambi, which had a very different climate from their homeland, with no obvious way of making a living. Eventually the British government offered the islanders £500,000 as a final settlement for all the effects of the mining and their forced exile. The islanders rejected the offer and took the British government to court in the 1970s in what was then the longest civil case ever heard. They failed in the main part of their case because the court held that the purchase of the mining rights for £50 in 1900 was a legally binding contract. However, the court did find that the British government had failed in its duty of care to the islanders but refused to make any award of compensation. Eventually the soon-to-be-moribund British Phosphate Commission offered a sum that just covered the costs the islanders had incurred in bringing the protracted legal case. By 1980 Ocean Island had been destroyed by phosphate mining and the deposits were exhausted. The islanders had lost their home and had effectively received no compensation for their loss. That was the real price of cheap fertilisers for Australia and New Zealand and cheap food imports for Britain.

Underdevelopment and Inequality

The fate of the Banabians was symbolic of much that had happened to the colonial and dependent world created by the expansion of Western Europe. The creation of a world economy dominated by western Europe and North America in the period after 1500 should have produced, according to the doctrines of liberal, free-market economics, a world-wide division and specialisation of labour. This should have allowed each country and area to concentrate on growing or making the commodities it was best suited to produce. As a result of this specialisation every area should, according to these theories, have benefited from the most efficient allocation of resources. The problem with this theory is that it completely ignores political factors – countries and areas were not equally powerful and they were not left free to decide what to produce. Political control enabled the colonial powers to ensure that the commodities they required were produced and allowed them to enforce a highly asymmetrical series of exchanges. The words of Cecil Rhodes, one of the driving forces behind British expansion in Africa in the late nineteenth century, reveal the realities behind the theories of liberal economics:

> We must find new lands from which we can easily obtain raw materials and at the same time exploit the cheap slave labour that is available from the natives of the colonies. The colonies would also provide a dumping ground for the surplus goods produced in our factories.

The way in which one part of the world – western Europe, North America and the white settlement colonies – became 'developed' and the way in which another part became 'underdeveloped' are not separate phenomena.

In the world economy that was created after 1500 one region was able to extract a large surplus of products and natural resources from the dependent area. The dominant economies of the industrialised core were characterised by the production of capital-intensive goods and relatively high wages and profits together with rising levels of consumption and wealth. The subordinate, peripheral economies were characterised by crop, raw material and mineral production that were of low capital intensity and linked to low wages and the repatriation of profits to the developed world. Although development took place in the subordinate colonial economies, it was almost entirely geared to the needs of the colonial or economically dominant power. Railways were largely confined to links between inland regions and a few key ports and their purpose was to facilitate the export of crops and raw materials. The achievement of political independence did not bring economic independence because the structure of the world economy had already been established. Only a few countries avoided this trap

– those that retained their political independence such as Japan, those that escaped European colonialism (and had huge US support during the Cold War) such as South Korea and Taiwan, the oil-rich states of the Middle East and the trade-based economies of Hong Kong and Singapore. In the late twentieth century China, once it had recovered from the disasters of the century between 1850 and 1950, industrialised at a rapid rate and other countries such as Brazil and India made more modest steps in that direction. For most countries in the developing world, particularly those in Africa, Latin America and much of Asia, these options were not available. All they could do was increase production of a few cash crops or minerals in an attempt to raise income and exports. The problem was that this approach tended to lower prices, lower incomes, increase dependence on a few commodities and create greater vulnerability.

The consequence of this unbalanced development was a world characterised by increasing inequality. The industrialised world was able to live beyond the constraints of its immediate resource base. Raw materials were available for industrial production and food could be imported to support a rapidly rising population. This formed the basis for a vast increase in consumption and the highest material standard of living ever achieved in the world. Much of the price of that achievement was paid by the population of the rest of the world in the form of exploitation, poverty and human suffering. The environmental problems produced by this growing inequality in the world were different for the rich and the poor. The current environmental problems in the world can only be understood in the context of the nature of the world economy produced since 1500.

10
DISEASE AND DEATH

The way in which humans have lived and the environments they have created have had a profound effect on their health, life expectancy and the causes from which they have died. It is yet another example of the influence of the environment on human history. Diseases have made their impact on human history in three main ways. First, there have been the highly lethal outbreaks of disease such as the 'Black Death', which killed almost a third of the population of Europe during its first outbreak in the mid-fourteenth century. Second, there have been the persistent, endemic diseases such as trypanosomiasis (sleeping sickness), malaria and schistosomiasis, that have either ruled out areas for settlement or produced long-term debilitation. Third, the inadequate diet on which much of the world's population has had to survive has resulted in deficiency diseases and general poor health, making people more vulnerable to other diseases.

The pattern of disease and death has been affected by four major developments in human history. First, the creation of farming communities opened humans to a whole range of diseases that originated with animals. In addition, permanent settlements made it difficult to provide clean drinking water and a number of diseases became endemic. Second, the development of cities brought humans together in numbers sufficient to allow the major epidemic diseases to develop and eventually spread. Third, the gradual drawing together of human communities around the globe, first within Eurasia and then when the Europeans reached the Americas and Oceania, spread new diseases to peoples who had no natural resistance. Finally, in the industrialised world in the late nineteenth century improvements in public health significantly altered the impact of diseases long before effective medical treatment became available in the twentieth century. Medical treatment had a significant but limited impact and, by the late twentieth century, it faced a new threat from the changing pattern of disease.

The diseases of agriculture

Although the available information about the health of gathering and hunting groups is somewhat sketchy, a very clear pattern does emerge. Their use of a wide range of foods meant that malnutrition was rare and deficiency diseases unknown. Their main diseases are likely to have a low level of intestinal parasites from poorly cooked food and unclean drinking water. The fact that numbers were low meant that the main infectious diseases could not take hold. Death in childbirth and infant mortality would have been high – probably about 200 per thousand, which is comparable to seventeenth-century France and lower than Washington DC in the late nineteenth century, which had an infant mortality rate of 300 per thousand. Overall life expectancy was not, therefore, high – about 22–25 years, the same as India in the 1920s. Studies of modern hunting and gathering groups have found that about a tenth of the population is aged over sixty and it is unlikely that the early agricultural societies exceeded this figure.

There is little doubt that the adoption of farming and the subsequent rise of settled societies led to a significant reduction in human health in a number of ways. The domestication of animals for the first time brought humans and various animals into close contact – animals often lived alongside humans inside the primitive houses. Many diseases that were endemic to particular animal species were able to adapt, mutate and jump to humans. Many of the common human diseases are close relatives of animal diseases. Smallpox is very similar to cowpox and measles is related to rinderpest (another cattle disease) and canine distemper. Tuberculosis originated in cattle as did diphtheria. Influenza is common to humans, pigs and birds and the common cold came from the horse. Leprosy came from the water buffalo. The result is that after living for almost ten thousand years in close proximity with animals, humans now share sixty-five diseases with dogs, fifty with cattle, forty-six with sheep and goats, forty-two with pigs, thirty-five with horses and twenty-six with poultry. Although many of these diseases took time to develop because of the low population densities, they were to have a huge impact on human history.

The first impact of disease in the early settled societies would have been caused by poor or non-existent sanitation. The mixing of human and food waste with drinking water would have led to an increase in intestinal worms. More important would have been the emergence of typhoid caused by a member of the salmonella family of bacteria and spread through polluted water, direct contact and flies. It remained common across the world until well into the twentieth century, especially in towns which often lacked any sanitary system. It died out in the industrialised world in the early twentieth century and also in the richer parts of the developing world later in the century. Nevertheless in the late twentieth century there were still fifteen million cases of typhoid a year worldwide and over one million deaths a year. Alongside typhoid there would also have been other similar diseases such as dysentery.

Second, the development of irrigation in the first farming communities led to the spread of schistosomiasis, a fluke that causes extreme debilitation and listlessness. The blood fluke has an elaborate life cycle that involves humans and water snails as hosts at different stages. Irrigation ditches, where people stood for long periods of time, turned out to be prime breeding grounds for the water snail. The disease appears to have been widespread in Mesopotamia and Egypt from the earliest agricultural communities, in China as rice growing in paddy-fields developed but not, for some unknown reason, in the Indus valley civilisation. The disease has been endemic in these regions ever since. By the mid-twentieth century probably ten million people in China were affected and nearly half the Egyptian population suffered from the disease as cotton grown in irrigated fields became a prime commodity for export in the nineteenth century. The huge increase in the world's irrigated area in the twentieth century (about five-fold) has spread the disease to even more people. It is now endemic in seventy-four countries and 200 million people are infected with about a tenth of that number suffering severely from the disease.

Third, the adoption of agriculture would have produced a number of deficiency diseases through the much reduced variety of food available. Overall the quality of food in general is likely to have declined because storage for long periods reduces the level of a number of key vitamins and nutrients. The first deficiency disease to emerge was anaemia caused by iron deficiency. This disease was unknown among gathering and hunting groups but is found on a rapidly increasing scale in the first farming communities. It was the result of cereals making up too high a percentage of the diet and inhibiting iron absorption. In the Americas the disease emerged as soon as maize formed a large part of the diet. In China beriberi caused by thiamine deficiency and leading to nerve problems and a swelling of the face and limbs was common by 2000 BCE. It was caused by a diet too high in rice (usually over 80 per cent) when too much of the outer husk of the rice was removed. There was a major outbreak across south-east Asia in the late nineteenth and early twentieth centuries when the Europeans introduced mechanical rice mills which removed all the husk. The impact of the disease was reduced by adjusting the mills to leave some of the husk on the rice and after thiamine was made synthetically in 1936.

Scurvy, caused by a lack of vitamin C, was endemic across the world wherever there was a lack of fresh fruit and vegetables in the diet. It can first be identified in Egypt about 1500 BCE. The disease has a slow but cumulative impact, leading to the re-opening of old wounds, soft gums and loose teeth. The disease was particularly prevalent on European ships once they began making long sea voyages after 1600 and it probably killed about one million sailors in the next two hundred years. It was prevalent among children in late nineteenth-century Europe and North America as mothers shifted from breast milk to condensed and evaporated milk that contained no vitamin C. On the other hand pellagra (from the Italian for 'rough skin', one of the early symptoms of the disease which eventually causes dementia) was related to eating too much maize. This produced a niacin deficiency which could easily be cured by eating a well-balanced diet. This was not possible for the poor who often had no other source of food than maize. The disease was first recognised in Asturias in 1735 and then spread to France and Italy when maize (in the form of polenta in northern Italy) became the main crop for poor peasants. It was also common among poor African-Americans in the south of the United States in the late nineteenth and early twentieth centuries and the deprived African mine workers in South Africa who ate almost nothing but mealie-meal made from maize. Although the native Americans, especially in Mesoamerica, had a very high percentage of maize in their diet they avoided pellagra by treating their maize with lime and cooking it as tortillas.

Perhaps the strangest of the food diseases was ergotism which was common in Europe from around 600 CE until about 150 years ago. It was caused by the ergot fungus on grains, especially rye, and developed because of damp conditions especially during planting and later in storage. It produced reddening and blistering of the skin ('St Anthony's Fire') and in acute cases affected the central nervous system resulting in convulsions, fits and hallucinations. There was a major outbreak in France in 922 when 40,000 people died and again in Paris during the winter of 1128–29 when 14,000 died. In nineteenth-century Russia there were ten epidemics with an average death rate of 40 per cent of those infected. It returned during the collectivisation drive of the 1930s when there were four epidemics. Ergotism was almost certainly responsible for many of the outbreaks of religious mania in the Anabaptist movement that was common across Germany in the 1530s, and especially for the major outburst in Münster in 1534. Britain seems to have escaped the worst effects of ergotism because of the high proportion of dairy products in the diet which helped break down the poisons in the rye.

Overall the health and life expectancy of these agricultural societies was very poor. In the Levant in the early Islamic period (around 700 CE) the average age of death was the same as it had been in the earliest farming villages about 9000 BCE. In Scandinavia around 1200 CE the average lifespan was about eighteen years and half of all individuals died as children. A detailed study of over a hundred cemeteries north of the Alps dating to between 1000 and 1200 CE shows that, apart from the very high infant mortality rate, a third of males and a quarter of females died between fourteen and twenty (mainly from malaria, smallpox, dysentery and tuberculosis). Overall less than a quarter of the population survived until they were over forty. The result was that these societies had a very different population structure from modern societies. Only about 5 per cent of the population were aged over sixty-five and children, not the old, were the main class of dependents.

The 'Great Plagues' and their spread

After the emergence of settled societies it was the development of the first cities with their large and concentrated populations that began to expose humans on a major scale to the diseases that had originated with animals. Infectious diseases such as smallpox and measles that do not depend on transmission through water or another host species require a minimum number of human hosts in order to survive. Recent studies on islands show that if the population is less than about 250,000 measles will die out. There would have been no more than short-lived, local outbreaks of these infectious diseases until relatively large cities began to develop. It was not until after 500 BCE that any city reached this size and another five hundred years before Rome and the Chinese capital, Loyang, did so. It is not known where either smallpox or measles originated or how many early outbreaks there were simply because surviving accounts of disease outbreaks are not accurate enough to identify specific diseases and, in addition, the nature of some diseases has changed over time. The impact of these diseases was affected by another key element in the history of disease – the growth of trade and travel enabling diseases to spread relatively quickly from one area to another. The early civilisations traded with each other but it was not until the opening up of the two major cross-Eurasia trade routes between the Mediterranean and China – the Silk Road and the sea routes via India and south-east Asia – in the period between 200 BCE and 200 CE that diseases could spread across the whole continent. It seems likely that smallpox developed somewhere in the east of Eurasia and it almost certainly arrived in the Mediterranean in 165 CE in what is known as the 'Antonine plague', the first of the two great plagues of imperial Rome. Troops who had been campaigning in Mesopotamia brought the disease back with them and the outbreak of smallpox lasted for fifteen years with a death rate of a third to a half of the population in the affected areas. There was a major outbreak of smallpox in China in 161–162 (the likely origin of the outbreak in the Roman empire) and again in 310–312 when the death toll reached 40 per cent in many areas. This was followed in 251–256 by the first outbreak of measles which killed about 5,000 people a day in Rome at the height of the outbreak. From this time both measles and smallpox became endemic across Eurasia, although measles may have taken some time to become established in western and northern Europe because of the low and dispersed population.

Measles, although derived from animals, is a purely human disease that was much more virulent in its early history. The disease spreads before the characteristic rash becomes apparent but it also suppresses the immune system and increases vulnerability to other diseases and this is one of the major causes of death from measles. The impact the disease could have on populations that had not previously been exposed to it was demonstrated by two events at opposite ends of the earth in the mid-nineteenth century. Measles arrived in the Faroe Islands in March 1846 and by July over 6,100 people out of a total population of 7,864 were infected. In this case there was a limited amount of immunity in the islands after a previous outbreak in 1781. What happened to a totally unprotected population is demonstrated by events in Fiji some thirty years later. In 1875 King Cakobau visited New South Wales and travelled back to the islands on *HMS Dido*, accompanied by two other ships. All three vessels had people infected with measles on them. The King met all his sixty-nine chieftains and their retinues on his return which ensured that the disease spread across the islands. The minimum death toll from measles was 40,000, about a quarter of the population of the islands.

It was not until the late nineteenth and early twentieth century that death rates from measles declined in western Europe and North America as children became better fed and lived in less crowded conditions. Although the virus that causes the disease was identified in 1911 there was no protection until a vaccine was developed in 1963 and the only solution was isolation of the infected. The measles vaccine is not as effective as those for other diseases and cannot be administered during the first nine months of life. Nevertheless the impact of the vaccine was substantial. In the early 1970s the death toll worldwide was about eight million a year and this fell to about one million a year by the late 1980s as a major immunisation campaign took hold. However, take-up rates of the vaccine have declined (to around 70 per cent in most industrialised countries) and this means that measles still infects about forty million people a year and kills about two million. The disease is contained but not eliminated and if vaccination rates continue to fall a major outbreak is likely.

Like measles, smallpox (so-called because syphilis was the 'great pox') originated with animals but was a purely human disease needing large populations to survive and spread. It was endemic across Eurasia by 1000 CE and spread into Africa via the Trans-Saharan camel routes. Travel remained an important way of spreading the disease because the incubation period was about twelve days before the first symptoms were apparent. Death could be rapid through internal haemorrhages but if this was survived the characteristic pimples and pustules emerged and when they fell off they left a pock-marked skin. The disease could be spread via the pus or scabs but the main route of infection was airborne. From about 1500 in Eurasia the disease seems to become slightly less deadly and transformed itself into a major disease of childhood. In the seventeenth and eighteenth centuries it probably killed about 400,000 people a year in Europe and was the cause of a third of all the cases of blindness. Overall the death rate was about a third of those infected and of those who survived about three-quarters had pock-marked faces. (A survey in Vietnam in the 1890s discovered that 95 per cent of adolescent children had pock-marked faces caused by smallpox.)

The first preventive measures were adopted in China in the first century CE (strongly suggesting that this was where the disease originated). This was variolation – using the pus and scabs of victims to protect others. It gave people a mild dose of smallpox with a death rate of only 1 or 2 per cent. By about 1000 CE the use of variolation had spread westwards and was common in the Islamic world and Africa. It was not adopted in western Europe until the early eighteenth century after the wife of the British ambassador in Constantinople saw it being used. It was also introduced into the Americas by the slaves brought from Africa. About a century later in 1798 a British doctor, Edward Jenner, noticed that cowpox provided immunity to smallpox and began the practice of vaccination. Within little more than a decade nearly two million people across Europe had been vaccinated and the disease was brought under control.

However, progress outside of the industrialised world was much slower. In 1950 there were still fifty million cases of smallpox every year and over the course of the twentieth century about 300 million people died from the disease. The newly founded World Health Organisation proposed a world-wide eradication programme in 1953 but this was rejected by the industrialised countries on grounds of expense. The Soviet Union and China undertook their own programmes and eradicated the disease within a few years. The WHO's aim was accepted in 1958 but no money was provided and it was not until the newly independent countries of Africa and Asia had greater influence that the eradication programme was finally adopted in 1966, against the continued objections of the industrialised world. The programme was a huge success – smallpox was eliminated in west and central Africa by 1970, Brazil a year later and in Asia by 1975. The last natural case occurred in Somalia in 1977. The smallpox virus now only exists in a frozen state in special disease control centres in the United States and Russia. This success was possible because smallpox has no animal host and therefore vaccinating the human population was sufficient – all that was required was enough will and money to organise an effective programme. However, most countries stopped vaccination in the mid-1980s and the effectiveness of that vaccination declines rapidly after about 20 years. The pool of people who are now highly vulnerable to smallpox is therefore huge and any accidental or deliberate release of the virus could be catastrophic.

The third of the 'Great Plagues' of human history was bubonic plague. The disease is caused by a bacterium that usually infects rodents (particularly rats) and their fleas. Humans can be infected by flea bites and this usually happens after a major outbreak of the disease among rats, causing a high death toll which forces the fleas to shift to feeding on human blood. Humans have little immunity to the bacterium and cells in the area of the bite die quickly, creating a blackening pustule and blister. The disease spreads through the lymphatic system, causing swelling of the nearest lymph node (usually the groin or the armpit), which swells to the size of an orange about four to six days after the original bite. Within ten days of the bite about 60 per cent of those infected die. Some develop a secondary lung infection and produce an airborne form of the disease which is almost always fatal.

It is unclear where bubonic plague originated but the first known outbreak occurred in the eastern part of the late Roman empire in the early 540s. It seems to have spread from India to Egypt and then, probably on a grain ship, to the capital Constantinople. Here it killed about 10,000 people a day at the height of the outbreak. Overall the death toll seems to have been between a quarter and a third of the population, more in the rat-infested and crowded port cities of the empire. The disease spread into north-west Europe but the isolated, small peasant communities in this region are likely to have suffered least. The plague arrived in China in 610 (on a boat from India) through the port of Canton where it killed about a quarter of the population. There were more outbreaks of the disease in Europe until the 770s but then, for reasons which are unclear, it seems to have died out for almost 600 years.

Much more is known about the next and most serious outbreak of plague commonly known as 'The Black Death'. Its spread demonstrates how trade and travel is central to the spread of diseases. It seems to have begun somewhere in central Asia and spread into China in 1331. It was then spread along the overland trade routes and by the Mongol armies to reach the Genoan trading colony of Caffa in the Crimea in 1346. From there it was transmitted to the Mediterranean and southern France by late 1347 and northern France and southern England during the next year. Then it travelled north, reaching even Greenland before moving back eastwards, reaching Moscow in December 1350. Everywhere the disease struck, death rates rose rapidly, especially once the pnuemonic form took hold. Many tried to flee (as did the ten wealthy young men and women from Florence who tell the stories in Boccaccio's *Decameron*), thereby helping to spread the disease still further. Medical treatment was either non-existent or ineffective and public health measures were limited to disposing of the bodies (sometimes even this broke down) and boarding up the infected houses together with their inhabitants. The impact of the plague in Europe, Egypt and the Levant was terrible and far worse than in China. (It also spread to India and Africa but so little information is available that it is impossible to assess its impact.) In Europe and other areas around the Mediterranean about a third of the population died, in many cities it was higher and only a few remote areas escaped altogether. It produced human suffering and social dislocation on a huge scale as families were destroyed, communities collapsed and villages were abandoned.

The plague regularly returned to Europe for the next 350 years. Between 1350 and 1536 there was a major outbreak somewhere in Europe on average every eleven years and even in the next 150 years the rate only fell to once every fifteen years. During the seventeenth century about two million people in France died of the plague, including 750,000 in a single outbreak in 1628–32. The last outbreak of the disease in England was the Great Plague of London in 1665 – it had spread from Amsterdam. At its height in September about 6,000 people a week were dying and the King, Court and government moved to Oxford. Samuel Pepys, a senior official in the Admiralty, also left the city for the safer areas of Greenwich and Woolwich and in his diary he describes the impact of the disease during a visit to London on 14 September when he was horrified to see:

> dead corps's of the plague, carried to be buried close to me at noonday through the City in Fanchurch street – to see a person sick of the sores carried close by me by Grace church in a hackney-coach – to hear that poor Payne my water[man] hath buried a child and is dying himself – to hear that a labourer I sent the other day to Dagenhams to know how they did there is dead of the plague; and that one of my own watermen, that carried me daily, fell sick as soon as he landed me on Friday morning last . . . is now dead of the plague . . . to hear that Mr Lewes has another daughter sick – and lastly that both my servants have lost their fathers of the plague this week – doth put me into great apprehensions of melancholy, and with good reason.

The plague began to die out in north-west Europe in the late seventeenth century and the last major outbreak was in Marseilles in 1720–21 when the death rate was 50 per cent. In Eastern Europe the last major outbreak was in Moscow in 1770–71 where, again, about half the population died.

The disease remained endemic in south-west Asia, Egypt and India for far longer. There was no effective medical treatment and the only public health measure that worked was the imposition of a quarantine (first introduced in Ragusa – modern Dubrovnik – in 1377). The modern outbreak of plague began in Yunnan in the mid-nineteenth century during the disruption caused by a series of major civil wars in China. It reached Hong Kong and Canton in 1894 and spread to Bombay (and the rest of India) two years later. In 1900 it reached Sydney and San Francisco. The result was that, for the first time, the plague bacillus spread to the rodents of Australia, the Americas and southern Africa, where it is now endemic. Europe is now the only area of the world which has no naturally occurring plague in rodents. (In the Americas thirty-four species of burrowing rodents and thirty-five species of fleas now carry the disease.) There is no effective vaccine but modern medical treatment can reduce the death toll among those infected. The main area where it affects humans is now Africa, which has all but 1 per cent of the world's cases – there were 2118 known cases in 2003 and 182 deaths. Bubonic plague is only kept as a minor disease through public health controls – if these ever broke down it could return with devastating effect.

Other Eurasian diseases

One of the oldest human diseases is leprosy, which, although it originated with the water buffalo, only affects humans. It is a bacterial infection which produces severe disfiguring, especially of the upper lip and nose, and often a very hoarse, rasping voice. It was common in all early societies and although the first clear case can be dated to 600 BCE it had been present for much longer. The disease first appeared in Europe in the sixth century CE but was not widespread until the eleventh century. The reaction to the disease was typical of most societies – stigmatisation and isolation. In 1179 the Lateran council of the Catholic Church set out the rules for Christians to follow. The diagnosis of leprosy was equivalent to death – the leper was made to stand in an open grave as the priest said the ritual for the dead and threw earth over the leper's head. Lepers lost all rights over their property, could not attend church or be buried in a Christian cemetery and were regularly tortured during the Inquisition. In the early thirteenth century they were banned from walking in the streets of both London and Paris. By this time there were 19,000 lazarets across Europe to confine and isolate lepers.

From the fifteenth century the disease was in decline in Europe although it remained endemic in other parts of Eurasia. The reasons for this decline are obscure but may be linked to the rise of tuberculosis as a disease – if tuberculosis is caught first it seems to give immunity to leprosy. Lepers continued to cause fear and loathing (in late nineteenth-century Europe it was portrayed as part of the 'Yellow Peril'). Although the bacterium responsible for the disease was identified in 1873 there was little treatment, other than segregation, until the first drug treatments became available in the 1940s. However, drug-resistant strains of leprosy appeared in the 1960s but a major WHO programme (with free treatment with more effective drugs) has had a major impact on the disease since the mid-1980s. In the last twenty years over fourteen million patients have been cured and the number of cases has fallen from over five million. Leprosy has been eliminated in 113 countries and is now found mainly in Brazil, India, Madagascar, Mozambique and Nepal, where there are about 400,000 new cases a year.

One of the consequences of the rise of tuberculosis in Europe, mainly caused by a diet high in milk from tubercular cows, was the development of scrofula or 'The King's Evil'. This was an inflammatory kind of tumour in the lymph glands, especially the neck. It was rarely fatal, unlike the pulmonary tuberculosis that became common later. Scrofula could be passed from mother to child and before the nineteenth century it seems likely that between a third and a half of the population of western Europe suffered from the disease. It is unclear when the idea that the touch of an anointed monarch could cure the disease took hold. Louis IX of France seems to have been the first to begin the practice of touching in 1254 and English kings did so shortly afterwards. The practice died out in England after 1714 – Samuel Johnson was 'touched' by Queen Anne in 1712 but without being cured. In France Louis XV touched 2,000 children at his coronation in 1722 and although the practice ceased with the revolution in 1789 it was revived briefly by Charles X in the late 1820s.

A disease which arrived relatively late in Europe, although it had probably existed for a long time elsewhere in Eurasia, was typhus. It is another disease confined to humans and is caused by a microorganism 'Rickettsia' which is neither a virus nor a becterium and is spread mainly by the human body louse. The major outbreaks were found in crowded, unhygienic conditions when people were wearing layers of clothes where the louse could breed. It was therefore often known as 'war fever' and 'jail fever' and it was not until the mid-nineteenth century that it was clearly distinguished from typhoid (the symptoms of the two diseases are very similar). It produced a steady death toll (the fatality rate was about 10 to 40 per cent) but not a demographic catastrophe. It was brought to western Europe in 1490 by Spanish soldiers who had been fighting in Cyprus. It had an immediate impact. Four years later the French had to abandon the siege of Naples as it ravaged their army and had to do so again in 1528 when 30,000 soldiers died. It was rampant across central Europe during the Thirty Years War of the first half of the seventeenth century.

Typhus was one of the major factors that ensured that in every war until the early twentieth century more soldiers died of disease than from combat with the enemy. For example, in the Crimean war, of the 95,000 French casualties, 75,000 were from disease. Of the British death toll of 22,000 only 4,000 resulted from combat. The first army in which more soldiers died as a result of fighting the enemy rather than from disease was the Japanese in the war with Russia in 1904–05. The major western European powers learnt from the hygienic precautions taken by the Japanese and the role of the louse in spreading typhus was finally identified in 1909. It was only the establishment of 'delousing stations' and other measures that avoided a major outbreak of typhus and enabled the slaughter on the western front to continue for four years during the First World War. However, typhus was rampant on the eastern front and spread widely during the chaos brought about by the Russian revolution – about thirty million people were infected and over three million died of the disease. There were further outbreaks during the Second World War (for example in Naples in 1943–44) but these were controlled. There is still no effective vaccine for typhus although the disease can be controlled using antibiotics. It has retreated to poorer, colder areas where the louse can survive in clothing. The disease can remain latent in patients for decades after recovery and a new epidemic could easily develop if louse numbers ever multiplied.

The second great transmission of diseases

Until the end of the fifteenth century the Americas remained isolated from Eurasia and Africa. The large cities such as Teotihuacan and the Aztec capital of Tenochtitlan are likely to have produced endemic intestinal diseases but it is clear that the lack of domesticated animals, in particular the complete absence of cattle, meant that the diseases of Eurasia (in particular smallpox and measles) were unknown. After several millennia of exposure to these diseases Europeans had developed a considerable degree of natural immunity. The native Americans had no such immunity and the impact of these European diseases was therefore catastrophic.

The first disease to strike the Americas was smallpox, which reached the Antilles in 1519. That it had not arrived earlier was mere chance – the long voyage and the lack of disease carriers on the ships. It reached the mainland and Tenochtitlan with the relief expedition sent to help Cortés after he and his group of adventurers were forced to flee the capital. The outbreak of disease caused massive disruption in society – there were so many dead that the survivors were unable to bury them all and the bodies rotted in the streets. Whole families were wiped out. Those that did survive would usually have been pockmarked, blind or both. Even more disturbing, the Spanish would have been largely immune to the disease. The Spaniards were left with only a demoralised wreck of a society to conquer. The same impact of disease occurred when the Spaniards moved on to the conquest of the Incas in the 1520s. The impact of smallpox was compounded by the arrival of measles in 1530–31 and then typhus in 1546.

The overall scale of destruction brought about by these Eurasian diseases is almost unimaginable. The population of central Mexico (roughly equivalent to the Aztec empire) was twenty million in 1500, equivalent to a quarter of the European population and four times as big as that of Britain. Within a century there were less than one million people in the area. The population of the Inca area fell from eleven million in 1500 to less than a million in 1600. The population of the Caribbean (about six million in 1500) was almost entirely exterminated. A population collapse of about 90 per cent may seem unbelievable but it is consistent with other evidence of the impact of disease on populations with no natural resistance. How many people actually died? It is unlikely that the figure is much less than 100 million during the sixteenth century. Death on this scale is the greatest any society has ever had to bear. In addition it was combined with immense cultural shock as the Europeans looted the area, imposed Christianity and effectively enslaved the remnants of the native population.

The other major disease to be taken to the Americas was yellow fever. This disease had long been endemic in west Africa where it had passed to humans from monkeys and had established itself in a mature form of low death rate and high levels of immunity among the local population. It is spread by mosquitoes but is difficult to transmit over long distances because the disease only lasts for seven to ten days and the mosquito has to live long enough for the virus to mature in its body before it infects another human host. The disease affects the liver, producing jaundice and the characteristic yellowing of the skin before extensive internal bleeding and a black vomit. Death rates in human populations newly exposed to the disease are around 60 per cent, often higher. Its impact meant that Europeans could establish no more than a foothold in west Africa for centuries – a series of small trading posts along the coast. Records kept by the British Royal African Company show that for every ten soldiers sent to west Africa between 1695 and 1722 six died in the first year, two more died between years two and seven and only one man survived to be discharged in Britain. Until well into the nineteenth century the Royal African Corps was composed of military criminals who exchanged their sentence for service (and a likely death sentence) in Africa – in some years death rates approached 80 per cent.

The transmission of the disease to the Americas did not take place for more than a century and was the direct result of the European slave trade which developed in the early seventeenth century. It first reached Barbados on a slave ship in 1647. Although the Africans taken as slaves to the Americas had considerable immunity (which increased demand for them as slaves) it ravaged the European populations of the Caribbean. In 1655 the French sent an expedition of 1,500 soldiers to conquer St Lucia – only 89 survived. A similar expedition to reconquer Haiti in 1801 produced 27,000 deaths. The disease also spread into the southern United States – Florida had major outbreaks in the 1820s and 1830s. When it reached Memphis in 1878–79 the death rate was over 50 per cent and the city almost ceased to function. Occasionally it reached European ports such as Lisbon, Barcelona, St Nazaire and Swansea but did not take hold because the mosquito could not survive.

The way in which the disease was transmitted was first established by a United States army doctor, Walter Reed, in 1900 and it was control of mosquitoes through drainage and insecticides (at first pyrethrum) that made the construction of the Panama Canal possible. An effective vaccine was developed in the 1930s but it is difficult to eliminate the disease because it survives in the monkeys of the tropical forests. There were major outbreaks throughout the twentieth century in Central America and Africa (30,000 people died in Ethiopia in 1960–62) and it is still endemic in Brazil. By the early twenty-first century there were 200,000 cases a year with a death toll of 30,000 but this is likely to be a severe under-estimate because of large-scale under-reporting. The disease now seems to be increasing again and the much greater speed of travel now makes it far more likely that it could spread from one area to another even though an outbreak ought to be containable with an effective vaccination campaign.

A disease related to yellow fever that was also taken to the Americas was dengue fever (also known as 'breakbone'). In its haemorrhagic form (which mainly affects children) the death rate is about 30 per cent. In its common form it is rarely fatal but produces fever and intense pain in the joints and bones. It is spread by mosquitoes and because the disease exists in four subtypes and immunity is only obtained after all the types have been contracted it is very difficult to produce a vaccine. The only treatment is therefore 'supportive' – pain relief and fever reduction. The disease seems to have originated in south-east Asia and then spread to India before reaching Africa. The first major outbreak in the Americas occurred in Philadelphia in 1779–80. Puerto Rico had nine major outbreaks in the decade after 1977. Dengue fever seems to be spreading as mosquitoes become immune to pesticides, urbanisation continues to increase and air travel becomes more frequent. About one and a half billion people live in the areas affected by the disease and the number of cases per year must run into several millions.

The transmission of diseases between the Americas and Europe in the period after 1500 as the Atlantic economy constructed by the Europeans developed is unlikely to have been in just one direction. Syphilis is closely related to other treponematoses diseases – yaws (known as bejel in south-east Asia) and pinta. It is caused by a spirochaetal bacterium and humans appear to be the only natural host. Syphilis can be contracted from the open lesions but is normally transmitted sexually. The disease has three phases with latent periods in between. The first occurs about three weeks after infection with a small lesion usually at the point where the bacterium entered the body. After a month or so this disappears and there a latent period of several weeks before multiple lesions appear all over the body (and often internally as well), hence the original name 'great pox'. The third phase may not occur for several years when the disease attacks the internal organs, skeleton and often the central nervous system, producing insanity and death.

The first outbreak of syphilis in Europe is unusual among diseases in that it can be dated precisely to the early 1490s. The first cases seem to have been in Barcelona in 1493 and it was then spread, as with so many diseases, by armies. The French army of Charles VIII took it to Italy in 1494 and when the army was disbanded after the failed expedition it spread further. It reached England and the Low Countries in 1496 and Hungary and Russia in 1499. It then spread along the Eurasian trade routes and the speed at which it did so demonstrates how closely integrated Eurasia was becoming. By 1498 it was in Egypt and India and by 1505 it arrived in Canton and from there spread to Japan. Every society saw it as a foreign disease and the names chosen relate to how it spread. To the English and the Italians it was the 'French disease'. The French called it the 'Naples disease' and to the Poles it was the 'German disease'. The Indians called it 'the disease of the Franks' and to the Chinese it was 'the ulcer of Canton'. The Japanese called it both the 'T'ang disease' (suggesting a Chinese origin) and 'the Portuguese disease'.

The crucial question is whether the disease was brought back by those on Columbus's voyages (the first was in 1492). Certainly contemporaries were absolutely convinced that this was how it arrived in Europe. That opinion is reinforced by the fact that that there is no unequivocal description of the disease in Europe before this time and if it did exist it must have been in a very mild form. There is little doubt that syphilis was a new disease in the 1490s because of its extreme virulence. If syphilis did not come from the Americas then the existing disease, yaws, would, coincidentally, have had to have changed its nature around 1490 so that it was transmitted sexually and became much more severe. This might have happened if it crossed with another, perhaps mild, treponematoses disease from the Americas. However, it seems far more likely that syphilis was the price that the Europeans paid for their 'discovery' of the Americas.

The initial impact of syphilis in the early sixteenth century in a population with no resistance was terrifying and it usually reached its tertiary and terminal form very quickly. By the 1520s Europeans knew that it was transmitted sexually and within a another couple of decades the severity of the disease was already in decline as it became endemic rather than epidemic. By the seventeenth century it was a dangerous infection but far less violent in the symptoms it produced. However, there was no known cure until the mid-twentieth century. There were two popular remedies. The first combined mercury or arsenic with sweating. This seemed to slow up the disease but because both metals are toxic the treatment probably killed as many people as the disease. The second, guaiacum, made from the wood of a tree found in the West Indies, was, as early as the 1530s, reckoned to be ineffective but at least it did not kill the patient. With no alternative available it remained a British-approved drug until 1932.

The bacterium that causes syphilis was first identified in 1905 but there was no effective treatment until the development of penicillin in the early 1940s and even then treatment was still patchy because of the strong moral overtones associated with the disease. In the second half of the twentieth century tertiary syphilis and death could be avoided by treatment. But the disease has not been eliminated and by the late twentieth century was on the increase. It remained an endemic and debilitating disease although it could be contained given reasonable public health systems.

The perennial diseases: malaria

Malaria is one of the longest-established human diseases and has proved to be one of the most difficult to conquer. It is first known in China by 2700 BCE and by 1600 BCE it can be identified in Mesopotamia and the Indus valley although it was probably present long before this date. In the early twenty-first century it remains the major cause of death in the developing world and about 40 per cent of the world's population is at risk from the disease. The name comes from the Italian 'mal' aria' or 'evil air' from the notorious marshes around Rome where the disease was endemic. The disease is caused by a plasmodium (probably related to a type of algae) and is spread by mosquito bites. Four types affect humans, of which falciparum is the most severe and can cause death. Others are highly debilitating, causing continual fevers and lassitude – in the vivax form of the disease these occur every other day. Malaria almost certainly originated in monkeys and first spread to humans in south-east Asia. From there it spread everywhere that its main carrier, the anopheles mosquito, could survive. Falciparum malaria is limited to the tropics but vivax is more tolerant of the cold and can survive as far north as England and southern Canada. For several millennia malaria has been endemic across much of Eurasia and Africa. It was probably spread to the Americas by the early European settlers and the African slaves they took to the continent. Only the relative in-efficiency of the American mosquito kept the Caribbean relatively free of the disease.

There was no cure for the disease although the Chinese used 'Qinghuasu' from the sweet wormwood tree to lessen the symptoms.

The Europeans discovered that the bark of the cinchona tree that grew in the Andes provided a partially effective remedy. Why it did so was unknown until two chemists in the 1820s isolated the active element in the bark – quinine. By the 1830s this was manufactured in large quantities and could be used as a cure for vivax malaria. By the 1840s it was being used in large doses as a prophylactic before infection against falciparum malaria. It was these discoveries that made European settlement of Africa possible in the later nineteenth century even though death rates remained high. The route by which the disease was transmitted was finally established by a British army doctor in India, Ronald Ross, in 1897 although he relied on a large amount of earlier work by others. Control measures – mosquito nets, quinine and the draining of swamps (for example the Pontine marshes around Rome) – reduced the impact of the disease in the early twentieth century.

It was the development of two new chemicals that seemed to promise the elimination of the disease. The pesticide DDT, first used in the 1940s, killed mosquitoes and could be sprayed on to walls where it remained active for up to three months and therefore broke the cycle of infection. At the same time the development of the highly effective chloroquinine attacked the disease once it had been contracted. In 1955 the WHO declared that it was possible to eradicate malaria. Almost immediately it became apparent that this was an impossible aim even though the goal was not formally abandoned until 1992. By the 1960s the highly toxic nature of DDT became apparent and its use had to be abandoned. At the same time strains of malaria that were resistant to chloroquinine began to emerge. The situation worsened in the 1980s when strains of multiple-drug-resistant malaria emerged along the Cambodia–Thailand border. By the early twenty-first century there was only one drug that could cope with all types of malaria and the WHO warned that if it continued to be used on its own then it would not be long before resistant strains emerged. In these circumstances the world would be back in the situation found before the 1940s of having no effective drug treatment.

The situation would, however, be far more serious than eighty years ago. Rapid population growth, resistant strains of mosquitoes and climate change mean that whereas 10 per cent of the world's population was at risk from malaria in 1960 that figure has now risen to 40 per cent. Malaria is now the main cause of death in the developing world with over 90 per cent of infections taking place in sub-Saharan Africa. Between 300 million and 500 million people are infected with the disease and the death rate is between 1.5 million and 2.7 million people a year, of whom one million are children. In the long interaction between humans, the environment and disease, malaria remains one of the most significant human failures.

The endemic diseases: influenza

One of the most persistent diseases in the world has been influenza. Like so many other diseases it originated with animals, and now develops new strains in domesticated pigs and chickens; it can also be spread by wild birds. The virus responsible mutates easily and rapidly and new strains constantly arise. As an airborne disease it is easily spread between humans through sneezing and coughing. In most years the death rate is low – about 1 per cent – and it is usually the young and the old who are most affected. There have, however, been numerous pandemics when a new strain emerges because previous infection provides only limited immunity against the new variants of the disease.

There is no unambiguous mention of influenza before 1510 although it must have existed for a long time before then. There were a series of pandemics in Europe in the sixteenth century in 1510, 1557 and 1580. Its incidence seems to have declined in the seventeenth century before re-appearing regularly in the eighteenth century with a major pandemic in 1781–82. In the nineteenth century there were three great pandemics across the world: in 1830–31, 1833 and 1889–90. By far the most serious of all the influenza pandemics was that at the end of the First World War. It first appeared in the spring of 1918 in the United States and appeared to be very mild before mutating in August 1918 into a much more virulent form. It appeared almost simultaneously in three ports – Boston (Massachusetts), Freetown in west Africa and Brest – and was clearly linked to troop movements in the latter stages of the war. From these centres it spread rapidly across the world in the winter of 1918–19 and reappeared in a milder form during the next winter. This influenza variant seemed to lead very easily to pneumonia and uniquely half the victims were healthy adults between the ages of twenty and forty. (Eighty per cent of the casualities suffered by the US army in the First World War were from influenza.) This may have been related to conditions at the end of the First World War – widespread malnutrition (especially in Europe) and social and economic breakdown in many countries. The death toll across the world was between thirty and forty million. India may have accounted for almost half this total with 500,000 dying in the United States and 250,000 in Britain. In other areas the impact was even greater – in Western Samoa about a fifth of the population (7,542 people) died of the disease. Overall, however, the death toll among those infected was only about twice the normal rate with influenza – the problem was that about a fifth of the people in the world contracted the disease.

The influenza virus was first isolated in 1933 and all but one of the world pandemics in the rest of the twentieth century were caused by mutations in the virus. There were major outbreaks of the disease caused by mutation in 1933, 1957 and 1968 (the 1977 outbreak was caused by the reappearance of the 1933 strain). Death rates though remained low – usually about 50,000 overall. The problem in dealing with the disease is that vaccines have only limited effectiveness and their use is usually restricted to vulnerable groups in the population. Because the influenza virus can mutate so easily a new pandemic is inevitable at some time.

The diseases of industrialisation and urbanisation

From the late eighteenth century there was a wave of rapid population growth, industrialisation and urbanisation in western Europe and North America. The environment that these changes produced resulted in a wave of new diseases. One of the most important was tuberculosis, which seems to have shifted from its lymphatic form, common in the previous centuries, into the much more deadly tubercular form. This latter form had existed for several thousand years but it was not a major problem until the nineteenth century when there was growth of cities and the development of large slums with people existing on poor diets and living in grossly overcrowded conditions. It is caused by a bacillus that is spread through the air by coughing and spitting. It gradually destroys the lungs but also causes a general wasting away of the body, hence its other name 'consumption'. Not all who are infected develop the disease and its main impact is on the malnourished living in damp conditions. It was the main cause of urban death in nineteenth-century Europe and North America. For most people there was no effective treatment but the affluent were able to spend several years in a sanatorium in the mountains resting and eating a diet rich in dairy products – a world described in Thomas Mann's *The Magic Mountain*.

The bacillus that caused tuberculosis was identified in 1882 but there was no vaccine until 1921 and even then it was only partially effective. The first drug treatment became available in 1952 but, as with so many treatments, resistant strains of the TB bacteria emerged rapidly. Strains that are resistant to a single drug now exist throughout the world and in Russia there are some that are resistant to all drugs. Tuberculosis remains a disease of poverty – currently there are about forty million TB cases diagnosed every year with a death toll of almost three million a year. Both figures began to rise steadily from the early 1990s. Nine out of ten cases are in developing countries (a third of all cases are in India) but rates elsewhere are rising too. In New York City rates doubled in the 1980s, primarily among the poor and homeless. There is still no fully effective vaccine.

The effects of poor diet available to most people who lived in the growing urban conurbations of the emerging industrialised world were compounded by the huge palls of smoke from industry and the burning of coal in factories and in the home that cut out much of the sunlight. This produced an epidemic of a deficiency disease known as rickets. This is caused by a lack of calcium and also vitamin D, which comes mainly from sunlight but can be found in foods such as liver, egg yolks and fish oils. The main victims of the disease are children who need calcium and vitamin D for bone growth and without it grow up with deformed skeletons. Rickets can first be identified in China around 1000 BCE, in the Mediterranean about a millennium later and in a mild form in medieval Europe. In most areas of the pre-industrial world sunlight partially compensated for the poor diet. In England rickets increased from the early seventeenth century as coal-burning became common in the main cities. By the nineteenth century it was known as the 'English disease' with nine out of ten children in the main cities suffering some symptoms. The other group that suffered from rickets was African-Americans. Their black skin absorbed only one-third of the sunlight of a white skin and so the slaves and their descendants in the less sunny climes of North America were badly affected. The impact of the disease slowly lessened during the early twentieth century as diet improved and the skies over the industrial cities of Europe and North America gradually began to clear.

It was the insanitary conditions of the cities in the industrialising world that produced a formidable epidemic that caused fear amongst both rich and poor – cholera. The disease is caused by a bacillus that only infects humans but can survive outside the human body without an animal host. It spreads through polluted drinking water, causing major intestinal problems and dehydration. In its original form the death rate was about 50 per cent for healthy adults but much higher for children and the old. The disease had been long-established in the lower reaches of the Ganges and its spread to the rest of the world in the early nineteenth century is another example of the impact of transportation improvements on the relationship between humans and disease.

There was a major outbreak of cholera in Bengal in 1817 and the movement of troops, both British and Indian, spread the disease rapidly to other parts of India such as Mumbai, most of which had never suffered from the disease. In the next hundred years cholera killed about thirty-eight million Indians. From India cholera was spread by ships and war. It reached the Gulf in 1821 and from there travelled along the trade routes to east Africa. In 1826 it infected the Russian army fighting in the Caucasus and in 1830–31 that army brought it westwards into the Balkans. Within a few months most of the cities of western Europe had the disease and by 1832 it had reached the Americas. Death rates were high because of the lack of effective sanitation systems – sewage and drinking water intermingled freely. There was a long controversy over how the disease was caused and spread. Many thought that it was caused by a miasma of foul air and it was not until an English doctor, John Snow, disconnected the handle of the pump at the well at Broad Street in London during September 1854 and stopped an outbreak of cholera that opinion gradually shifted towards the correct cause of the disease. It was the impact of cholera that slowly forced governments across Europe and North America to construct proper water-supply and sewage-treatment systems. This occurred long before the bacillus was identified in the 1890s.

Ever since the early nineteenth century there have been waves of cholera outbreaks that have swept across the world. Luckily, since the mid-twentieth century it is the milder 'El Tor' strain of the disease that has been dominant. The seventh and latest cholera pandemic began in Indonesia in 1961. By the mid-1960s it had spread to India, Iran, Iraq and the Soviet Union. By 1970 it had reached west Africa (for the first time in a hundred years) and by 1991 could be found in Latin America. It remains prevalent in areas with poor sanitation and where public health systems break down under the strain of war and civil war. It has largely been eliminated in the wealthy, industrialised world.

The role of environmental and medical improvements

In the last century and a half there has been a remarkable transformation in the pattern of human disease that had been prevalent for several millennia. For most of human history a majority of children died within a few years of birth. Now in the rich, developed world only about 1 per cent of children fail to reach the age of five and those that do die usually suffer from rare, inherited disabilities. Life expectancy at birth has increased dramatically from between thirty and forty years at the end of the eighteenth century to well into the seventies today. Death rates have fallen steadily. In the 1840s in England about twenty per 1,000 of the population died every year – current rates are about a quarter of that figure. Although the timing of these changes varied from country to country (and there are still differences between them) the overall pattern is similar.

The main reason that people are living longer is that far fewer are killed by the infectious diseases that were the scourge of human history for thousands of years. How to explain this decline is a subject of considerable debate. Some diseases appear to have evolved over time into less virulent forms – this is certainly the case with scarlet fever and syphilis and it is a result of the way in which viruses and bacteria can survive (if they kill too many of their hosts they will die out). Another key factor is advances in medical knowledge and treatments. Vaccination eliminated smallpox as a disease and controlled diseases such as measles and yellow fever even if it could not eliminate them. A clear illustration of the impact of vaccination can be seen with polio which, for reasons that are unclear, became an important disease in the early twentieth century after having been endemic at a very low level for thousands of years. People such as President Franklin D. Roosevelt suffered from the disease and at its height in the 1950s in Europe and North America it left tens of thousands of people a year badly paralysed. Once a vaccine had been developed in the mid-1950s governments could organise mass vaccination programmes that eliminated the disease – the Soviet Union was the first to do so. By the mid-1990s the WHO had ensured that about half the world's children under the age of five (a total of 400 million) were vaccinated. The world death rate fell from 350,000 a year in the mid-1980s to about 3,000 at the beginning of the twenty-first century and the number of countries that had the disease was reduced from 125 to seven. However, eradication of the disease is unlikely because this would require a near-universal vaccination programme – only 1 per cent of those infected with polio show any sign of the disease. Even so, medical advances can only go so far and the development of a vaccine is only part of the story. Equally important is the will of governments and the international community and the provision of enough money and resources to create an effective vaccination programme.

Apart from vaccination the most important medical advance was the development of antibiotic drugs in the mid-twentieth century. The first, penicillin, was discovered accidentally by Alexander Fleming in 1928. However, it was more than a decade before it could be made on a substantial scale and until the mid-1940s its use was mainly restricted to the Allied armies fighting in the Second World War. By the late twentieth century there were over 25,000 different types of antibiotics. They played a crucial role in curing diseases and saving lives in the last fifty years of the twentieth century. However, bacteria were able to evolve new, resistant strains. Medical treatment became a race to develop new drugs as resistance to existing drugs developed. The situation was worsened by the misuse of antibiotics – over-prescribing (often for diseases that did not need such treatment) and because they were fed as part of the routine diet of animals kept in increasingly unhealthy conditions in the industrialised world as agriculture became ever more intensive. (In the United States animals are given thirty times more antibiotics than the human population.) The first bacteria resistant to penicillin developed as early as 1946 but these could usually be killed by other antibiotics. By the mid-1970s the inevitable happened – multiple-drug-resistant bacteria developed. In 1977 the first incurable strains of TB were found in South Africa and have since spread to much of the former Soviet Union and eastern Europe. By the late twentieth century over fifty million people in the world were infected with this form of TB. Similar MDR strains of dysentery and cholera also emerged (the latter in Bangladesh in 1992). By the early twenty-first century it seemed increasingly likely that the mid-twentieth century development of antibiotics might be no more than a temporary gain in the long human battle with disease.

The impact of medical advances played an important but limited part in the reduction of disease in the industrialised world. Of far greater importance were a better diet and environmental improvements. Public health measures were crucial – effective sewage systems and treatment of drinking water drastically reduced the impact of water-borne intestinal diseases such as dysentery and cholera. Overall about a fifth of the reduction in mortality levels in the nineteenth century came from these programmes. A better diet in Europe as more food became available, and better housing (which reduced over-crowding, damp and poor ventilation) also improved resistance to disease. All of these improvements were crucial in the reduction of tuberculosis as was the prevention of spitting in public places, the construction of sanitoria to isolate patients and the slaughter of infected cattle. The pasteurisation of milk (first undertaken in Chicago in 1908) was another factor. The death rate from TB in England was reduced by half between 1838 and 1882 before the bacillus that caused the disease had even been identified. (A similar pattern can be seen with cholera where the bacillus was not identified until the 1890s.)

The mortality pattern of the industrialised world in the nineteenth and early twentieth centuries was not repeated in the developing world in the twentieth century. Here the reduction in mortality rates was far greater and quicker. This partly reflected the appalling conditions of the early twentieth century when average life expectancy was about twenty-five years at birth (about half that of western Europe and North America). By the late twentieth century this had more than doubled to sixty-three years. This was the result of the importation of advanced medical techniques, vaccination, antibiotic drugs and the chemical spraying of mosquito breeding grounds. This had an imme-diate impact. For example, in Mauritius the death rate fell by 80 per cent in little more than a decade after the mid-1940s – a similar decline took over 150 years to achieve in Europe. However, by the last third of the twentieth century the fall in the death rate slowed and in some countries went into reverse. This reflected the limits of medical intervention in the face of major social, economic and en-vironmental problems – malnutrition, poor water supply and sanitation and the rapid growth of slums as the urban population rose swiftly. The result was a huge disparity in the impact of disease and death between the rich and poor countries. In the late twentieth century the chances of a child born in Sierra Leone dying before the age of five was thirty-three times greater than for a child born in Iceland. Deaths before the age of five accounted for half of all deaths in Bangladesh. Child mortality in Togo and Benin was at the same level as Sweden in 1800.

The impact of medical improvements was also limited in dealing with some of the endemic diseases of the poorest countries as we have seen in the cases of schistosomiasis and malaria. Similarly trypanosomiasis (African sleeping sickness), which is spread by the tsetse fly, still rules out large areas of sub-Saharan Africa to human settlement. In the twentieth century there were major outbreaks of the disease in Uganda and the Congo basin (1896–1906) and across much of sub-Saharan Africa in the 1920s. By the late twentieth century the WHO estimated that there were about 500,000 cases a year but with under-reporting of the disease rife an accurate estimate is impossible. What is known is that in parts of Angola and the Congo up to half the population suffers from this debilitating disease.

The new infectious diseases

In the late twentieth century a range of 'new' infectious diseases emerged, primarily in Africa. The rapidly rising population and the consequent demand for new agricultural land meant that settlements pushed further into the tropical forests. This had long been a source of human diseases (for example yellow fever) as infections once confined to monkeys and other primates could jump across the species (as so many diseases had done in the past) to infect humans. It is here that the most devastating of the new diseases – AIDS – spread from chimpanzees (where it had long been endemic) to humans somewhere in West and Central Africa probably in the 1920s – the first cases can be identified in Kinshasa in 1930. It did so almost certainly through hunting and eating of the animals. It became more widespread during the Angolan civil war of the 1970s and 1980s, reaching South Africa and eventually the whole of the world.

AIDS begins as HIV (human immunodeficiency virus), where the infection by the virus produces a vigorous immune response but which, unusually, does not kill off the virus. There can be a very long period (often up to a decade) without any major symptoms before full-blown AIDS develops and the immune system collapses, leading to infection by other diseases. The slow progress of HIV (and the lack of symptoms) gives the virus many opportunities to spread but unusually only about 5 per cent of those who are exposed to the disease are infected (with smallpox the rate was about 99 per cent). HIV spreads through sexual contact, intravenous drug use and the increased use of blood and blood products – in Japan half of all haemophiliacs were infected by contaminated blood and in many countries, especially France, blood products were not treated for years even though the route of transmission was known. Between 30 and 50 per cent of children are infected if their mother has the disease – in the womb, during birth or through breast-feeding.

AIDS has become the greatest of the modern plagues and only the fact that it kills people slowly has lessened its impact on public consciousness – in some places death rates are as high as during the Black Death in Europe. In the early twenty-first century about forty million people in the world were infected with HIV/AIDS (twice the number only a decade earlier), there were five million new cases every year and the death rate was just over three million a year. The disease is, like so many others, a disease of poverty. It is overwhelmingly concentrated in Africa where there are over 25 million people infected – over 7 per cent of the population of sub-Saharan Africa. In Botswana and Swaziland about 40 per cent of adults have the disease. (In western Europe infection rates are about a third of 1 per cent of the population.) Life expectancy is falling dramatically in these countries and by 2010 there will be about twenty million children in sub-Saharan Africa who have lost both parents to the disease. There is no vaccine available and aggressive drug therapy which can slow up the disease to a remarkable extent is very expensive – about $15,000 per person per year. The average medical expenditure per person in Africa is $6 a year and so, not surprisingly, only 30,000 people receive drug treatment.

The other new diseases that emerged in the late twentieth century were the group known as haemorrhagic fevers although they are caused by different viruses. The first to emerge in 1969 (though it had existed for much longer) was Lassa fever (named after the town where it was first identified in Nigeria). Its natural carrier is a mouse but the virus is highly virulent when it crosses to humans, with a death rate of up to 60 per cent through internal bleeding. Only the fact that it is difficult to spread directly from human to human (the only known route is contaminated blood) has restricted the spread of the disease. A similar disease that causes death through massive internal bleeding is Ebola, first identified in 1976 in western Sudan and the Congo. In the last thirty years there have been regular outbreaks across central Africa. In the 1,850 known cases the death toll was 1,200 but far more people have been infected; because they have never been treated in the medical system the cases are not recorded. There is no vaccine against the disease and, like Lassa fever, only the fact that is contracted through direct contact with contaminated blood has restricted its spread. The third disease is hantavirus which was probably first identified in China a thousand years ago and spread to the southern United States in the early 1990s where it took on a new form. It is carried by the deer mouse but if passed to humans the death rate is about 50 per cent. The disease has spread from New Mexico to Florida and as far north as Rhode Island.

The diseases of affluence

The major decline in infectious diseases in the industrialised world in the last two centuries has been paralleled by the rise of new diseases that have radically altered the way of death for this minority of the world's population. High infant mortality and early death from infectious diseases have been replaced by cancer and cardiovascular disease which together now account for two-thirds of deaths in the affluent societies of the industrialised world. Part of the explanation for this change is that people are now much more likely to live into old age and become susceptible to degenerative diseases, especially if, as with some cancers, they have a genetic origin. However, some of the explanation, particularly with other types of cancers, lies in environmental factors – the increased levels of pollution, in particular from the highly toxic artificial chemicals produced in the second half of the twentieth century.

Another factor, especially in the rise of cardiovascular diseases, has been the major change in the diet over the last two hundred years. Many of the dietary changes have been beneficial in eliminating deficiency diseases. People are also much taller than in the past because they are better fed. Although it is difficult to generalise, the average height of the population of medieval Europe was about 30 cm less than now. Contemporary British children are about 20 per cent taller than they were two hundred years ago. However, many of the dietary changes have been harmful, in particular a reduction in fibre intake, a rise in sugar consumption, much higher levels of fat intake and a higher proportion of processed foods. White bread is an example of processing that reduces the amount of fibre and nutrients in food. The removal of bran and germ from flour was probably first adopted in the fourteenth century but white bread remained a luxury item for the rich because the frequent shortages of grain meant that all of the available flour had to be used so as to maximise bread production. (At times of acute grain shortage governments often banned the making of white bread.) Before 1750 only about a twentieth of the population of France ate white bread and it was not until the late nineteenth century in Europe, as grain shortages disappeared, that white bread became common, eventually dominating the bread market by the mid-twentieth century. The reduction in the fibre content of food in the industrialised world has been associated with the rise of many complaints such as constipation and intestinal disease together with colon and bowel cancer.

The rise in sugar consumption has had an even more deleterious effect on health. Until the rise of the great slave-worked sugar plantations of Brazil and the West Indies in the seventeenth century the use of sugar was almost unknown in western Europe. Foods were sweetened using honey (or maple syrup in North America). By 1750 sugar intake in Europe and North America had risen to about two kilograms per person per year. It is now over thirty times that rate. The most immediate effect was a rapid rise in dental caries. Prehistoric skeletons suggest that less than 3 per cent of teeth were affected and in many societies with low sugar consumption tooth decay is almost unknown. The rise in sugar consumption is also directly linked to the increase in the number of people suffering from diabetes. This was a very rare disease before the eighteenth century when it first became noticeable among the English aristocracy who had a high consumption of sugar. Diabetes rates have risen sharply since the early twentieth century – it now affects 1.8 million people in Britain (3 per cent of the population) and in the United States (where sugar consumption is even higher) it is twenty-one million people (7 per cent of the population). World-wide the figure is about 150 million but this is expected to double in the next twenty years.

Fat intake has increased throughout human history. The first major step was the 'secondary products revolution' – the use of goat, sheep and cow milk to create dairy products. However, the poor grazing and lack of fodder in the early agricultural systems meant that these animals had only a low output of milk and poor transportation meant that it was difficult to move products with a short life. All these factors kept dairy consumption at relatively low levels. Technological changes in the late nineteenth and early twentieth centuries – refrigeration, pasteurisation, canning and faster transport via railways – made dairy products regularly available to the rapidly rising urban population. The greater consumption of meat also raised fat levels in the diet. These technological changes were also central to the rise of a new phenomenon – the food industry, which concentrated on selling processed food rather than distributing fresh food. Over the course of the twentieth century the consumption of processed food by the average American tripled and the consumption of fresh fruit and vegetables fell by over a third. Processing food not only removes many of the nutrients and important trace minerals but also introduces additives such as antioxidants, emulsifiers, thickeners, flavour enhancers, dyes, artificial sweeteners and bleaching agents. Many of these are needed to disguise the poor quality of the initial ingredients. The average person in Britain now consumes about two kilograms of these chemical additives every year.

All of these changes in the diet, linked to the much greater consumption of food, have had a major impact on human health in the affluent world. The rate of obesity increased dramatically in the late twentieth century. In Britain one in five adults is now clinically obese – a rate double that of the 1970s. In the United States the situation is far worse. About sixty million adults are obese – about a third of the population – with the rate having doubled since the 1970s. By 2010 it is expected that a third of British adults will be obese. In the early 1960s only 4 per cent of American children were seriously overweight – in the next forty years that rate quadrupled. Obesity, excess food consumption and a diet high in fat significantly increases death rates, especially from cardiovascular diseases. Heart disease was almost unknown a century ago except among the rich who could afford a diet high in fat and sugar and ate too much food. Even in 1930 coronary heart disease was responsible for only 1 per cent of British deaths. By the mid-1990s this had increased to just over 30 per cent and it is continuing to rise.

Part of the rise in heart disease can be attributed to the rise in smoking. The large increase in tobacco consumption since the seventeenth century in the form of snuff, cigarettes, pipe tobacco and cigars is also directly related to increased cancer rates. Smoking increases the risk of contracting cancer by about a third and it also increases the risk of heart disease, bronchitis and other lung diseases. About half of all regular smokers are killed by the habit and it is the major preventable cause of cancer in the world. In the twentieth century tobacco products probably killed 100 million people. Lung cancer rates in the industrialised world rose by 80 per cent in the two decades after 1960 and similar increases are now occuring in the developing world as tobacco companies increasingly concentrate their marketing activities in these countries. Overall cancer is now the second most common form of death in the industrialised world – one in three Americans contracts cancer (compared with one in twenty-seven in 1900) and in one in four dies of the disease. European cancer rates are ten times higher than those in west Africa but half of all the world's cancers now occur in developing countries and here treatment is poor – about 80 per cent of patients will die of the disease compared with 50 per cent in the industrialised world. Overall treatment remains patchy and the chances of a cure remain remote. Huge amounts of money have spent over the last fifty years in an attempt to find cures and treatments for different types of cancers but overall the results have been disappointing except in the cases of some rare types of the disease. Most of the effort has gone into high-technology medical research and very little has been done to reduce the environmental factors in cancer apart from anti-smoking campaigns, which have had limited, though increasing, success in some countries.

11
THE WEIGHT OF NUMBERS

One of the greatest changes in human history has been the unprecedented and rapid increase in population over the last 250 years. The
most graphic way to convey the accelerating rate of growth is to
compare the length of time it has taken to increase the world's population by one billion. The total number of people in the world first
reached one billion in about 1825, and it had taken about two million
years to reach this level. The next billion was added in only a hundred
years. A further billion (taking the total to three billion) took about
thirty-five years from 1925 to 1960. The next billion was added in
only fifteen years (by 1975) and the next billion took about twelve
years until the late 1980s. In the next twelve years another billion was
added, bringing the world's population to six billion just before 2000.
The number of people in the world is now about six and a half billion.

World population, 1750–2000

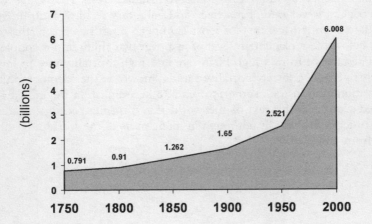

The rise in population was a world-wide phenomenon that began in the mid-eighteenth century and the exact causes are unclear. In China the long period of stability and prosperity that lasted from the late seventeenth until the mid-nineteenth centuries was clearly important. In Europe, particularly in the Netherlands and Britain, agriculture was slowly becoming more productive as a multitude of small improvements from better rotations to improved fodder crops had their impact. In addition the new crops brought from the Americas, in particular maize and potatoes, raised food output. It was not until well into the nineteenth century that the decline in the impact of infectious diseases had much effect. At the same time the import of food from abroad was also important in raising both the quantity and variety of the diet.

The pattern of population growth varied greatly from continent to continent. In Europe the population was about 160 million in 1750 but rose by over 70 per cent in the next century to about 275 million (the fastest period of growth the continent had ever experienced). In the next fifty years, under the impact of the reduction in infectious diseases and better food supply, it rose even faster to reach just over 400 million by 1900. In the first half of the twentieth century the rate of growth fell steadily, mainly as a result of a significant fall in the birth rate. In England the number of children born to each woman fell on average from about four and a half in 1850 to less than two. This pattern, eventually repeated in all the industrialised countries, was a radical departure from the pattern that had prevailed throughout human history up until this point. It resulted from a later age of marriage, a lower rate of marriage and voluntary fertility control. By the late twentieth century in most European countries the birthrate was below the replacement level of just over two children per couple.

The change from a high birth rate and high mortality rate to low birth rate and a lower mortality rate is known as the 'demographic transition' and in took some 150 years to accomplish in the industrialised world. In the rest of the world this transition occurred later but much more rapidly, although in many parts of the world it is still far from complete.

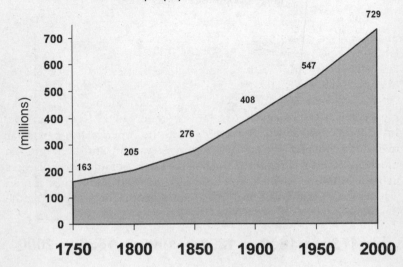

Europe population, 1750–2000

In Asia the rate of population growth from 1750 to 1900 was about half that of Europe so that numbers only grew from about 500 million to about 950 million. The huge growth in population took place in the twentieth century when numbers more than tripled whereas Europe's did not even double.

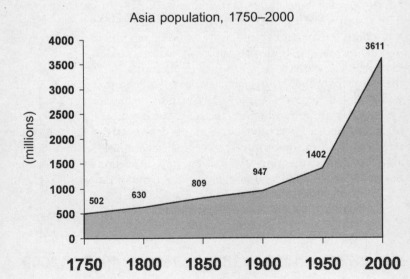

Asia population, 1750–2000

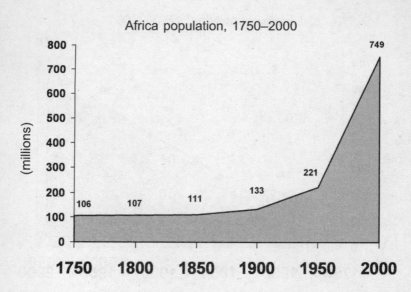

Africa population, 1750–2000

In Africa the pattern was even more extreme. The population hardly increased at all between 1750 and 1900 and then rose more than five-fold in the twentieth century. The pattern of growth in the Americas (and Oceania) was strongly influenced by immigration but in Latin America the population rose seven-fold in the twentieth century.

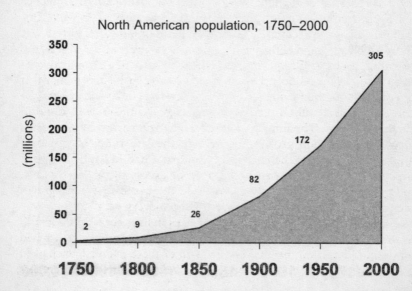

North American population, 1750–2000

Latin America population, 1750–2000

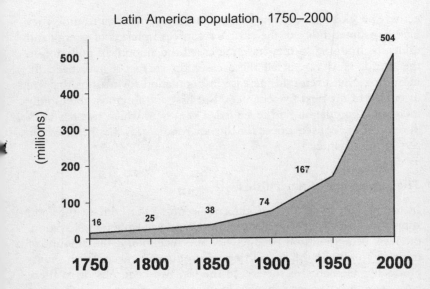

Much of the increase in these areas of the world was the result of the very rapid decline in mortality in the mid-twentieth century. In Egypt the mortality rate fell by half in the twenty-five years after 1946 – in Sweden a similar fall took from 1800 to 1920. There was then a time lag before the shift to a lower fertility rate occurred – overall the average number of children per woman in the developing countries fell from over six to about three in the half century after 1950. But the decline was patchy. In China the demographic transition was completed by the end of the twentieth century – the number of children per woman was at European levels partly because of a strong government programme of incentives and penalties to achieve this goal. Most of Asia has followed a similar pattern. However, in Africa on average five children are still born to each woman – about the level found in England in 1750. The impact of all these changes meant that the peak of world population growth occurred in the late 1960s at just over 2 per cent a year. By the end of the century it had fallen back to just over half this level. However, the peak of excess births over deaths did not occur until the early 1990s when the world's population was increasing at just over ninety million people every year.

There are now more than seven times as many people in the world as there were only two centuries ago. This unprecedented rise has had a profound effect on the environment. All of these people have had to be housed and therefore the number and size of human settlements across the globe has increased dramatically. At the same time they have consumed more of the earth's resources in terms of energy and minerals. In doing so they have increased the amount of pollution in the world. However, in all these cases the increases have been far more than the seven-fold increase in population for reasons that are discussed in the next five chapters. But first and foremost the rapidly increasing population had to be fed. The way in which this has been done, and the consequences for the environment, is the theme of the rest of this chapter.

The expansion of agriculture

In the period up until the eighteenth century the limit to the food supply was mainly set by the amount of land available for cultivation, the level of agricultural productivity and technology, the amount of trade in food and the proportion of production taken by the non-productive elements in society. In the last two centuries or so there have been revolutionary changes in most of these areas. The fact that the earth now supports seven times as many people as it did only two hundred years ago seems, at first glance, to be a triumph of human ingenuity in getting round the limitations on food supply that had prevailed for thousands of years and restricted the growth in human numbers to very low levels. However, many fundamental problems have remained unresolved, many have worsened, the impact on the environment has been profound and it is far from clear that the advances of the last two centuries can be continued.

In Europe and China the traditional response to rising population was to bring more marginal land into production even if the yields from the new areas (which usually had poorer soils and climate) were lower than those from long-established regions. This process has continued across the world since 1700. In the last 300 years the amount of pasture in the world has increased by 680 per cent and the amount of cropland by 560 per cent. This has been paralleled by a 43 per cent drop in the grassland area and a 22 per cent fall in the amount of forest and woodland. By the end of the twentieth century (after ten thousand years of farming) about a third of the world's vegetated area was covered in domesticated plants and pasture.

One of the most important trends was, as we have seen, the increasing political and economic control that Europe exercised over the rest of the world. This was particularly apparent in the European settlement colonies of the Americas, Australia and New Zealand. Here the climate

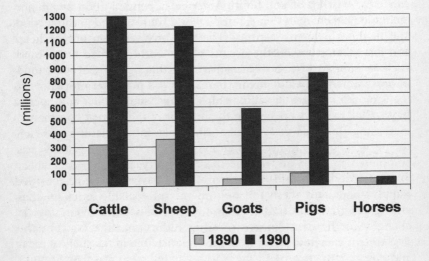

World domesticated animal numbers, 1890–1990

was generally benign and European crops could be grown relatively easily. There were waves of emigration from Europe – in the ninety years after 1846 about seventy million Europeans left the continent. This not only relieved population pressure in their countries of origin but also led to the opening up of huge frontiers of settlement that swept across the new continents, creating new farm land. In 1800 European settlement in North America was still restricted to a small area along the eastern coast yet within a hundred years it had reached the Pacific. Across the world in the sixty years after 1860 about half a billion hectares of new land was brought into cultivation, mainly in the United States and Russia. Overall in the period 1860–1960 the area under crops in the United States increased two-and-a-half-fold, in Russia/Soviet Union it quadrupled, in Canada it rose eight-fold and in Australia it was twenty-seven times larger at the end of the period than at the beginning. The expansion of cropland stopped in Europe around 1920, in the United States about a decade later, in Japan in 1960 and in the Soviet Union in 1966. For most of the twentieth century the expansion of cropland took place away from the temperate areas of the world and was concentrated in the tropical area – west Africa, the interior of South America and Indonesia. For example, the area under crops in Brazil rose six-fold between 1930 and 1970 and then accelerated still further as the Amazon area was opened up.

However, this new agricultural land, in particular in North America, Australasia and South America, depended upon two major technological changes that created a world trade in food in the second half of the nineteenth century. The first involved transport. Cheap and fast steamships on the oceans of the world enabled bulky food such as grains to be moved easily. Until this period Europe's food imports were high-value luxury items that did not deteriorate during the long sea voyages in sailing ships – sugar, coffee, tea and cocoa. In parallel, railways opened up the interior of these great continents, increasing the pace of settlement and enabling foodstuffs to reach the ports. The first transcontinental railway in the United States was completed in 1869 and in the next twenty years American wheat exports rose seven-fold. In Argentina the British owned all of the railways and used them to open up the interior to create ranches and farms that exported their produce. The second technological change was the development of refrigeration and freezing. For the first time it was now possible to move perishable items such as meat and dairy products around the world. Chilled meat was first brought from New York to England in 1875 and from Buenos Aires to France two years later. Ships with freezer compartments brought meat from Australia to England in 1879 and from New Zealand three years later. By the 1890s butter and cheese from New Zealand were also being imported. In 1901 the first refrigerated banana boat brought fruit from Jamaica to England. In parallel with these developments in the European settlement colonies the European states in their other colonies began to remake the economies to produce the goods that Europe wanted.

The result of all these changes was a vast increase in the amount of food traded in the world. In the 1850s total exports were no more than four million tonnes; by the 1880s they had increased to eighteen million tonnes and at the outbreak of the First World War had reached forty million tonnes. They stayed at this level until the late 1950s when there was another period of very rapid growth – by the late twentieth century it had reached nearly 250 million tonnes (a more than sixty-fold increase in the space of 150 years). The nineteenth century therefore marked the beginning of the end of several thousand years of largely self-sufficient agriculture with only limited trading in a few luxury items. European countries, especially Britain, became dependent on imported food in the late nineteenth century and it was one of the major factors that enabled them to industrialise on such a large scale and sustain highly urbanised populations. In the early twentieth century Britain imported 80 per cent of its wheat consumption, 65 per cent of its fruit and 40 per cent of its meat.

High input farming

The huge increase in the area under crops and pasture in the last three centuries and the development of a world trade in food would not have been sufficient on their own to sustain the seven-fold increase in the world's population in this period. This depended on a series of technological changes that radically transformed agriculture in the industrialised world into a high-input, high-energy business with much higher levels of output.

Until the mid-nineteenth century there had been only a very slow increase in agricultural productivity across the world as the result of the accumulation of a series of small-scale improvements – improved rotations, new fodder crops, better drainage, new equipment and some new crops. In the six hundred years before 1800 crop yields in Europe doubled as a result of these changes and the same happened in China with the rice crop between 1450 and 1800. The first major change came with the huge increase in fertiliser use. Until the nineteenth century farms were almost entirely dependent upon manures and composts produced on the farm itself in order to maintain soil fertility. (In China and Japan the farms around the cities could use the 'night soil' collected there by the highly efficient cleaning organisations.) Mixed farming (combining arable and animals) was therefore the norm. The first shift came with the importation into Europe in the 1820s of guano from the huge artificial islands off the Pacific coast of Latin America.

The first major departure from the use of natural fertilisers came in 1842 when an English farmer, John Lawes, applied sulphuric acid to phosphate rock and produced the first artificial fertiliser – superphosphate. There was little suitable rock in Europe and so it was imported from Florida (1888), then Morocco (1921) and later from Thailand. The Soviet Union relied on the slave workers of the gulags in the Kola peninsula and the Australians wrecked the landscapes of Ocean Island and Nauru to find the rock. Nitrogen had, for thousands of years, been obtained through planting legumes which had bacteria on their roots that could 'fix' atmospheric nitrogen. An artificial process to duplicate this based on ammonia synthesis was first devised by a German chemist, Fritz Haber, in the late nineteenth century and developed on an industrial scale by Karl Bosch. It became commercially feasible after the First World War but took up large amounts of energy to produce the artificial nitrogen fertilisers and it was not until the second half of the twentieth century that their use expanded dramatically.

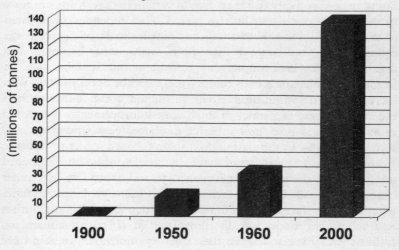

World inorganic fertiliser use, 1900–2000

Overall the world's use of artificial fertilisers rose from 360,000 tonnes in 1900 to 137 million tonnes in 2000 with a peak usage of 150 million tonnes in 1990. Most of this increase came after 1950 when fertiliser use was still only ten million tonnes a year. However, much of the fertiliser was wasted. At least half of it simply ran off the soil into water courses where it caused numerous environmental problems. Excessive use of these fertilisers also reduces the uptake of key trace minerals in the soil, which can damage plant growth and lower nutritional levels. Some crops such as maize also respond better to high fertiliser use and therefore they have expanded in area. The massive use of artificial fertilisers has had only a limited impact on crop yields. In western Europe fertiliser use rose ten-fold between 1910 and 2000 yet crop production only doubled. By the end of the twentieth century across the world it was clear that fertiliser use had reached its effective maximum – adding more to the soil no longer increased yields and merely increased other problems.

In parallel with the increasing use of artificial fertilisers there was a rise in the mechanisation of agriculture replacing some of the back-breaking labour of the previous thousands of years. Horse-drawn threshers and reapers were used from the 1830s but the large-scale use of steam power on the farm was uneconomic. The real leap forward came with the development of the internal combustion engine and the tractor in the early twentieth century. In the United States, where there was a labour shortage, tractors were used from the early 1920s – there were 2.3 million by 1945. The second state to use them on a large scale was the Soviet Union when most of farming was collectivised in the 1930s – these huge farms needed tractors and their production and use became a fetish in the Stalinist state. In Europe tractor use only became common after the Second World War. (In 1950, Germany still used over two million horses on its farms.) Overall the number of tractors in the world increased from 300,000 in 1920 (two-thirds of these were in the United States) to twenty-six million by the late twentieth century.

Other forms of mechanisation spread rapidly in the second half of the twentieth century. Although the United States had combine harvesters in 1920 they were rare in Europe before the Second World War – in 1944 Denmark had none but in 1970 had 40,000. The first electric milking machines were introduced in 1895 and 70 per cent of farms in New Zealand used them by 1920. As late as 1950 only 3 per cent of cows in western Europe were machine milked; by the 1980s only 3 per cent were hand-milked. Other crops were also mech-anised. In 1965 just 1 per cent of the Californian tomato crop was picked by machines; within three years only 5 per cent was picked by hand.

One of the consequences of increased agricultural productivity and high levels of mechanisation was the rapid decline in the agricultural labour force and an increase in the size of farms. Until the early nine-teenth century nowhere in the world could more than about 10 per cent of the population be employed in nonagricultural activities (in many areas it was much lower than this) because agricultural produc-tivity was so low. As late as 1851 half of the British population was still rural and in 1920 half of the American labour force still worked on farms. By the late twentieth century in the industrialised world no more than about 2 or 3 per cent of the workforce were working on farms. In parallel the number of farms declined rapidly – in the United States there were seven million farms in the 1930s but less than three million fifty years later when over half of all agricultural produce came from 5 per cent of the farms. The average size of the American farm tripled in the fifty years after the mid-1930s and in the Soviet Union the average state farm covered over 40,000 hectares.

Until the twentieth century traditional methods of raising domesticated animals were extensive and the numbers that could be supported were limited by the availability of pasture and fodder crops. Improvements were largely restricted to selective breeding such as the increasing use of the Merino sheep for their fine wool. Intensive systems did exist on a small scale. In sixteenth-century England, one observer reported that many pigs were kept 'in so close a room that they cannot turn themselves round about whereby they are forced always to lie on their bellies'. For centuries poultry and game birds were kept in the dark and had their feet nailed to the floor or cut off because it was thought this made their meat more tender. In France geese were force-fed to produce foie gras. In 1686 in England a large landowner, Sir Robert Southwell, reported that he had seen a 'new invention of an ox-house, where the cattle are . . . to eat and drink in the same crib and not stir until they be fitted for the slaughter'. In the twentieth century these methods were applied on a semi-industrial scale with the use of increasingly intensive, high-energy systems. Instead of feeding outdoors on natural foods such as grass, animals were brought indoors and fed on artificial feeds. Chickens are kept in grossly over-crowded battery cages, cattle in small stalls and pigs chained to walls in sties small enough to ensure they cannot move. Animals which are herbivores are fed on a diet which includes a high percentage of dead animals, recycled manure, growth hormones and even newspaper and cement dust. In addition they are routinely given high levels of antibiotics to counter the diseases that would otherwise be rife in such conditions.

Agricultural output in the industrialised world in the last sixty years has also been raised artificially by the use of huge subsidies. Ironically this has occurred at exactly the time when agricultural output and employment has played an increasingly marginal part in national economies. Across the industrialised world subsidies in various forms now amount to about $500 billion a year. For example the European Union subsidises the sugar industry at a cost of £1.34 billion a year through inflated price guarantees, export subsidies and high tariffs against imports. The EU guarantees sugar producers a price of £423 a tonne against a world price of just over £100 a tonne. Imports from developing countries are blocked by a tariff of 324 per cent and the surplus EU sugar is dumped on the world market with an export subsidy of £352 a tonne, which further depresses the price obtained by farmers in the poorest parts of the world. In the United States the government provided income support for rice farmers of $1.3 billion a year at the end of the twentieth century yet the total value of rice production was only $1.2 billion. In Japan government subsidies to farmers amount to 1.4 per cent of GDP yet agriculture only contributes 1.1 per cent of Japanese GDP. Cows in the United States and the European Union are each subsidised to the extent of $2.7 a day, which is twice the average income of a farmer in the developing world. At the same time India, the largest milk producer in the world, is forbidden by the World Trade Organisation from subsidising milk production.

Overall modern industrialised agriculture is highly energy-inefficient. Although output is high this relies on huge inputs that consume large amounts of energy. Machines have to made, and then consume fuel when being used. Huge sheds containing animals have to be heated and lit. Animal feeds have to be produced in factories. Huge quantities of artificial fertilisers, pesticides and herbicides have to be used and these require large amounts of energy to produce. Then the products of the farms have to be transported over large distances and stored. Overall agriculture in the United States now consumes more energy than it produces.

Agriculture in the developing world

The recent agricultural history of the developing world has been very different from that of the industrialised world. These countries had only limited access to the products of the new lands opened up by the Europeans after 1800 and most of the newly cultivated land in Latin America was used to provide food for export. The amount of new land in Asia, particularly in places such as China, was extremely limited because production was already relatively intense and took up the most productive land. These countries faced two additional problems. European control of their empires meant that increasing amounts of land were devoted to growing crops for export that the imperial power wanted. In addition, land was very unequally distributed. In Latin America two-thirds of the land is now owned by 1.5 per cent of landowners and a third of the population own just 1 per cent of the land. (Many of the rest are landless labourers.) In Africa three-quarters of the agricultural population own just 4 per cent of the land. Apart from the appalling social problems caused by this inequality, food output was lower than it might have been because the large landholdings tended to concentrate on export crops.

Given this background the rapidly rising populations of Asia, Africa and Latin America have caused acute agricultural problems. The amount of arable land per head of the population began to fall – in China from the 1870s, in India, South Korea, the Philippines and Vietnam after 1900 and in Java after 1920. Production had to become more intensive in order to produce enough food. After the Second World War this was achieved mainly through the so-called 'Green Revolution' – the use of new high-yield varieties of wheat and rice. These were developed in Mexico (wheat) in the 1940s and the Philippines (rice) a decade later. Their impact in raising production was immediate. Mexican agricultural production rose at 5 per cent a year in the twenty years after their introduction and after they were first used in India and Pakistan in 1965 wheat yields doubled in a decade. The new rice varieties had the same effect.

Part of the motivation behind the 'Green Revolution' was political. The United States, which funded much of the research through the Rockefeller Foundation, saw it as a way of countering social discontent and the threat of Communism in the newly independent developing countries. In practice the social, economic and environmental impact of the 'Green Revolution' has been disastrous for the majority of the population. Some countries did benefit, in particular South Korea, China and India, but sub-Saharan Africa was hardly affected by the new crops. Overall the revolution did not give developing countries food independence. The new crops were also highly energy-inefficient – only a quarter as efficient as hand and hoe agriculture in paddy fields. This was because although the new varieties increased yields they required large amounts of fertiliser, water and pesticides. (In the forty years after 1950 fertiliser use in Asia rose thirty-eight-fold.) Farmers were also tied to the seed companies (located in the industrialised world) for new seeds because the new varieties were sterile and covered by patents. Only farmers who could afford the higher inputs could hope to benefit from the 'Green Revolution'. Small peasant farmers did not have enough land and capital to benefit from these new varieties. Large landowners became richer, expanded their holdings, bought up peasant land and turned the peasants into landless labourers. In Mexico 80 per cent of the extra production from the 'Green Revolution' came from just 3 per cent of farms, the number of days worked per year by landless labourers fell from 194 to 100 and their real income fell by a fifth.

In many cases, for example Ethiopia in the 1980s, the effects of these changes actually increased social and economic tensions and led to revolution.

World agriculture at the end of the twentieth century

Despite the large increases in production during the twentieth century there were signs of increasing fragility in the world's agricultural system by the end of the century. Most of the increase in production had been achieved through a massive increase in inputs. The agriculture of the industrialised world is not any more 'efficient' than that of the rest of the world – what it is able to do is purchase more inputs and therefore obtain higher output. That process is now near to its theoretical limits. The level of fertiliser use is such that increased application no longer produces a bigger crop because most of the cultivated varieties are near to their maximum possible output. Production also depends on a very limited number of crops and varieties. During the twentieth century about three-quarters of the world's crop plants were lost and no longer cultivated. Ninety per cent of the world's calories now come from just twenty species and half the world's food intake comes from just four – rice, maize, wheat and potatoes. Of these four, 60 per cent of the output comes from specialised high-yield varieties. This makes the world potentially very vulnerable to any disease that affected one of these varieties.

The rise in agricultural productivity began to slow at the end of the twentieth century. From 1950 to 1990 productivity rose at about 2 per cent a year – since 1990 it has only been at half this rate, which is lower than the growth in population. The amount of land devoted to grain production reached a peak in 1981 but has since fallen by over 11 per cent as land has gone out of production as the result of environmental degradation. From 1950 until 1984 world grain production per person rose by 37 per cent. However, 1984 was the peak and in the last two decades grain production per head has fallen by 18 per cent. World cereal production has been on a plateau since 1994 and since 1999 grain stocks have fallen every year because consumption has been greater than production.

In some countries the situation was far worse. By the 1980s fifty countries in the developing world which had been self-sufficient in food in the 1930s were net importers of food. In the thirty years after the Second World War, in thirty-four countries, affecting a quarter of the population in the developing world, food output did not keep pace with population. The worst affected area was Africa where food supply per head fell steadily after 1967.

World food in the twentieth century

Although the world's agricultural system now supports just over six billion people it does so very unequally. The result is hunger, malnutrition and even famine on a major scale. Overall there is enough food in the world to feed everybody at an adequate level – the problem is its unequal distribution. Put simply, the people of the industrialised countries eat half the world's food although they are only a quarter of the world's population. Grain consumption per head in the United States is five times higher than in sub-Saharan Africa. More food is sent from the poorest countries to the richest than goes in the opposite direction. Most developing countries are net exporters of food. For example between 1995 and 2005 the food output per head in Nigeria, Ethiopia, Sudan, Kenya, Tanzania and DR Congo (countries which make up 60 per cent of the population of sub-Saharan Africa) fell by a fifth yet agricultural exports per head rose. A large proportion of this trade is, as it was in the past, in luxury items to provide more variety in the diet for those who are already well fed.

The result is a world where there is growing obesity and diabetes in the rich countries of the world caused by too much food and too much sugar and continuing malnutrition across much of the rest of the world. However, although the average American eats 50 per cent more food than the average person in the developing world there have always been marked inequalities within the industrialised world. In the late 1980s about fifteen million people in the United States did not earn enough to consume a well-balanced diet and another fifteen million were constantly hungry (a total of over 10 per cent of the population). The situation was far worse in the early twentieth century, particularly during the depression of the 1930s. Then people starved to death on the streets of New York and in Britain it was officially estimated that a third of the population was too poor to buy the minimum diet the League of Nations thought essential to maintain health.

Much of the food in the rich countries is wasted. The latest estimate is that in the United States food waste in the production, distribution and consumption chain is equal to 85 per cent of the food actually eaten. Overall 40 per cent of the world's grain harvest (over 70 per cent in the industrialised world) is used inefficiently by being fed to animals to produce meat rather than being eaten directly by humans. Pets in the richest countries of the world eat more meat than the people of the poorest countries in the world. Overall about a fifth of the world's grain harvest is eaten by rodents.

There are no statistics available for the scale of world hunger before the 1930s, but the earliest figures show a pattern that must have prevailed for several centuries with the worst problems being concentrated in Africa, Latin America and parts of Asia. In the 1930s about one billion people (half the population of the world) suffered from malnutrition. By the early 1950s this had risen to about one and a half billion people or six out of ten people in the world. During the 1950s the proportion of the world's population that was malnourished fell slightly but because of the rapidly rising population the numbers affected rose to nearly two billion. The situation improved more rapidly from the early 1960s so that by the early 1980s about a quarter of the world's people were malnourished. However, because of the continuing rise in population the number of people affected was still over one billion, about the same as in 1950. After 1980 the situation began to deteriorate again. By the end of the twentieth century about one billion people in the world suffered from chronic malnutrition and another billion had to live on a grossly inadequate diet. Overall a third of the people in the world do not get enough food to live a healthy life and another substantial group live on the margins of subsistence, facing the constant threat of food shortage. About forty million people a year die from hunger and its related diseases – equivalent to 300 Jumbo jets crashing every day, with half of the passengers being children.

In Europe the spectre of famine that had haunted the continent for thousands of years disappeared in the eighteenth and nineteenth centuries as production increased and more food was imported from abroad. (It only returned, in the form of an absolute shortage of food, because of the social and economic dislocation brought on by war – Belgium and Germany in the First World War, Russia during the revolution and civil war, Greece, Leningrad and the Netherlands during the Second World War.) In the poorest countries in the world it remained a constant threat. The number who died in the famines of the first half of the twentieth century is unknown: often the colonial authorities took little interest. In 1908 there was a major famine in northern Nigeria, but the government in the capital Lagos knew nothing until they read the annual report of the officer in Kano several months later. He wrote that mortality had been 'considerable' but he hoped 'not so great as the natives allege' and continued that 'we had no remedy at the time and therefore as little was said about it as possible'. In French West Africa there were major famines across the Sahel in 1913–14 and again in 1931. The French laid the blame on African 'idleness', 'apathy' and 'fatalism' and the government continued to requisition grain throughout both famines.

The requisitioning of grain illustrates a central fact in all famines – there is no absolute shortage of food. The people affected by famines are those who are too poor to buy food at the highly inflated prices that prevail during a shortage. In 1943–44 about three million people died in Bengal after rice prices quadrupled in two years and fishing was disrupted by the war. The British government did little to help and the famine occurred after the largest ever rice crop, when food stocks were at record levels. In Ethiopia in 1972–74 about 200,000 people died in the provinces of Tigre and Wollo after only a small fall in the harvest but at a time when food was still being exported from the two provinces and from the country as a whole. In Bangladesh in 1974 when rice prices doubled in three months after severe flooding, one and a half million people died of starvation. But there was no shortage of food – production, both in absolute and per head terms, was the highest ever. Once again it was those who lacked the resources to buy food who starved. The number who died in the famines of the twentieth century is unknown but a conservative estimate would suggest at least 100 million.

Under increasing public pressure a number of attempts have been made to deal with famine, malnutrition and hunger. In 1963 President John Kennedy stated: 'We have the means . . . to wipe hunger and poverty from the face of the earth in our lifetime.' In 1974 the World Food Conference agreed to eradicate hunger within a decade. In 1996 the World Food Summit agreed a much more modest aim – to halve the number of people suffering from hunger (to 400 million) by 2015. Efforts so far have failed abysmally. They have failed because of a failure to correctly analyse the problems and the use of food aid as a political weapon. There is no shortage of food in the world, only a maldistribution. This unequal access to food is made worse by the agricultural subsidies of the rich world. There is a strong suspicion that much food aid is simply a way of getting rid of some of the huge agricultural surpluses generated by the policy of high subsidies. In addition the provision of food aid by the United States is tied to the attitudes towards the United States adopted by the recipient country. Over 90 per cent of food aid (99 per cent in the US) is tied to purchases in the donor country and the United States also insists that the overwhelming majority of the aid is carried in US ships. Overall food aid costs the recipient half as much again as buying food locally and a third more than buying from other countries.

Agriculture and the environment

The huge increase in the amount of land under cultivation, the extension of pasture land into new areas and the intensification of agriculture have all led to increased environmental degradation. Natural ecosystems have been destroyed through deforestation, the ploughing up of grasslands, and the extension of the cultivated area on to marginal land and steep slopes with the consequent increase of soil erosion, degradation and desertification. The increasing use of irrigation has led to a huge increase in the demand for water and often the building of dams that have themselves further damaged ecosystems and many human communities.

Even in the long-established agricultural systems of the industrialised world where there has been no expansion in the amount of agricultural land (in Britain in the early twenty-first century it covered the same amount of land as in 1860) there was considerable environmental destruction in the second half of the twentieth century. During that period in Britain almost all its lowland meadows, two-thirds of its lowland heaths, half its ancient lowland woods, bogs and wetlands and a quarter of its hedgerows (amounting to about 225,000 kilometres) were destroyed. Elsewhere the drainage of wetlands to provide agricultural land led to a steady loss in natural ecosystems – in the last century the United States lost half its wetlands. The biggest loss was in the Florida Everglades where drainage started in 1883, partly to provide land for urban development (the population of the area has risen from 11,000 to over four million) but mainly for sugar cane cultivation. Rivers were dredged, new canals dug and the natural drainage of the area was destroyed. The result was that sea water flowed in, the main lake was affected by eutrophication (lack of oxygen leading to the death of animal life), the peat dried out and the level of the land fell by a third of a metre a year. Most of the wildlife, including 90 per cent of the two and a half million wading birds, died out.

Deforestation

The clearing of natural forests has been a continuous process throughout human history as the easiest way of obtaining new agricultural land. Before agriculture about 45 per cent of the earth's surface was covered by forest – about a third of all those forests have been destroyed in the last ten thousand years. In many places this has been a long, slow, steady progress as for example in the Mediterranean. In China too natural forests originally covered about three-quarters of the land – by the early twentieth century forests were restricted to the inaccessible and mountainous areas and by the end of the twentieth century covered no more than 5 per cent of the land. In India what is now the Thar desert in Rajasthan and Punjab was still impenetrable jungle two thousand years ago. What is clear, however, is that the pace of destruction accelerated after 1700 and increased still further in the second half of the twentieth century.

European settlement of the new lands of the Americas and Australasia triggered off forest destruction on a vast scale. For example on Haiti (which means 'green island' in the local language) less than 10 per cent of the original forest remains. In the eastern United States and Canada half of all the forests were cleared in the 300 years after European settlement. Australia was only settled at the end of the eighteenth century but half of the original forests no longer exist and three-quarters of the rain forest has been destroyed. New Zealand officially became part of the British empire in 1840 and within 150 years half the natural forests were destroyed. European control elsewhere destroyed forests for exotic timbers for furniture, specialist hardwoods to make ships and to provide land to grow the crops that the Europeans wanted – sugar, cotton, coffee and tobacco in particular, followed by rubber and palm oil in the early twentieth century.

In the twentieth century a clear pattern of destruction emerged. In the industrialised world forest cover began to increase (especially in the second half of the century). In the nineteenth century industry and the railways of the United States were fuelled by timber – it was cheaper and easier to use than coal. However, timber use per head peaked in 1907 and it was replaced by new forms of domestic heating and in industry and construction by iron, steel and plastics. Forest clearance in the industrial world is now primarily for logging and not for agricultural land. The increase in the forest area across the industrialised world (it grew by 36 million hectares in the 1990s alone) has been achieved through the importation of timber from the developing world – for furniture and as pulp for paper.

The impact of demands from the rich world, combined with the pressure for agricultural land from a rapidly rising population, have produced a swath of destruction in the tropical forests of the world since 1950. Overall the area covered by tropical forests has fallen by almost half – from 2.8 billion hectares to 1.5 billion in the last half century. Some of the earliest destruction took place in the Philippines and Indonesia (mainly to meet Japanese demands) so that in the former country only 3 per cent of the original forest cover is now left. Overall in Asia about nine-tenths of the original forests have been destroyed. The scale of destruction in west Africa was even faster with clearance rates of around 15 per cent a year in some countries. In Ivory Coast only a fortieth of the original forests still survive and in Nigeria only a seventh are left.

From the 1970s much of the destruction was concentrated in Amazonia, which contains about 40 per cent of the remaining tropical forest in the world. For the Brazilian government, which strongly supported this programme, the 'opening up' and development of Amazonia served a number of purposes. Given the highly unequal land-holdings in the settled part of the country and the political power of the landowners it was a way of defusing social tensions by giving land to landless labourers – it was 'a land without men for men without land'. It was also a way of settling the vast interior of the country and securing its remote frontiers. Roads were built into the forest (the population along the Brasilia–Belem road rose from 200,000 when it was opened in 1960 to two million within a decade) so that logging could begin and land be cleared for settlement and plantations.

The type of development was, however, determined by the soils of the area. In tropical forests most of the nutrients are held not in the soil (which is poor) but in the trees and plants. The land is usually cleared by peasant farmers and the burning of the forest cover helps provide nutrients for a few years. The soil is soon exhausted and the peasants sell out to large landowners and move on to new land. The large landowners create large cattle ranches on the poor grassland which is all the soil will then support. However, even these often cannot be sustained for long – nearly all the ranches established in the Amazon area before 1978 had been abandoned by the mid-1980s. The rate of destruction of the Amazon forests is difficult to estimate but by the early twenty-first century at least a fifth of the original forest had disappeared. However, the results of a satellite survey published in *Nature* in late 2005 suggested that because 'selective logging' (the removal of a few key trees such as mahogany) badly degraded the surrounding area the level of destruction each year could be twice that previously estimated.

Overall the rate of tropical forest destruction across the world in the early twenty-first century was estimated to be on average 86,000 hectares a day (an area larger than New York City) or thirty-one million hectares a year (an area larger than Poland). At this rate the world's tropical forests will be destroyed by the middle of the century at the latest. Tropical forest clearance involves the destruction of an entire ecosystem and these forests contain about half of all the plant and animal species on the planet (about thirty million different species). The actual rate of species loss is unknown but the best estimate would suggest about 50,000 a year, most of whom were unknown to science when they became extinct. Tropical forest destruction can also have a significant impact on the climate. When vegetation cover is removed, solar energy, instead of being partly absorbed by the trees, is reflected from the nearly bare ground, increasing temperatures, drying the soil, creating dust in the atmosphere and helping to stop rain clouds forming. Estimates suggest that about 260,000 square kilometres of forest needs to be cleared before these effects are noticeable on a significant scale. In the last century about four times this area of forest has been cleared in west Africa and the consequences have become apparent. Rainfall in the Sahel and parts of sub-Saharan Africa is now a third less than it was a century ago, with over three-quarters of the years being highly arid. Similar effects are also now being felt across Amazonia.

Since the 1980s there have been a number of attempts to control deforestation but they have been almost totally unsuccessful. The fundamental problem is that for many countries timber exports are one of their few sources of export earnings in a world trade system that is already structured against them. Governments argue that they are entitled to use their natural resources as they see fit, industrialised countries did the same in the past and they require compensation from the industralised world if they are to restrict logging. (The situation is made worse by high levels of corruption.) Effective monitoring of logging is difficult – about a third of Britain's timber imports come from illegal logging. Even when it is claimed that 'sustainable' logging takes place (where one tree is replanted for every tree cut down) the richness of a tropical forest is usually replaced by a monoculture of quick-growing trees such as eucalyptus.

Soil erosion

In almost every part of the world modern agriculture has led to severe soil erosion in the wake of deforestation, ploughing up of grasslands and the cultivation of steep slopes. Soil erosion has led to dust storms, flooding, loss of fertility and the abandonment of cultivation. Soil degradation now affects about a third of the world's land surface and a third of the world's cropland is losing soil faster than it is created. In India nearly 800,000 square kilometres of land is affected by soil erosion, Haiti has no quality topsoil left, and in Turkey about three-quarters of the land is affected, half of it seriously. The United Nations Environment Programme estimates that since 1945 human activities have degraded two billion hectares of land, of which 430 million hectares have been irreversibly destroyed.

The experience of the United States provides a particularly clear illustration of the problem of soil erosion. To the early settlers the amount of land available seemed inexhaustible and they paid little attention to maintaining soil quality. Cultivation was extensive and it was easy to abandon ruined land and move to new areas. The worst problems were associated with the cultivation of tobacco and cotton – the former requires eleven times the nitrogen and thirty-six times the phosporus of a food crop and therefore exhausts the soil very quickly. Tobacco farmers found that the second crop on new land was the best but that after a couple more seasons maize or wheat was all that could be grown. This could only be done for a few years before the soil was completely exhausted. Land was abandoned and the ruined soil was easily eroded away by the wind and rain. Settlers moved steadily westwards from the Tidewater to the Piedmont region, clearing the forests to find new land. Within less than a century from the first settlement in 1607 Virginia was suffering from severe flooding brought on by deforestation. In the eighteenth century Georgia had the same experience with soil erosion gullies over forty metres deep in places. Continued cotton growing had the same impact on the soil and this was one of the main pressures behind the continued westward movement of the frontier, cotton cultivation and slavery. By 1817 in North Carolina the amount of abandoned land equalled the area under cultivation.

Until the latter half of the nineteenth century the Great Plains area was avoided by settlers because the ploughs they had available could not break up the tough, compacted grass. The development of the heavy steel plough, pulled by teams of six to twelve oxen, enabled farmers to move into an area that had previously supported large herds of bison and the Native Americans who hunted them. The Great Plains were ploughed up to grow wheat even though this was a marginal area for cereal cultivation – the climate was semi-arid with only about fifty centimetres of rain a year and the thin topsoil was only held together by the grass. Despite the experience of previous centuries (and in defiance of all ecological principles) the US Bureau of Soils claimed in 1909 that: 'the soil is the one indestructible, immutable, asset that the nation possesses. It is the one resource that cannot be exhausted; that cannot be used up.'

When they made that claim the early stages of one of the world's greatest ecological disasters were being enacted on the Great Plains. The last large territory, Oklahoma, was opened for settlement in 1889. In the next forty years about ten million hectares of virgin land were ploughed up and cultivated with drought-resistant wheat. After 1919, when Russian wheat exports ceased, American production expanded. Another 1.25 million hectares were ploughed up and output rose to two-and-a-half times the 1914 level. However, the history of American over-exploitation of the land was about to meet its nemesis.

In the early 1930s one of the periodic droughts that regularly affect the Great Plains occurred. The loose, fragile and dry soil, which had lost its protective grass cover, was blown away in the high winds. Huge dust storms were created across a region covering west Kansas, south-east Colorado, north-west Oklahoma, north Texas, north-east New Mexico and parts of Nebraska and the Dakotas. The first big storm in May 1934 picked up about 350 million tonnes of topsoil and deposited it over the eastern United States (an estimated twelve million tonnes fell on Chicago alone) and dust was even detected on ships 400 kilometres out into the Atlantic. In March 1935, over one million hectares of wheat were destroyed by dust storms and by 1938 over four million hectares of land had lost the top twelve centimetres of soil and more than another five million hectares the top six centimetres. The social and economic effects were terrible (and immortalised in John Steinbeck's novel *The Grapes of Wrath*) – respiratory diseases across the region rose by a quarter and the infant mortality rate by a third. By 1938, when 850 million tonnes of soil were still being lost every year over three and a half million people had abandoned farms in the area. Oklahoma lost almost a fifth of its population – in some counties almost half the population left. Better conservation practices improved the situation in the 1940s but the periodic droughts returned in 1952–57 when wind erosion affected twice the area damaged in the 1930s and again in the 1970s when another four million hectares suffered from severe erosion.

The creation of the 'dust bowl' acted as a catalyst in forcing the US government to carry out a full survey of the extent of soil erosion in the country. The results, published in 1938, revealed an alarming situation. Even in a prime agricultural area such as Illinois a third of the land was damaged by erosion. In total an area as big as South Carolina had been eroded away, an area the size of Oklahoma and Alabama had been seriously damaged and the amount of sand and gravel washed down the rivers would have covered an area as big as Maryland. The situation worsened in the second half of the twentieth century. By the 1970s a third of the topsoil of the United States had been lost and fifty million hectares of cropland had been either ruined or made very marginal for cultivation. Another 430,000 square kilometres were suffering unacceptable rates of erosion. Topsoil was still being eroded away at the rate of five billion tonnes a year – six times the rate a century earlier. This widespread devastation was the result of extensive and continual monocropping on marginal land and the practice of regarding the soil as no more than a medium to prop up plants while large quantities of artificial fertilisers, herbicides and pesticides were poured on to the crops.

A similar disaster took place in the Soviet Union in the 1950s and 1960s. The first Five Year Plan of 1929 proclaimed: 'We must discover and conquer the country in which we live . . . Our steppe will truly become ours only when we come with columns of tractors and ploughs to break the thousand-year-old virgin soil.' The tractors arrived on the marginal grasslands of Kazakhstan in the 1950s as part of the 'virgin lands' programme. In total about forty million hectares of land were ploughed up between 1954 and 1960. Yields peaked in 1956 and then fell steadily. Practices such as deep ploughing and leaving the soil bare during fallow periods led to severe soil erosion following the severe drought of 1963. Even before 1963 land was being lost at a rate of almost a million hectares a year and and in 1964 and 1965 a total of over ten million hectares was badly damaged. Overall within a decade of the start of the ambitious programme almost half the newly ploughed-up land had been severely affected by erosion. From the mid-1960s, on average another million hectares of land were abandoned every four years. The amount of cultivated land in Kazakhstan fell by over a half in the last two decades of the twentieth century.

In Australia, the first proclamation restricting tree-felling because of increasing soil erosion was issued in 1803, little more than a decade after the first settlers arrived. Most of Australia consists of desert or highly marginal land (nearly three-quarters of the country is semi-arid or worse) and the remaining land is easily eroded if ploughed or over-grazed. By 1878 there was extensive soil erosion in Victoria, in the 1920s Sydney was suffering from heavy dust storms and twenty years later it was estimated that half of New South Wales was affected by soil erosion. By the mid-1980s over half of all Australia's agricultural land was affected by soil erosion.

In China, the growth of population and the extension of the agricultural area over several millennia have produced extensive deforestation and rising soil erosion. The situation was bad in the nineteenth century but worsened considerably in the twentieth. In just twenty years in the middle of the century over eight million hectares of agricultural land were lost to soil erosion. In total over a third of what was once arable land has now been abandoned. The dust caused by spring ploughing in China can now be detected thousands of miles away in Hawaii.

A major secondary problem caused by soil erosion is that rivers carry away large amounts of the soil and this causes river beds to rise (increasing the risk of flooding), it silts up any downstrean dams and eventually extends deltas at river mouths. In the Himalayas about 40 per cent of the forests were destroyed in the thirty years before 1985 and since then the rate of destruction has increased. The beds of the great rivers that flow out of the mountains rose at almost fifteen centimetres a year, causing extensive flooding and the rapid expansion of the deltas in countries such as Bangladesh. Here land is in short supply and with a rapidly rising population people have little choice but to settle on the new land created in the deltas. Here they are very vulnerable to further flooding, and cyclones moving in from the Bay of Bengal cause major loss of life and livelihoods from so-called 'natural disasters'. Almost an eighth of the world's population now lives in areas affected by flooding caused by deforestation and soil erosion in the Himalayan region.

Another factor in increasing the rates of deforestation and soil erosion has been over-grazing by pastoral groups who live in the marginal areas of Africa, the Near East and Central Asia. They need to travel with their animal herds over considerable distances in a carefully integrated cycle of activities during the year in order to maintain their way of life and avoid over-stressing the environment. In the twentieth century these pastoral groups (about forty million people across the world) came under pressure from two sources – the expansion of the area under cultivation and a continuing rise in their own numbers. The rising number of farmers forced pastoral groups away from the better land and into ever more marginal areas. Rising population in the pastoral groups led to more animals being kept, further stressing a very vulnerable environment. In the Sahel area of west Africa the number of animals has more than doubled since 1940 and in Rajasthan the area available for grazing in the 1950s fell by 15 per cent but the number of animals increased by more than half. Aid programmes designed to turn the pastoralists into sedentary herders have worsened the situation by concentrating the destructive effects of grazing.

The most extreme form of soil erosion and loss is desertification, which now affects the south-western parts of the United States, northern Mexico, north Africa, the Sahel, large parts of southern Africa, parts of Australia and an increasing amount of land in China. The best estimate is that the world's deserts are expanding by about 70,000 square kilometres a year. Throughout the twentieth century the Sahara moved steadily northwards (the great grain-growing lands of the Roman empire in Libya had long been lost) at about 100,000 hectares a year and the speed of advance in the south, particularly in Sudan, was far greater. In Chile the Atacama desert is advancing at over three kilometres a year. Overall about 700 million people (just over 10 per cent of the world's population) live in the arid and the semi-arid areas of the world threatened by the encroachment of deserts. In some areas the figure is far higher – in Kenya a third of the population is affected by desertification.

Water

The intensification of food production and the huge increases in output in the last two centuries have been partially achieved by a massive increase in the amount of irrigated land. In 1800 there were about eight million hectares under irrigation across the world. In the nineteenth century this area quintupled to about forty million hectares. In the twentieth century there was another seven-fold increase to about 275 million hectares. Overall the growth was just over thirty-four-fold. In total slightly over 15 per cent of the world's arable land is now irrigated, with the highest proportion in Asia. Irrigation allows crops to be grown where otherwise it would not possible and it does (as the earliest farmers seven thousand years ago discovered) increase yields dramatically – in south Asia only a third of the rice-growing area is irrigated yet it produces almost two-thirds of the total crop.

World irrigated area, 1800–2000

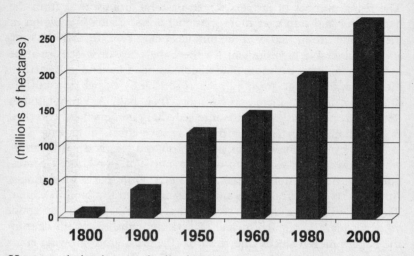

However, irrigation can badly degrade the land and lead to waterlogging and salinisation as the early societies in Sumer discovered over four thousand years ago. These effects are now found in half of the irrigated land in Syria and Iraq, a quarter of the irrigated land in the United States and four-fifths of the irrigated area in the Punjab. In the last thirty years of its existence the Soviet Union lost more previously cultivated land to waterlogging and salinisation than the cultivated area of Ireland and Belgium combined. The main growth in the irrigated area came in the middle of the twentieth century when it increased at about 3 per cent a year. After 1978 this slowed to about 1 per cent a year. However, in the last quarter of the twentieth century the loss of irrigated land was also at about 1 per cent a year so that there was little net increase in the area under irrigation.

The expansion of the irrigated area has, alongside a rising population, the demands of industry and increasing consumption per head, placed severe pressure on the world's water resources. The world depends on a very limited amount of fresh water – 97 per cent of the world's water is salt water in the oceans and two-thirds of the fresh water is frozen, mainly in Antarctica. Most of the rest is in underground aquifers that are too deep to access. Only a quarter of 1 per cent of the world's fresh water is in lakes and rivers and a quarter of that is in Lake Baikal in Siberia. The world's annual water consumption has risen from roughly 110 cubic kilometres in 1700 to about 5,200 cubic kilometres now – a forty-seven-fold increase but with most of that increase concentrated

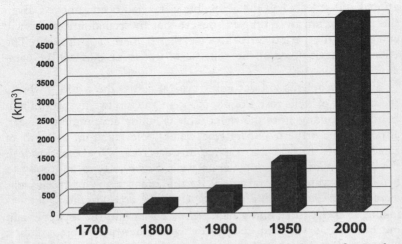

World water use, 1700–2000

in the last fifty years. Throughout history most human use of water has been for irrigation – well over three-quarters of total usage until about 1970 and it is still about two-thirds of the total (nearly all the rest is used by industry). The problem with irrigation is that the vast majority of the water is wasted – in India and China about two-thirds of irrigation water is lost through evaporation and seepage out of irrigation canals and in the United States half the water is lost.

By the early twenty-first century the scale of the world's water problems was becoming clear. World water use rose almost ten-fold in the twentieth century with most of the increase concentrated in the second half of the century. However, most of this increase was a 'free ride' – using up the water resources of underground aquifers far faster than the rate at which they are replenished. (Fossil aquifers such as the Ogallala under the Great Plains of the United States, that of the North China Plain and the one under Saudi Arabia cannot be replaced at all in any useful timescale.) The Ogallala aquifer is an underground river stretching from Texas to South Dakota that has been accumulated over about 25,000 years. The first settlers in the area used windmills to pump water but these could only reach a depth of about ten metres and irrigate an area of about three hectares. From the 1930s the use of oil-drilling technology enabled deep wells to reach the aquifer. In the thirty years after 1950 extraction rates quadrupled and water was being drawn out at ten times the rate of replenishment. As wells went deeper to reach the remaining water costs rose until they became uneconomic – irrigation began to fall from about 1980. In the 1990s there were still 150,000 pumps operational and at least half the accessible water had been removed. Even the most optimistic estimates suggest that the remaining water will run out in the period 2010–2020, earlier if there is a drought. The effects on agriculture of the collapse of water supplies will be catastrophic.

Similar problems are becoming critical across the world. In India the number of wells rose from 800,000 in 1975 to twenty-two million in 2002. In many areas the water table is falling at one metre a year, although in northern Gujarat the water table fell from a depth of fifteen metres to over 400 metres in thirty years. In Uttar Pradesh the government dug 2,700 village wells between 1970 and 1985 to bring fresh water to rural areas. By the 1990s 2,300 of these wells were dry. In Pakistan the water table in the area around Quetta is falling at about three and a half metres a year and the area will run out of water in the next decade. In Yemen the water table is falling at two metres a year and with the population of the capital Sana'a growing at over 3 per cent a year the city is expected to have no water within the next few years. Other cities have suffered different problems from the over-extraction of water. Mexico City is built on a lake and its water use grew thirty-five-fold in the twentieth century, causing the city to sink over seven metres into the lake. In the forty years after 1920 Tokyo sank five metres because of water extraction but the introduction of strict controls managed to stabilise the situation.

In Saudi Arabia the government decided to use its oil wealth to become self-sufficient in wheat. This involved subsidising domestic production at five times the world price and extracting huge amounts of water from the underground aquifer. Half the reserves of water available when extraction started in 1984 have now been used and some wells need to be a kilometre deep in order to find water. The policy is economically and environmentally unsustainable. The situation in China is equally grim. Over-extraction of water from the 4,000-kilometre-long Yellow River caused it to run dry in 1972 and since 1985 for part of every year it fails to reach the sea. The water level in the aquifer under the North China Plain dropped three metres in just one year (2000) and around Beijing wells need to be a kilometre deep to reach fresh water. The cities face a water crisis and farm output is falling because of a shortage of water. The wheat harvest in the area fell by 30 per cent between 1997 and 2003. This is one of the major grain-producing areas in China and as a result the Chinese grain harvest fell by almost 14 per cent in the period – an amount equal to the total Canadian grain harvest. The World Bank has warned of 'catastrophic consequences for future generations'.

The increasing shortage of water in the world has led to a growing number of disputes over this key resource. This is nothing new. About 2500 BCE Lagash and Umma fought a 150-year-long war over control of irrigation water. In 1953 Israel tried to divert part of the Jordan river to irrigate the Negev but had to alter its plans following widespread condemnation. When Syria and Jordan drew up a plan for water diversion, Israel destroyed the preliminary works during the 1967 war. There have been two disputes over the Euphrates – in the mid-1970s between Syria and Iraq and again after 1990 when Turkey proposed to build a huge dam that would seriously affect both countries. Egypt and Sudan have often been in dispute over the Nile. As water shortages grow more severe over the next few decades the number of conflicts is likely to rise.

In the drive to provide irrigation water (and hydroelectric power for industry) numerous huge dams have been built across the world – in the last half of the twentieth century the number rose from 5,000 to 40,000. Overall the area covered by the world's reservoirs is now twice the size of Italy and two-thirds of the world's river flow passes through dams. Many of these projects have been highly damaging both environmentally and socially. The flooding caused by the reservoirs displaces large numbers of people – probably as many as forty million in the last sixty years (half of them in India) and the current rate is about four million people a year. Most receive no compensation. The lakes that are created cause a major loss of land – the Volta dam flooded 800,000 hectares and the Kariba dam about half this amount. Often this is good agricultural land. In addition the fisheries that have often been established on the lakes have failed. Evaporation rates in tropical areas are very high so that much of the water is lost and deforestation of the surrounding slopes produces very high run-off and siltation rates. In India the Nizamsagar dam in Andra Pradesh lost two-thirds of its capacity to silt accumulation and in China the Sanmenxia dam, completed in 1960, was abandoned four years later because the reservoir had completely silted up. The record was, however, set by the Laoying dam project which had to be abandoned before it was completed because of siltation.

A clear illustration of the problems caused by large dams can be found in the Nile where the low-level irrigation system that was the basis of the great civilisations of Egypt and which sustained the population for several thousand years has been destroyed in the last two centuries. In the traditional system, adjusted to the annual flood of the Nile, barley and wheat were harvested in the spring so that the late summer flood did not damage the crops and instead provided the basis for new planting in the early autumn. In the mid-nineteenth century the semi-independent ruler of Egypt, Mehmet Ali, directed a programme of extensive cotton growing as a basis for exports (and also the income of the social elite). Cotton production rose six-fold in less than thirty years after 1855. However, cotton was harvested in August and September and the crop could be ruined by the summer flood. New controls over the Nile were thought to be essential. After the British took over control of Egypt in 1882 and then the Sudan in 1898 a scheme became feasible. In 1902 the low Aswan dam was built – it was heightened in 1912 and 1934. This was a modest control mechanism that only caught about a fifth of the annual flood. A Greek-Egyptian engineer, Adrian Daninos, proposed a much higher dam at Aswan in 1912 and again in the late 1940s. This was rejected by the British, who favoured a dam in Uganda or Sudan. After the 1952 revolution the new leader of Egypt, Colonel Nasser, rejected these ideas (he wanted the dam to be under Egyptian control) and the high dam at Aswan was eventually built with Soviet help using money from the newly nationalised Suez canal.

The lake created (named after Nasser) now stores about two to three years' worth of the flow of the Nile. The controlled release of water allows two to three crops a year to be grown – cotton, maize and rice can be grown in the summer – and the dam generates about a third of Egypt's electricty. However, these gains have been achieved at a very high price. The lake is in the area of the highest evaporation and about a sixth of the water in the lake is lost every year. The dam also stops the flow of silt down the river – the basis of the fertility of Egyptian agriculture for several millennia. This has to be replaced by chemical fertiliser production, which takes up much of the electricity generated by the dam. The lack of silt has caused the Nile delta to shrink – sea water has infiltrated up to fifty kilometres inland and in places the delta is retreating at eighty metres a year. The discharge of Nile water into the Mediterranean is now one-sixteenth of the rate fifty years ago and this has increased the salinity of the Mediterranean significantly. Permanent irrigation has caused schistosomiasis rates to rise ten-fold and in some areas all of the population suffers from the disease. Many of the great monuments of ancient Egypt were drowned by the lake and elsewhere as water levels downstream have risen salt and mineral damage to others has increased dramatically. The dam has not solved Egypt's water problems – at the beginning of the twenty-first century Egypt was facing one of the most acute water shortages in the world.

The Aral Sea

The greatest single water and irrigation disaster in world history took place in the last third of the twentieth century in what was then the Soviet Union. The penchant for large-scale schemes and modifications of the natural environment characterised the USSR for all of its history. The Aral Sea was the fourth-largest inland sea in the world and was unusual in that two rivers flowed into it (the Syr Dar'ya and the Amu Dar'ya) but none flowed out – it was maintained in size by the high evaporation rate. Small-scale irrigation schemes had long existed in the area. In the early 1970s the Soviet authorities finally adopted a long-considered scheme to divert both rivers to irrigate over seven million hectares of land (equivalent to the whole of Ireland) to grow cotton in a highly unsuitable climate with poor soils. The consequences of this diversion were well understood and accepted. In the late 1950s one of the advocates of the scheme, A. Babayev, the president of the Turkmenistan Academy of Sciences, said:

> I . . . consider that drying up of the Aral is far more advantageous than preserving it. First, in its zone, good fertile land will be obtained . . . Cultivation of cotton alone will pay for the existing Aral Sea, with all its fisheries, shipping and other industries. Second, the disappearance of the Sea will not affect the region's landscapes.

What happened in the next decades was not an unexpected or unplanned disaster.

The diversion of water reduced the inflow to the Aral to 10 per cent of its usual amount. The sea level fell by nineteen metres and the area covered by the sea shrank by over three-quarters – by 1990 it had split into two separate lakes. As the water evaporated the salinity of the Aral increased by over 50 per cent and it is now two and a half times more salty than sea water. The shrinking sea exposed over 35,000 square kilometres of the former sea bed. Over 100 million tonnes of salt were blown away every year in the dust storms that raged on average for fifty days a year. The remains of the old sea bed and some of the badly waterlogged adjacent land became a new desert of solid salt marsh – the 'Aralkum'. The shrinking of the sea significantly affected the climate of the region – winters are now longer and colder but summers are drier and far hotter with temperatures up to 49°C.

The fishing industry collapsed in 1982 but the Mayanak cannery survived until the end of the Soviet Union by airlifting frozen fish from the Baltic and transporting the rest along the Trans-Siberian railway from the Pacific. It was the economics of the lunatic asylum and ended in 1992. As well as all the fish in the sea half the mammal species and three-quarters of the birds in the area became extinct. Human health was badly affected too as drinking water became polluted with salt and the fertilisers and pesticides poured on to the cotton. The mortality rate rose fifteen-fold in a decade, cancer and lung disease rates rose thirty-fold and the region now has the highest child mortality rate of any area of the former USSR. The state of the Aral Sea is a monument to the idea (particularly prevalent in the Soviet Union) that humans can manipulate the environment with impunity. Attempts at reclaiming the Aral Sea have had little impact and the problem appears to be insoluble. Any attempt at reclaiming the southern lake has been abandoned and most of it is expected to disappear by about 2010 as evaporation rates increase still further in the increasingly shallow sea.

Modern agriculture has, like its predecessors, produced a mixture of achievements, problems and environmental disasters. What has changed is the scale of the problems. Although more people have been fed than ever before this has been done on a highly unequal basis. Most of the population of the industrialised world has a rich (in many cases over-rich) and varied diet while a large part of the world's population suffers from a poor diet, malnutrition and the diseases associated with these conditions. The need to bring more land into production and the intensification of that production has produced numerous environmental problems – deforestation, soil erosion, desertification, salinisation and the over-loading of land and water with fertilisers, pesticides and herbicides. The problem that has affected agricultural societies throughout history – ensuring an adequate supply of food for all – has still not been solved on a global scale.

12

THE SECOND GREAT TRANSITION

The second great transition in human history involved the exploitation of the earth's vast (but limited) stocks of fossil fuels. It led to the creation of societies dependent on high energy use. This was a fundamental change – until the nineteenth century every society across the globe had very few sources of energy and the total amount of energy they could generate was small. This transformation was at least as important as the development of agriculture and the rise of settled societies. In its impact on the environment its effects were far greater and took place over a far shorter period of time. Until this transition all the forms of energy used by human societies were renewable (although trees were normally exploited in a non-renewable way). The last two centuries have been characterised not just by the use of non-renewable fossil fuels (coal, oil and natural gas) but by a vast increase in energy consumption.

Societies need energy for a number of fundamental tasks – lighting, cooking and heating in houses and power for agriculture, construction, industry and to provide mobility. Obtaining energy always involves a cost that limits what energy sources can be exploited. This cost may be quite low as in the case of gathering wood or very high as in the case of offshore oil. Until the last hundred years the availability of energy has also been constrained by transport problems. Wood and coal are heavy and bulky to move in any quantity (except by water) and water and wind power can only be provided in certain places. It was not until the development of electricity in the late nineteenth century (and then the building of high-power transmission systems) that a flexible, easily distributed form of energy capable of providing power, lighting and heating was available.

Human power

Until the nineteenth century about three-quarters of the world's mechanical energy came from humans and nearly all the rest was provided by animals – wind and water power were only marginal sources of energy. Humans are more efficient energy converters than animals. The amount of food they need is far less than the pasture and fodder requirements of animals and so in societies dependent on low-productivity agriculture there was little choice but to use people as the main source of power. For thousands of years it was the vast amounts of human toil and effort, with its cost in terms of early death, injury and suffering, that was the foundation of every society. Humans provided the main energy input into farming, carrying out a multitude of tasks, such as clearing land, sowing, weeding, digging, harvesting, constructing terraces and irrigation ditches, with limited assistance from animal power and with no more than primitive tools. With 90 per cent of the population living as peasants this was the reality of human life for all but a fraction of the last eight to ten thousand years. As late as 1806 one French agricultural writer could still advocate abandoning the plough and returning to digging fields by hand which, he argued, although slower, was cheaper (there was usually plenty of surplus labour available) and more thorough. Humans also provided much of the power for industry. The Great Crane in the market place at Bruges, regarded as the technological marvel of the fifteenth century, was powered by a human treadmill. In the nineteenth century prisons in Britain operated a treadmill which could be hired by local industrialists. The highest lock on the Grand Canal in China was worked by teams of several hundred men using capstans and ropes. Human power was also the main form of energy in the house until the invention of a range of labour-saving household appliances in the twentieth century. A hundred years ago two and a half million people (over 80 per cent of them women) were employed as domestic servants in Britain and they constituted the largest single occupational category. However, the new household appliances were only available for the relatively prosperous. Even in the early twenty-first century hard work around the home, especially gathering wood and collecting water, was the norm for hundreds of millions of women around the world.

One of the major problems societies faced was mobilising enough labour to undertake the projects the ruling elite wanted. Most of that labour was provided by compulsion either from the people of their own society or by slaves (often from prisoners of war and conquered peoples). Most of the great monuments of the ancient world from the temple mounds and palaces of Mesopotamia to the pyramids of Egypt and Mesoamerica, the Grand Canal and the Great Wall of China were built using forced labour. During the Fourth Dynasty in Egypt (2575–2465 BCE) the building of tombs in the provinces and simple memorials for the bulk of the population ceased. Instead, effort was concentrated on building the huge pyramids devoted to the Pharaohs and monuments for their kinsmen and high officials. The vast amounts of labour involved were controlled centrally by the state and there is no evidence that it was given willingly by the bulk of the population. In China the Grand Canal (which brought food from the south of the country to the capital) involved over five million conscripted peasants guarded by 50,000 police. Probably about half the workers died before the project was finished.

Some form of forced labour was common throughout European history too. For more than a thousand years agricultural workers were required to work a certain number of days a year on their landlord's estate in return for the land they cultivated themselves. Only in this way could the landlord obtain enough labour to cultivate his land and the peasants were usually tied to their land and unable to move away. Serfdom lasted in some form into the eighteenth century in western Europe and for almost another century in eastern Europe and Russia. Forced labour was required for other purposes too – as late as the eighteenth century the corveé to build and maintain roads still functioned in France. The totalitarian states of the twentieth century, in the gulags of the Soviet Union and China and the camps of Nazi Germany, resorted to forced labour for their major construction projects.

In addition to forced labour from their own people many societies relied on slaves to provide an unpaid energy source that only needed feeding. Slavery was common in the first settled societies and remained so across the world until the last century. There were, however, three societies in particular that relied on large-scale slavery. In ancient Greece Sparta treated the conquered peasants of Messenia as state slaves (or helots); forced to wear dog-skin hats, they could be killed with impunity by any Spartan. In Athens a third of the population (about 100,000 people) were slaves who worked in the state-owned silver mines at Laurion and on the small farms of the majority of Athenian citizens. If the latter had time to participate in the politics of Athens it was because of the slaves that they employed. Both of these societies were dwarfed by the slavery of the Roman empire. By the first century CE Italy had about two million slaves and they made up about a third of the population. Slavery on this scale required a continuing influx from war (prisoners and the populations of captured cities) and a well-developed slave trade encompassing western Europe, the Black Sea area and Africa. Probably about 250,000 people were enslaved every year. The third society that depended on large-scale slavery were the European colonial empires, mainly in the Americas. In total about twelve million Africans were enslaved to work on the plantations growing the luxury crops – sugar, tobacco, cotton, rice and indigo – that the Europeans wanted. Many also worked as domestic servants and in industry. In addition many of the indigenous population were conscripted as forced labour to work on farms and in terrible conditions in the silver mines of Bolivia that funded much of European trade for centuries.

Human power was also important in providing transport. On land, porters provided one of the commonest methods of carrying loads for thousands of years and still do in many parts of the world. In the Inca empire the main method of communication, in the absence of domesticated animals, was a highly efficient network involving teams of runners to convey messages along the roads built by the state (using conscripted labour). At sea many ships relied on human rather than wind power (or a combination of the two). The great ancient Greek fighting ship, the trireme, was human powered and in the Mediterranean the galley remained the main warship until the late sixteenth century. Galleys, both commercial and military, were powered by slave or penal labour in many societies from the Roman empire to nineteenth-century France. In China even their far more technologically advanced ships were powered by paddle wheels driven by human-powered treadmills. In Brazil from the sixteenth century, a shortage of animal power and a surplus of slave labour led to the slaves rather than oxen or horses being used to pull carts. In nearly every society people have been used to transport other people and this has acted as a powerful status symbol. The methods used have included seats and baskets on the back of a porter (still found in many parts of the world), palanquins carried by four or six people (common in the Near East, India and China until the nineteenth century), the sedan chair carried by two people (widely used in the cities of seventeenth- and eighteenth-century Europe), the rickshaw and its modern variant – the bicycle rickshaw.

Animal power

Apart from people, the other readily available power source (in Eurasia and Africa) was animals and they remained vital, even in the industrialised world, until well into the twentieth century. The main constraint on their use was the problem of growing enough feed. A horse needs about two hectares of land to obtain sufficent food, although an ox needs slightly less. Given the low level of agricultural productivity, most of the land was needed to feed humans and the number of animals that could be kept was therefore limited.

Animals were first used to carry loads as pack animals – loads were strapped over the back. Some of the earliest domesticated animals such as asses and onagers remained important in this role for thousands of years. Mules were also important, particularly in the Mediterranean area (Spain had twice as many mules as horses in the nineteenth century) and after the Europeans conquered South America mules were introduced as the main pack animal – there were over two million of them in the eighteenth century. Wheeled vehicles were first used in Mesopotamia about 3500 BCE (slightly later in the Indus valley and Egypt). They were drawn by oxen or onagers. Use of the onager died out and in Europe and the Near East the ox was the primary draught animal until the eighteenth century, when there were still twenty-four million oxen, compared with fourteen million horses, engaged in agricultural work in Europe. In India in the early twentieth century there were over five million ox-carts and they remain an important form of transport in many countries.

The horse was first domesticated about 3200 BCE somewhere on the plains of southern Ukraine although where the bridle and bit to control the animal were developed is unknown. However, for about three thousand years its use (apart from pulling lightweight chariots) was limited mainly to carrying individual riders or packs because of the problems involved in harnessing it to pull loads. Horses were first controlled through a throat and girth harness derived from those used for oxen. This was effective on oxen, cows and buffalo because it pulled against the vertebrae joining the neck to the body but on horses it produced the main point of traction on the throat and windpipe, which tended to choke the animal if it was pulling heavy weights, for example a plough. An effective harness required the pulling point to be low, along the sides of the animal, but this required a different vehicle design with two side shafts not a central pole. These problems were first solved in China with the breast-strap harness in the third century BCE, which enabled horses to pull heavy loads – it was not known in Europe for almost another thousand years. The modern horse collar (a padded frame round the neck with the shafts attached at shoulder level for the most effective pulling power) was first used in China around 500 CE but not for another five centuries in Europe. In northern and western Europe horses needed to be shod in order to counter the effects of damp. Iron horseshoes spread from Central Asia to China and then westwards to the Islamic world and finally to western Europe around 900 CE.

These technological developments enabled horses to be used as draught animals and, in western Europe, allowed the heavy plough capable of breaking up the soils of the region to come into use in the late first millennium CE. It was a long time before horses replaced oxen – they needed more feed and they had little value when they were too old to work whereas the ox could be sold for meat. In Europe the horse was little used for ploughing before 1100 and in England the ox remained the dominant draught animal for centuries after this. By the eighteenth century horses were the main power source for all agricultural work and remained so until the middle of the twentieth century when they were replaced by tractors. The agricultural horse population probably reached its peak in England and France about 1920 and as late as 1939 Germany still had over three million horses on its farms.

Animals were also important in warfare. Some of the earliest vehicles were solid-wheeled chariots on which the leaders of Mesopotamian armies rode to battle (they fought on foot). The development of cavalry and mounted warfare was limited until the invention of the stirrup. It was probably developed by nomadic groups somewhere in Central Asia and was known in India in the third century CE, reaching China where by 477 it had developed to enclose the whole foot. It took two centuries to travel westwards and it was used in Iran by 694. Its first use in Europe was by the Frankish army in the 730s. The advantage of the stirrup was that it made the rider much more difficult to knock off the horse. Horses began to be a central part of European armies although numbers were limited by the scarcity of feed – by the fourteenth century armies could still not maintain more than 3–5,000 horses. The heyday of the cavalry was in the eighteenth and nineteenth centuries (when more fodder was available) but they were ineffectual in the increasingly technological warfare of the twentieth century. Nevertheless armies continued to rely on horses as the main means of transporting supplies. During the First World War the

British army used 1,200,000 horses and in the Second World War the German army had mechanised Panzer divisions but it also required the logistic support of 2,700,000 horses (twice as many as it used between 1914 and 1918).

The use of wheeled vehicles across Eurasia remained problematic and limited for centuries. It was uneconomic to move loads over long distances and the 'roads' were little more than rutted tracks that degenerated into deep mud whenever it rained. These problems lay at the root of what, at first glance, seems a strange development – the replacement of wheeled transport by camels across the Near East and later the Islamic world. The one-humped camel was probably first domesticated about 1500 BCE somewhere in the Arabian peninsula. Originally they were kept for their milk until the development around 100 BCE of the north Arabian saddle (an inverted V-frame over the hump with a pad for a seat and from which loads could be tied to either side). The camel had a number of key advantages – it was far cheaper to maintain than oxen and horses and there were no harnessing problems. Only limited loads could be pulled in carts and these were expensive to maintain. A camel needed only a small amount of equipment, could travel twice as far in a day, did not need roads and one person could control six animals when they were tied together. Use of the camel was, therefore, a major improvement and infrastructure investment could be concentrated on bridges and caravanserai. The camel was crucial in opening up the trans-Saharan trade routes and the two-humped camel (domesticated in Iran or Afghanistan shortly before 1500 BCE) was the basis of caravan traffic along the Silk Road to and from China although it was rarely used west of the Oxus river.

In every part of the world, for thousands of years, animals were an important source of power to drive machinery, in particular mills to grind grain and crush olives. In late eighteenth-century Britain in the early stages of the so-called 'industrial revolution' there was a rapid increase in the use of animal, particularly horse, power (hence the use of the term to measure power generation). Most of the early machines used in industry were small-scale and needed only a low power input – about two or three horsepower for a fulling mill and about one horsepower for every 100 spindles in the earliest textile machines. Joseph Arkwright, one of the earliest pioneers in textile machinery, used nine horses at his first factory in Nottingham to power 1,000 spindles. Horses were also used extensively in mine drainage and the brewing industry. They were cheap to buy and flexible and they were the normal source of power up to about ten horsepower. After about 1800 horses were gradually replaced by steam power although the changeover took several decades.

As a means of transport the horse was vital until displaced by the widespread use of the internal combustion engine. Most individual travel depended on horses and they provided the power for stage coaches and the barges on the canals that were built in increasing numbers across Europe from the mid-eighteenth century. The development of railways in the nineteenth century did not bring about the demise of horse-drawn travel (apart from the stage coach). By generating more traffic, railways actually increased the demand for horses and their numbers reached a peak around 1900. Britain had one of the earliest and densest railway networks yet horse numbers rose rapidly in the nineteenth century. In 1810 there were about 15,000 privately owned carriages but by 1870 there were about 120,000. The number of horses kept in towns for private and business use rose from about 350,000 in 1830 to 1,200,000 by 1900. In towns the developing public transport systems remained horse-drawn until the early twentieth century. In 1902 London had 3,700 horse-drawn buses (each one drawn by two horses and needing about ten horses a day to maintain the service), 7,500 hansom cabs and 3,900 hackney coaches. Railway companies depended upon horse-drawn vehicles to distribute goods from their depots – in London railway companies owned 6,000 horses and coal merchants used another 8,000. As late as 1913 nearly 90 per cent of London's goods traffic was horse-drawn.

In the early twentieth century Britain had a horse population of about three and a half million (twenty-five times the current level), France had roughly the same number and the United States had between twenty and thirty million. These numbers were about the maximum that the agricultural system could support. In 1900 Britain's horses consumed four million tonnes of oats and hay every year (taking up the production of about six million hectares of land) – without cheap grain imports Britain could not have supported this number of horses and fed the human population. The logistic problems in moving this amount of feed around the country, particularly in towns, were enormous. In the United States horse feed took up about 36 million hectares of land (a quarter of the total) and was near the maximum possible. The replacement of the horse (outside of agriculture) happened very quickly. In Britain horse-drawn buses had almost disappeared before the First World War and after the early 1920s the number of horses in the country fell by a half in a decade.

Water and wind power

Until well into the nineteenth century humans and animals remained the dominant sources of energy. Water and wind power were no more than supplements and almost entirely confined to industry. The first use of water power was about 100 BCE in Egypt with the development of an automatic irrigation wheel and a grain mill. These early mills were powered by horizontal wheels but, once the necessary cams and gears had been invented, vertical wheels, which gave much more power, were universally adopted. The spread of water power, which in Europe and the Mediterranean area was restricted almost entirely to grinding grain, was slow at a time when human labour (much of it from slaves) was plentiful. In China water power was developed separately but at about the same time as in the Mediterranean, and was used mainly for operating the double-action piston bellows in the iron and steel industry rather than grinding cereals.

Over a period of several centuries the use of water mills spread across western Europe. The first firm evidence for the scale of construction comes from the Domesday Book, compiled in England in 1086. It records 5,624 water mills serving about 3,000 settlements. In some areas there was a dense concentration of mills – the Wylye river in Wiltshire had thirty mills within about fifteen kilometres. In France along the Robec river near Rouen there were two mills in the tenth century but twelve three centuries later. In the fourteenth century there were sixty-eight mills along about a kilometre of the Seine near Paris and on the Grand Pont in the centre of the city there were thirteen mills under the arches of the bridge.

Water power saved labour and time but its use posed a number of problems. Water flow in the rivers was variable during the year and so for considerable periods the mills could not be used. The concentration of mills in the best parts of the rivers caused endless disputes over usage and access. At Toulouse, on the Garonne, floating mills under the bridges were demolished at the end of the twelfth century and a series of small dams were built to regulate water flow. This too caused endless disputes – if those downstream were built higher to increase the water flow this diminished the flow for those upstream. New uses for water power gradually spread across Europe. It was first used for fulling cloth in Normandy in 1086, for tanning leather in Paris in 1138 and for making paper in Valencia in 1238. It was also used for driving iron forges, sawing wood, operating bellows and grindstones, polishing precious stones, milling coins and making mash for beer. As the use of water power for fulling cloth spread in England the woollen industry relocated to the north-west of the country where plentiful water power was available.

The first tidal mills were operational at Basra in southern Mesopotamia in the early eleventh century and from there spread westwards to Europe where the first tidal mill was built at the head of the Adriatic near Venice. They gradually spread across Europe, especially in areas where the rivers were either not powerful enough or not regular enough to drive mills. They were largely restricted to grinding grain but they continued to be built into the nineteenth century when there were still twenty-five operational in Devon and Cornwall.

Despite its limitations, water provided a major source of power for machines throughout the world for centuries. In the early stages of the so-called 'industrial revolution' in the late eighteenth and early nineteenth centuries the use of water power was extended mainly because of the limited efficiency of early steam power. In Britain many of the early textile factories were built in Yorkshire and Lancashire along the banks of fast-flowing rivers. Water power was used in London to pump water from the Thames to provide part of the city's water supply. In 1900 Nuremberg still had 180 operational water mills. In the United States many industries depended on water power until the 1880s and steam power was usually only used when it was essential to locate a factory away from a river. Water power was capable of sustaining large industrial enterprises. The great textile centres of Lowell, Lawrence and Manchester along the Merrimack river in New Hampshire and Massachusetts had over 900 mills and factories powered by the river. The giant Mastodon Mill on the Mohawk river took water in 250-centimetre pipes to turbines that generated 1,200 horsepower and drove over three kilometres of shafting, turned sixteen kilometres of belts, 70,000 spindles and 1,500 looms, which produced about 60,000 metres of cotton a day. In Japan too, the first textile mills were water powered and steam was not used on a major scale until the 1890s.

The main supplement to water power in both Europe and China was provided by the wind. Windmills were first developed in Tibet and China as prayer wheels and their use spread slowly. By about 1300 they were a common source of industrial power and by 1600 they were used to pump water. In the Islamic world windmills developed around 1100 but they had vertical axles. Their invention and development in Europe was a separate phenomenon (one of the few areas in which Europe was technologically ahead of the rest of Eurasia at this time). European windmills had horizontal axles and could also turn their sails into the wind and these two features significantly increased their power output. The first windmills were operational in England sometime before 1200 and they then spread rapidly – there were 120 round Ypres within a few decades. They reached Poland and Russia in the fourteenth century and were adopted slightly later in the Mediterranean region. A vital source of power in areas where the rivers were too sluggish, their power output was even more unreliable than water-powered mills. They were mainly used for grinding grain but from the fifteenth century were used to pump water and, in the Netherlands, played a crucial role in draining land to extend the cultivated area. By the sixteenth century the Netherlands had over 8,000 windmills and they were also used to power saws, lift equipment from mines and in fulling cloth, dressing leather, making gunpowder, rolling copper plates and throwing silk.

Wood

The world's main source of fuel until well into the nineteenth century was wood. The only real alternative was dried animal dung but if this was used the level of agricultural fertility fell dramatically and so it was usually a fuel of last resort. Wood had many advantages – it was easy to collect, readily available, burned well when dried and it was often free. The problem was the level of demand for wood. Although it was possible to meet some requirements (particularly for charcoal) by methods short of felling (such as coppicing), the easy option of felling was usually taken. Wood can be a renewable energy source but while supplies seemed inexhaustible little attention seems to have been paid to replanting. Apart from fuel, wood was used in construction work of every kind, in industry to make casks, vats and machinery, in shipbuilding, in wagon, coach and wheel making, and, in the form of charcoal, as the primary fuel for iron smelting, brewing, glass making, brick production and as one of the key constituents of gunpowder.

All of these uses consumed large amounts of wood. A moderate-sized house in medieval England required a dozen oaks to be cut down and in the fourteenth century work on Windsor Castle resulted in the felling of over 4,000 oaks in ten years. Once the industrial use of wood developed, consumption rose even more dramatically. Charcoal production involved large amounts of labour and huge quantities of wood. In 1475 the small iron industry in the Oberpfalz district of the Rhineland created employment for 750 miners, 3,000 people to transport the wood and iron and over 5,000 woodcutters and charcoal makers to produce about 10,000 tonnes of charcoal a year. In the United States in the nineteenth century (when wood was the primary energy source for industry until the 1890s) an average blast furnace used about 100 hectares of woodland every year and the large Hopewell furnace in Pennsylvania was using three times this amount. The production of potash was just as destructive. One works near Archangel in Russia exported 1,000 tonnes every year and for each tonne of potash produced 1,000 tonnes of wood were consumed. In the late seventeenth century, Russian potash production was using three million tonnes of wood a year. Evaporating water to produce salt also used prodigious quantites of wood. In the Kama region of Russia, where there were over 1,200 salt works in the eighteenth century, all the local forests had been felled and wood had to be brought from over 300 kilometres away to fuel the boilers.

Over the centuries this steady but relentless destruction of the forests, with very little, if any, replanting, slowly exhausted the resources of India, China, the Near East and western Europe. The early industrialisation of China in the eleventh and twelfth centuries produced acute wood and charcoal shortages in the north of the country. The smaller population and the slower development of industry meant that the first signs of a timber shortage were not felt in western Europe until the fifteenth century. The first area to be affected was shipbuilding which required specialised types of timber. Venice, one of the major maritime states, exhausted local timber supplies and came to rely on imports from its colonies along the Dalmatian coast. By 1590 the Venetians had to import completed hulls and merely fit them out in the arsenal. In its rival Genoa the price of oak for shipbuilding rose eleven-fold in the century after 1460 (far faster than the rate of inflation). Portugal, the first European nation to reach the Indian Ocean, suffered from a timber shortage from the start of the voyages of exploration in the fifteenth century. By the sixteenth century nearly all Portuguese ships were built in their colonies, either in Goa, using the teak forests of the Indian west coast, or in Bahia, using Brazilian hardwoods. Spain suffered from similar shortages. When Philip II built the Armada to sail against Holland and England in 1588 he had to buy trees from Poland.

In England the shortage of shipbuilding timber was first noticed during the wars against the French in the 1620s and the shortage became acute in the 1650s, particularly for specialised requirements such as masts. A first-rate, 120-gun ship needed a mainmast forty metres long and over a metre in diameter. Until the mid-seventeenth century the navy could rely on the oak forests of England, especially those of Sussex. In the late seventeenth century the Admiralty belatedly introduced officially sponsored replanting schemes even though it would take a century before the trees could be used. In the meantime timber had to be imported. At first the main sources were Scandinavia and Russia. The first recorded imports from Norway were in 1230 but by the seventeenth century large-scale importing of firs and oaks from the Baltic was underway. The areas around the main ports of Norway, Sweden, Denmark and Russia, were rapidly denuded of suitable trees and supplies had to be floated down the rivers from inland. In 1756 the British bought the right to export 600,000 trees a year from Russia in order to supply the Royal Navy. Agents acting for the British would buy up whole estates, put an end to farming and set the serfs to work felling trees until there were none left, when the estate would be sold. The new colonies in North America became an important source of British naval timber. The first pine masts were felled in New England in 1652 and by the end of the century the economy of New Hampshire was totally dependent on the timber trade. In 1696 the first warships for the Royal Navy were built in North America because of the shortage of European timber and during the eighteenth century a third of British warships came from this source. The effects of this rapidly growing demand were soon felt. By 1700 most of the timber within thirty kilometres of the main rivers of New Hampshire had been felled and in another fifty years most of the eastern sides of the mountains had also been cleared. By 1772 the newer colony of Maine had overtaken New Hampshire as the main source of supply. By then the latter colony had been stripped of all its very tall pines and the Royal Navy had to use inferior composite mainmasts made in Riga. During the American War of Independence the Royal Navy suffered from an acute shortage of timber and throughout the Napoleonic wars it had to rely on imported timber from Canada. The ships of the Royal Navy that were the basis of British naval supremacy in the first two-thirds of the nineteenth century were not built from stout British oaks. Until the replacement of wooden ships by ironclads from the mid-1860s, the Royal Navy had to import what it regarded as inferior timber from all over the world to construct its warships.

The shortage of timber for ships was only one symptom of a

major problem affecting the whole of Europe. There was a serious and widespread shortage of wood which meant that the continent faced an energy crisis. Much of its early industry (small-scale as it was) depended on wood and charcoal. Local sources were becoming exhausted and the poor state of inland transport made it impractical to move supplies very far. The shortage was felt from the late sixteenth century. In 1560 the iron foundries in Slovakia were forced to cut back production as they ran short of charcoal. In 1595 the bakers of Montpellier in the south of France had to use bushes to heat their ovens as there were no trees left around the town. In 1715 in Burgundy wood was so scarce and prices were so high that the poor could no longer afford fires. A decade later at Wieliczka in Poland the salt evaporation works had to close because all the local wood was exhausted.

The crisis and its consequences can be traced in Britain, which was one of the first countries to experience the energy crisis and to begin the transition to new forms of energy. In the early seventeenth century one commentator, Edmund Howes, wrote:

> within man's memory, it was held impossible to have any want of wood in England. But contrary to former imaginations such hath been the great expence of timber for navigation, with infinite increase in the building of houses, with the great expence of wood to make household furniture, caskes, and other vessels not to be numbered, and of carts, wagons and coaches, beside the extreme waste of wood in making iron, burning of brick and tile, that at this present, through the great consuming of wood aforesaid, and the neglect of planting of woods, there is so great a scarcity of wood through the whole kingdom.

The scale of the problem first became apparent in the second half of the sixteenth century. Because of a shortage of fuel about two-thirds of England's salt had to be imported from France which could rely on the sun to evaporate the water. Government measures, including a prohibition on cutting large trees and an enquiry in 1548–49 into the large-scale cutting of timber by the iron industry in the Weald of Kent and Sussex, had no effect. The destruction of the forests reached unprecedented levels at this time. In 1560 Duffield forest had over 90,000 oaks but in less than thirty years only 6,000 were left. The shortage of charcoal meant that armaments production had to be cut back and iron guns had to be imported from Sweden. The steep rise in charcoal prices in the late seventeenth century – they increased by 150 per cent between 1630 and 1670 – reflected an acute shortage and the situation worsened still further in the early eighteenth century. In 1717 a newly constructed iron furnace in Wales could not begin production for four years until it had accumulated stocks of charcoal and even then it only had enough fuel to operate for thirty-six weeks before it was forced to close. In other areas of the country blast furnaces were only able to operate in short bursts of activity every few years.

The response to this increasingly severe energy shortage was a change to what was widely regarded as an inferior fuel – coal. As wood prices rose, first the poor, and later even the rich, were forced to use coal. As Stow's *Annals* for 1631 commented:

> there is so great a scarcity of wood throughout the whole kingdom ... the inhabitants in general are constrained to make their fires of sea-coal or pit coal, even in the chambers of honorable personages.

The demand for coal rose as wood and charcoal became scarce and more expensive. In 1550 English coal production was about 210,000 tonnes but by 1630 it had risen to 1.5 million tonnes. The bulk and weight of the coal meant that most of it had to be moved by sea and the long-established trade between Newcastle and London expanded rapidly. In 1550 it amounted to about 35,000 tonnes a year but by 1700 it had risen to 560,000 tonnes and rose again to 850,000 tonnes by 1800. Although, given the choice, people preferred to use wood for household heating and cooking, they could burn coal on existing fires and stoves. This was also the case for some industries such as smithing, brewing and soap-boiling, which could use any form of heat. However, in many industries the impurities in coal posed major technical problems because they damaged the end product. New processes had to be evolved and this took time. Coal was used for making glass after 1610 and for making bricks a decade later. By the 1640s coke was used to dry malt for brewing and by the 1680s to smelt lead, copper and tin. The last major industrial process to be adapted to coke was the production of pig iron. (The Chinese had solved the problem more than 700 years earlier but nobody in Europe was aware of this.) This was achieved by Abraham Darby in 1709 although it was not until the development of the puddling process by Henry Cort in 1784 that coke could be used to make wrought iron.

Until well into the eighteenth century there was no alternative to using charcoal in some industrial processes, hence the limitations on production.

Coal

The growing use of coal in place of the increasingly scarce and expensive wood and charcoal in western Europe (particularly in Britain) from the seventeenth century onwards was of far greater significance than the simple substitution of one energy source for another. All previous human societies had depended upon renewable sources of energy – humans, animals, water, wind and wood. The fact that they had usually 'mined' forests with little, if any, thought given to conservation or replanting meant that this energy crisis was self-inflicted – the result of a short-sighted approach repeated century after century. Only when the shortage became acute did societies have no alternative but to exploit coal on a large scale even though it was, in many respects, an inferior fuel. It was the start of the switch to dependence on non-renewable sources of energy. The first major exploitation of the world's fossil fuel reserves, created from the great tropical forests that existed over 200 million years earlier, began in the seventeenth century. It was the beginning of the second great transition in human history and its consequences were profound. It amounted to more than simply a shift to reliance on non-renewable energy sources. It marked a fundamental discontinuity in human history – a move from the energy shortage that had characterised human history up until this point to societies that depended upon rapidly growing and very high energy use. The consequences of that shift are only now becoming apparent.

All the main coalfields of Europe were being worked by the thirteenth and fourteenth centuries, albeit on a very small scale. The coal came from opencast mining or from shallow pits no more than about fifteen metres deep. Deep mining was not developed until the eighteenth century when the high cost of charcoal offset the extra costs involved and the development of relatively efficient pumping machinery made it possible to remove water from deep shafts and galleries. These pumps were some of the first machines to utilise steam power derived from coal. However, coal-powered industries developed slowly even after the initial technical problems in its use had been overcome. Much of the early industrialisation of the textile industry in Britain was based on water power because it was cheaper and more flexible than coal, which was still difficult to transport. This meant that industries dependent on coal had to be located on the coalfields of Europe – Yorkshire, South Wales, the Ruhr, north-east France, Kempen in Belgium and Limburg in the Netherlands. Countries such as Italy, Austria and the Scandinavian countries, which had few coal reserves, were some of the last to industrialise in Europe.

Nevertheless the nineteenth century witnessed an enormous rise in energy consumption. World energy use more than tripled between 1800 and 1900 and this was almost entirely based on the rapid increase in coal consumption. In 1800 world coal output was about ten million tonnes, by 1850 it had risen to seventy-six million tonnes and by 1900 it was just over 760 million tonnes (a seventy-six-fold increase). In the space of two years at the end of the nineteenth century the world used more coal than in the whole of the eighteenth century. From a negligible contribution, coal came to account for about 90 per cent of the world's greatly expanded energy consumption. The new, high rates of energy use could not have been sustained using wood – in 1900 the world's coal consumption was equivalent to destroying and transporting a forest three times the size of Britain every year. There were not enough forests in the world to sustain production on this scale for very long and it is doubtful if this amount of wood could have been moved around the world. It was the labour of millions of coal miners, usually

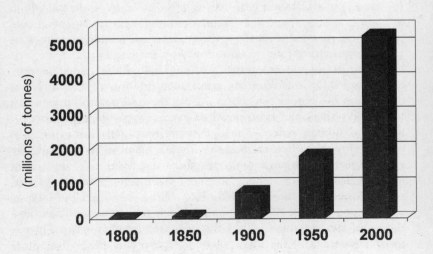

World coal production, 1800–2000

working in extremely difficult and dangerous conditions and without machinery (another example of the importance of human power), that supplied the energy that made the industrialisation of Europe possible.

Industrialisation in the United States depended upon a very different pattern of energy use from Europe. In such a recently settled area there were still large amounts of wood available until the late nineteenth century. In 1850 wood still accounted for 90 per cent of the United States' fuel supplies and half the country's iron was still produced using charcoal. Stoves and boilers still depended on wood as did the steamboats on the great rivers such as the Mississippi and railway locomotives were wood-fired (in Britain they were coal-fired from the start because of the shortage of wood). As late as 1870 wood still provided three-quarters of the fuel supply for industry and transport. Not until the mid-1880s did coal become the main source of energy in the United States as the easily available wood supplies became exhausted. The increased use of coal, when it came, was dramatic – consumption rose thirty-fold in the second half of the nineteenth century and by 1910 coal provided three-quarters of the country's energy. The experience of the United States shows, once again, that coal was not utilised until other more easily available and more flexible energy sources were on the point of exhaustion. The same pattern can be found in the later industrialisation of Japan. Water power sustained the initial rise of the textile industry and steam power was not adopted until the 1890s. As late as 1914 as much coal was used to fuel ships as in the whole of industry. Nevertheless as industrialisation spread Japanese coal production rose substantially – from less than a million tonnes in 1880 to over thirty-four million tonnes by the late 1920s.

The shift to non-renewable fossil fuels and the rise in energy consumption is illustrated by the transformation of the shipping industry in the nineteenth century. Until the mid-nineteenth century the world's ships had been powered by renewable energy sources – wind and human power. The first river steamboats came into use around 1810, they first crossed the English Channel in 1821 and by 1839 crossed the Atlantic (although these ships had sails too). Early technological developments – iron and steel hulls in the 1850s and 1860s – increased the effectiveness of sailing ships and the largest, about 2,000 tones, could still compete with the early, inefficient and expensive steam-powered ships that operated from the 1840s. It was the development of the high-pressure steam boiler made from steel that transformed the situation. By the late 1860s steamships could bring three times as much cargo from China to Europe in half the time taken by sailing ships. The amount of steam-powered shipping in the world rose from just 32,000 tonnes in 1831 to over three million tonnes by the mid-1870s and then rose exponentially as sailing ships gradually died out and the steamship took over the world's merchant and naval fleets. Britain built a chain of coaling stations across the globe to sustain the world-wide deployment of the Royal Navy. Not only had the amount of shipping in the world increased dramatically, its energy demands had increased even more.

The growing use of coal led to the rise of an important by-product – the use of the waste gases to provide the first non-natural source of lighting. Coal gas was first used to light a factory in Salford in 1807 and six years later a cotton mill in Rhode Island. The advantage for the factory owners was that artificial lighting enabled far longer hours to be worked. Coal gas was cheap to use for street lighting once the high installation costs (laying gas mains and installing new street lights) had been paid. It was cheaper than whale oil and available in much greater quantities. Street lighting could therefore spread on a much greater scale – by 1816 the first districts in London were lit by gas supplied from a central coal-burning plant through underground mains and by 1823 gas-lighting systems had been built in fifty-two towns. By the late 1820s gas lighting had been adopted in Boston, New York (which depended on imported British coal) and Berlin. The use of coal gas to light streets and houses (of those able to afford the installation and running costs) and eventually for cooking spread throughout the industrialised world in the nineteenth century.

The peak of the world's dependence on coal for its energy came in the first decades of the twentieth century. Its share of world energy consumption then fell from about 90 per cent in 1900 to less than 50 per cent by the early 1960s and to 25 per cent in the early twenty-first century. However, coal production continued to increase – from about 760 million tonnes in 1900 to just over 5,000 million tonnes in 2000 (a six-and-a-half-fold increase). The decline in the importance of coal occurred first in the United States because of its large oil reserves. In Europe the change came much later – in 1950 coal still provided over 80 per cent of the continent's energy. Yet by 1970 the proportion had fallen to less than a third as cheap imported oil replaced coal. Until the 1950s Europe's railways still depended on coal-fired steam engines as they had over a century earlier. Then in the space of a couple of decades they were replaced by diesel-powered locomotives and electrified systems. In 1900 Britain was the second largest coal producer in the world and the industry employed about 1,200,000 men. By the end of the twentieth century output was at 10 per cent of the level a century earlier and there were only 10,000 coal miners. Production also shifted from deep mines to opencast mining, which was much more destructive environmentally (but cheaper) – in the United States two-thirds of coal production now comes from opencast mines. However, coal is still the second most important energy source in the world. About 40 per cent of the world's electricity still comes from coal-fired plants and in many countries of the world it remains the primary fuel. It accounts for over 80 per cent of South Africa's energy consumption, two-thirds of China's and three-fifths of India's.

Electricity

One of the most significant energy developments of the last two centuries has been the use of fossil fuels to provide a highly convenient form of secondary energy – electricity. In 1821 Michael Faraday developed the first electric motor and a decade later the dynamo. Although the first electricity generators were made in London in 1834 it needed several decades of development before the new technology could be applied commercially. Its main use before the late nineteenth century was in communications, with the development of the telegraph and then the submarine cable (first laid between Britain and France in 1851). It was not until 1875 that the first large-scale lighting application was made at the Gare du Nord in Paris. Factories and shops across the industrialised world followed as new developments such as the carbon filament lamp (first used in England in the House of Commons in 1881) and the longer-lasting tungsten filament lamp introduced after 1911 made electric lighting more attractive. Soon, electric lighting began to replace gas in the homes of the wealthy.

The major increase in electricity production and consumption took place in the twentieth century and relied upon a number of linked developments. First, the construction of ever bigger power stations – in the 1920s the average capacity was about 30,000 kilowatts but within fifty years this had risen to 600,000 kilowatts. Second, the construction of high powered transmission lines and grids (first pioneered in Germany in 1885) to distribute power locally, nationally and eventually internationally. Although the first hydro-electric power station was built at Niagara Falls in 1886 the overwhelming majority of electricity has always been generated from fossil-fuel-fired power stations – at first coal and then oil and natural gas. This has been an important element in the ever-growing consumption of fossil fuels – in 1950 electricity generation took up 10 per cent of the world's fossil fuel production, and by the end of the twentieth century the figure had risen to 40 per cent.

The rise of a highly convenient form of energy, which, once the expensive infrastructure had been built, could be used in factories and homes for lighting, heating and power marked a fundamental shift in energy patterns. For the first time energy was easily available at the flick of a switch and, not surprisingly, this was one of the main driving forces behind much greater energy consumption. Electricity did not just replace steam power in industry, it made possible far greater automation of production processes through the use of machine tools and, in the late twentieth century, industrial robots. It also provided the energy for completely new industries such as aluminium production, which required massive amounts of electricity. It takes six and a half times more energy to produce a tonne of aluminium than it does to make a tonne of steel and world aluminium production rose almost a hundred-fold in the twentieth century. Electricity also provided the basis for the huge growth of domestic energy use and the industries that made the products that consumed it. The United States was the first country to be electrified on a major scale, in the urban areas in the 1920s and a decade later in the countryside. In Europe the countryside did not generally receive electricity until the 1950s. Nearly every household in the industrial world is now connected to a mains supply that provides lighting, heating, cooking and power for a vast array of household goods such as refrigerators, freezers, televisions, washing machines, dishwashers and computers. Overall electricity consumption has increased twice as fast as overall energy consumption. In Europe electricity consumption rose twenty-six-fold between 1920 and 1970. World electricity consumption has continued to rise at a very rapid rate – it doubled in the two decades after 1980.

Oil

Oil had been obtained for centuries across the world from places where it seeped on to the surface and was used mainly in the form of bitumen for caulking ships and for medicinal purposes. It was not until the second half of the nineteenth century that efforts to extract and exploit the product on a commercial scale began. The world's first commercial oil came from the Drake well in Pennsylvania in 1859 but for decades world production was small-scale. The pressure to overcome the technical problems in drilling, extracting, refining, storing and transporting oil came from two sources. First, lubricants such as whale and vegetable oils were increasingly inadequate for the demands of the new industrial machinery. Second, whale oil was in short supply and the price was rising steadily. It was these two uses that drove early oil production. In the late nineteenth century 85 per cent of crude oil was refined into kerosene for illuminating oil and the rest was made into industrial lubricants. The development of oil-burning furnaces in the early twentieth century created new uses – by 1909 fuel oil made up half the output of an increasingly important industry. The Royal Navy changed from coal to oil-fired warships at this time. The development of the internal combustion engine in the late nineteenth century and the enormous growth in the number of vehicles in the world transformed the oil industry. By 1930 gasoline became the main refined product produced by the oil industry followed in the next couple of decades by aviation fuel. Plentiful and cheap supplies of oil led to the production of new artificial materials such as nylon and rayon and then the development of a whole range of plastics for different uses.

The twentieth century witnessed the development of a range of oilfields across the world. In 1900 Russia was the largest producer (mainly from the fields around Baku in Azerbaijan). It was soon over-taken by the United States exploiting the huge oilfields of Texas, the surrounding area and California. By the 1920s and 1930s production in Venezuela, Mexico and Romania was also important. After the Second World War the development of the oilfields of the Near East provided much of the world's energy – production in the area rose forty-three-fold between 1938 and 1970. As new exploration and drilling technologies were developed, more difficult fields came into production – Alaska and then the offshore fields of the Gulf of Mexico, the North Sea and elsewhere. (By the late twentieth century a third of the world's oil output came from offshore fields.)

Just as coal proved to be the means to support an otherwise imprac-ticable increase in energy use in the nineteenth century, the availability of cheap oil was the motive force behind the even bigger increase in energy use in the twentieth. The world's consumption of oil has risen astronomically in the last century. In 1890 production was about ten million tonnes. This rose about ten-fold to reach ninety-five million tonnes in 1920, trebled to 294 million tonnes in 1940 and then

World oil production, 1890–2000

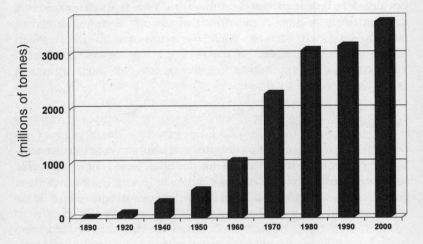

doubled every decade to reach 2,300 million tonnes by 1970. The rate of growth slowed in the last three decades of the twentieth century but consumption still rose to over 3,800 million tonnes by 2004 – a 380-fold increase in little more than a hundred years. In 1900 oil provided less than 1 per cent of the world's energy supplies but 40 per cent of the much greater level of energy use by the end of the twentieth century. Its impact on transportation has been even greater – it now provides over 90 per cent of the world's energy used in transport.

Natural gas

A major by-product of the exploitation of the world's oilfields has been the increasing importance of natural gas. Although it was used locally near American oilfields in the early twentieth century, its widespread use required major developments in high-pressure pipeline technology. This was first adopted in the United States in the 1930s but little use was made of natural gas until the 1960s – huge amounts were simply burnt off at the oil well heads. Major fields were developed off Groningen in the Netherlands (it now supplies about a third of the country's energy) and in the North Sea. Britain switched over from coal gas to natural gas in the 1970s and then in the 1990s it switched from coal-fired electricity to gas-fired. The use of natural gas in Britain rose eight-fold between 1970 and the early twenty-first century. The largest producer in the world now is Russia, exploiting the vast fields in Siberia, and much of this gas is now pumped to western Europe. In the early twentieth century natural gas provided no more than 1.5 per cent of the world's energy. A century later it was the third most important fuel (after oil and coal), making up just over 20 per cent of the total.

Non-fossil-fuel power

Only two technologies have contributed real alternatives to the increased consumption of fossil fuels – hydroelectric and nuclear. (Other renewable sources – wind, wave, geothermal and solar power – have not been developed on any major scale.) The use of hydroelectric power grew significantly in the first part of the twentieth century and in 1929 it provided 40 per cent of the world's electricity. However, by then most of the best sites had already been exploited and it became progressively less important over the rest of the century. It now provides less than 5 per cent of the world's energy. Nuclear power was a spin-off from programmes to manufacture nuclear weapons. Its history has been dogged by safety concerns, poor designs and a consistent failure to produce electricity cheaper than other sources. It survived only through massive government subsidies – over $100 billion in the United States in the forty years after 1950. In Britain when the electricity industry was privatised in the late 1980s no commercial enterprise would take on the nuclear power stations because they were so uneconomic. In most countries that operate the 450 or so nuclear plants across the world they provide about 20 per cent of the country's electricity; only France, motivated by a lack of oil, produces about 80 per cent of its electricity from nuclear power. In terms of total energy nuclear power only provides 7 per cent of world consumption.

The problems associated with nuclear power became increasingly apparent in the last third of the twentieth century. After 1970 two-thirds of the nuclear plants on order were cancelled (in the United States all were cancelled) and in many European countries it was decided to phase out nuclear power. Neither the World Bank nor the Asian Development Bank will fund new nuclear plants since they are not cost-effective. In the UK a government report in 2005 concluded: 'Nuclear power is . . . not currently competitive as a new investment option.' Part of the reason for this is that there is no solution to the problem of storing the high-level nuclear waste that remains at the end of a nuclear plant's life – the current estimate for decommissioning the UK's existing nuclear plants is over £70 billion. Despite this array of problems the phasing out of the existing nuclear plants will mean a major drop in electricity-generating capacity over the next two decades and this, combined with pressure from the nuclear industry, has caused many countries to consider building new plants – thirty-five were under construction in the early twenty-first century.

Energy in the early twenty-first century

World energy patterns are now completely different from those that prevailed for all but the last two centuries of human history. Until the early nineteenth century renewable resources – human, animal, wood, water and wind – provided nearly all the world's energy. Now, over 85 per cent comes from non-renewable fossil fuels (40 per cent from oil, 25 per cent from coal and 21 per cent from natural gas). That transition to fossil fuels has been accompanied by a spectacular rise in energy consumption. Until the nineteenth century all societies faced an energy shortage but that has been replaced by energy abundance. Modern societies could not function without high energy consumption. The major fossil fuels have seen huge rises in consumption. The world's annual consumption of coal is now about 500 times greater than it was in 1800. Oil was hardly used until 1900 but annual consumption is now about 380 times greater than it was a century ago. The consumption of natural gas rose 175-fold in the twentieth century. For most of the last two centuries fossil fuel has been cheap – spectacularly so in the case of oil in the mid-twentieth century. In real terms the price of oil actually fell for long periods in the twentieth century – 1919–39, 1948–73 and 1984–2000.

Percentage share of world energy consumption, 1900 and 2000

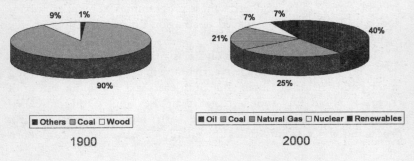

Energy efficiency

Low prices and seemingly inexhaustible supplies have not encouraged energy efficiency and therefore huge quantities of energy have been wasted. In some instances this was simply the result of new technologies necessarily being inefficient. In the nineteenth and early twentieth centuries most coal was burnt in open fireplaces where over 90 per cent of the heat was wasted (mainly by escaping up the chimney). Wood stoves were more efficient – about two-thirds of the heat was wasted. The earliest steam engines were hopelessly inefficient – at best 95 per cent of the energy was wasted and they could do the work of no more than 200 men. The development of high-pressure steam systems increased efficiencies roughly thirty-fold so that steam-driven turbines were about 20 per cent efficient by 1910 and further improvements doubled this figure by the 1950s. Nevertheless this still meant that more energy was wasted than performed useful work.

Although electricity provides a highly convenient form of energy it is a highly inefficient way of producing energy. Generating stations have to be built and operated and high-voltage transmission lines constructed, all of which consumes energy. The earliest generating stations were only about 4 per cent efficient. This rose to about 13 per cent in the mid-1920s and then to about 25 per cent by the 1950s. However, since then there has been almost no further improvement in efficiency. This means that although a third of the world's energy is used to produce electricity, at least two-thirds of it is wasted in generation and transmission. The United States wastes as much energy in electricity generation as the total energy consumption of Japan. These inefficiencies are compounded by inefficiencies in consumption – homes, factories and offices are poorly insulated and domestic appliances and light bulbs use more electricity than is necessary. The sharp rise in oil prices in 1973–74 and 1979–80 did lead to programmes designed to save energy in most of the industrialised world but once prices began to fall in real terms these programmes were quietly abandoned and the emphasis placed once again on making more energy available.

Modern industrialised agriculture is highly inefficient in energy terms. The most energy-efficient agriculture in the world is rice growing in the paddy fields of China and south-east Asia where the output of energy is about fifty times greater than the input. Other so-called primitive agricultural systems are also highly energy-efficient, producing about twenty times the energy they use. At best, modern cereal farming produces only about twice as much energy as it consumes in the form of fertilisers, pesticides, herbicides and machinery. Modern agriculture is also becoming less energy-efficient. In the twenty years after 1952 energy inputs rose by 70 per cent but production only increased by 30 per cent. Maize production in the United States shows an even worse situation. Energy inputs rose 400 per cent between 1945 and 1970 but yields only rose by 138 per cent. Overall the energy efficiency of American maize production has fallen by over a half since 1915. Modern animal rearing systems consume large amounts of energy in heating the sheds where the animals are kept and in producing the artificial feed and the antibiotics they eat. Meat production in the industrialised world now consumes between two and three times the energy it produces. The production of frozen fish is the most inefficient of all forms of food production – it consumes about twenty times as much energy as it produces. On top of these energy costs in food production it is necessary to add the energy cost of processing and distributing food. This takes about three times as much energy as producing the food itself.

In spite of these considerable inefficiencies modern industrial economies are far more energy-efficient (in terms of the amount of energy consumed per unit of GDP) than they were a century ago. The energy intensity of the British economy peaked as early as 1850–80 and it began falling slightly later elsewhere across the industrialised world – Canada in 1910, the United States and Germany in the 1920s, Japan in the 1970s and then in the newly industrialising countries such as China and Brazil in the 1980s. The United States is now about 60 per cent more energy-efficient than it was in 1920. Overall the world is about as energy-efficient now as it was in 1900 (reflecting the use of poor technologies by newly industrialising countries) but the peak of inefficiency came in 1970 and since then world energy efficiency has increased by about 20 per cent. There are, however, major differences between countries in their energy efficiencies. The United States is still 60 per cent less efficient than Italy and Japan and worse than India and China. Countries such as Ukraine, still saddled with hopelessly inefficient Soviet-age technology, are worse still – it is three times less efficient than the United States and five times worse than Japan.

There is, however, an important lesson to be learnt from the history of energy efficiency. Although the countries of the industrialised world are now more energy-efficient than they were a century ago this has not stopped a massive increase in energy consumption. Indeed, there is plenty of evidence that increasing efficiency tends to lower energy, particularly electricity, prices (especially in real terms) and this encourages the greater use of energy. The experience of the United States in the last two decades of the twentieth century illustrates this factor very well. The energy intensity of the US economy fell by 34 per cent in this period. Combined with a population rise of 22 per cent this should have produced a fall in energy use. However, GDP per head rose by 55 per cent and so the total amount of energy consumed rose by 26 per cent. Across the world it seems highly unlikely that this pattern will change in the future. Further increases in energy efficiencies through new technologies will not stop rising demand for energy and even higher energy consumption. Ultimately the impact of energy consumption on the environment comes from the total amount of resources used and the pollution that is produced. How 'efficiently' this is done is of little consequence.

Energy inequalities

The switch to fossil fuels and the development of high energy-use societies has heightened world inequalities. Fossil fuel consumption has been overwhelmingly the responsibility of the major industrialised countries. In the first half of the twentieth century the industrialised countries of western Europe and North America consumed over 90 per cent of all the fossil fuels used in the world. In the early twenty-first century the fifth of the world's population that live in the rich countries of the world still consume over 70 per cent of the world's energy. The United States makes up only 5 per cent of the world's population yet every year it uses 27 per cent of the world's energy. The majority of the world's people who live in the developing world use only 10 per cent of the world's energy. The poorest quarter of the world's population (just over one and a half billion people) use only 2.5 per cent of the world's energy. The average American now uses four times as much energy as their predecessors did a century ago, twice as much as the average European, thirty times more than the average Indian and almost a hundred times more than the average Bangladeshi. Indeed the US military now consumes as much energy every year as the total energy use of two-thirds of the countries in the world and it is more than the total energy use of a rich country such as Switzerland.

The energy problems of the majority of the world's population are closer to the conditions experienced in the whole world before the nineteenth century. Energy is still in short supply and both human and animal power are still crucial. Half the world's population (about three billion people) still depend on wood, charcoal and animal or crop residues for their fuel supplies. The rapidly rising population in the twentieth century placed a severe strain on these limited resources. At least 100 million people are unable to obtain enough fuel for even their minimum cooking and heating needs and nearly two billion people are depleting their stocks of wood faster than the replanting rate. Shortages of wood bring about a vicious circle in which dried animal dung is used for heating and cooking rather than as a manure, thus reducing soil fertility, crop yields and the ability to maintain animal numbers. This then exacerbates many of the other problems of people living on the edge of poverty and with a poor diet.

The last ten thousand years of human history have witnessed an enormous change in the pattern of energy consumption, from the minimal demands of gathering and hunting groups to modern American levels. However, nearly all that change has taken place in the last two centuries and by far the biggest increase in energy use has come in the last hundred years. Despite the increasing level of technological sophistication in obtaining and distributing energy over that period there has been a remarkable consistency of attitude towards energy. Human societies have rarely taken account of anything except short-term considerations and have treated all sources of energy as though they were inexhaustible. That cannot be the case with fossil fuels. Estimates of the point at which reserves will be exhausted are difficult to make because of the problem of estimating the size of undiscovered reserves and future consumption rates. However, most estimates agree that there is enough coal to last for several hundred years (even at increasing consumption rates) while reserves of oil and natural gas are likely to be exhausted during this century, perhaps within a few decades. Long before reserves are exhausted severe problems will be encountered as supplies become more difficult to obtain from remote fields and prices rise as demand comes up against supply constraints. However, before the world has to cope with a shortage of fossil fuels it is likely to have to face the far more severe environmental problems caused by their consumption over the last two hundred years.

13
THE RISE OF THE CITY

One of the greatest changes in the way people live and the environment they have experienced has been caused by the rise of the city and urban living in the last two centuries. Cities are the most artificial environment produced by humans. Cities emerged at an early stage in the development of settled societies and are generally seen as central to a civilised society. However, for thousands of years they played little part in the lives of the overwhelming majority of people. Until the early ninetenth century only a tiny minority – no more than 3 per cent of the world's population – lived in cities. With the increasing use of fossil fuels and the transition to high-energy, industrialised societies in Europe and North America truly urban societies emerged where about three-quarters of the population lived in cities. Overall though only about 14 per cent of the world's population lived in cities in 1900. The great transformation in the twentieth century, especially in its second half, was the rise of the city in the rest of the world, even in countries that were not industrialised. By the first decade of the twenty-first century a majority of the world's population lived in cities.

The scale of these changes can, because of the parallel phenomenal rise in world population, best be illustrated by the number of people living in cities rather than by percentages. In 1800 about twenty-seven million people lived in cities. By 1900 this had risen ten-fold to 225 million. By 1950 it had almost quadrupled to 810 million. In the next fifty years it more than tripled to 2,900 million. Overall the number of people living in cities has risen 107-fold in the last two centuries.

Early cities

The earliest agricultural groups lived in villages of at most a few hundred people. Within a few thousand years of this crucial step being

Percentage of people living in cities, 1800–2000

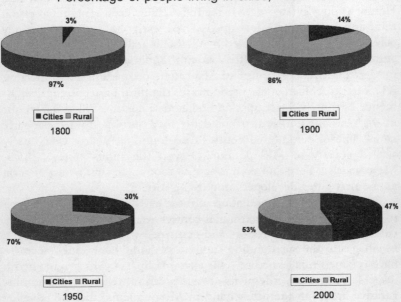

3%

97%

■ Cities ■ Rural

1800

14%

86%

■ Cities ■ Rural

1900

30%

70%

■ Cities ■ Rural

1950

47%

53%

■ Cities ■ Rural

2000

Number of people living in cities, 1800–2000

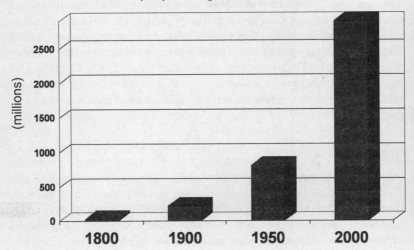

(millions)

2500

2000

1500

1000

500

0

1800   1900   1950   2000

taken, cities such as Uruk, Ur and Lagash in Mesopotamia, and Harappa and Mohenjo-daro in the Indus valley, each containing several thousand people, had developed. These early cities were essentially ceremonial centres dominated by their cult activities and with certain administrative functions. Often, as with Teotihuacan and Monte Alban in Mesoamerica, these cities were laid out to complex designs reflecting religious symbolism or oriented towards significant astronomical alignments. Normally the secular and religious elites controlled the labour of the peasants and redistributed the food surplus from these early cities. These 'ceremonial centres' are found in nearly every early settled society in Mesopotamia, Egypt, the Indus valley, China, Mesoamerica, Peru and even as late as 1000 CE in south-west Nigeria as the first cities developed among the Yoruba.

The nature of the population of these early cities varied from area to area. In Egypt the ceremonial centres were mainly populated by rulers, priests, administrators and craftsmen with the peasants living in villages in the surrounding rural areas. Some later 'cities' seem to have retained this structure – for example Sparta, which remained a dispersed series of settlements around a central religious and administrative area. On the other hand in Mesopotamia, the Indus valley and many of the cities of Mesoamerica such as Teotihuacan and those of the Maya most of the population seems eventually to have moved from the countryside into the cities and to have cultivated the surrounding area during the day and returned to their homes in the cities at night. In some the population was rigidly organised and strictly disciplined. In Teotihuacan the population lived in about 2,000 one-storey, windowless compounds (with underfloor drains) about fifty metres long. About sixty people seem to have lived in each of the compounds. They were probably occupied by kinship groups and the size of the compounds varied according to the status of the group and its occupations – potters for example all lived in the south-west part of the city.

'Pre-industrial' cities

The rise of the first empires after about 2000 BCE, the steady but slow increase in the population and the development of small-scale trade led to the development of 'pre-industrial' cities which had characteristics very different from the modern city. During a long period of over three thousand years most 'cities' were on the scale of a modern small town or village. Many had a population of no more than a thousand or so and many of the inhabitants were still engaged in farming in the surrounding fields. The only large population centres were the great imperial capitals such as Rome, Constantinople and Ch'ang-an or, on a smaller scale, the centres of seaborne trading empires such as Athens, Venice and Malacca. These cities depended on their empires for the import of food – Athens relied on the Black Sea area, Rome the Mediterranean. The various Chinese capitals relied on the transport of food along the Grand Canal from the great rice-growing areas in the south of the country. As the administrative centres of large empires these cities contained the elite and wealthy members of that society together with administrators and priests. The population was swelled by the servants and slaves of the elite and a miscellany of craftsmen, traders and shopkeepers who depended on the elite for a living. A large part of the population was also unemployed beggars living on the margin of existence. These large capital cities were parasitic on their societies. Not only did they import huge quantities of food, they were also extremely unhealthy places to live and depended upon a steady influx of people from the countryside to maintain numbers. People left the rural areas because making a living there was extremely difficult and begging in the cities often seemed a better alternative.

Despite their parasitic nature many of these cities grew to an enormous size. At its height around 150 CE Rome probably had a population of about 500,000. Lacking a firm economic base and dependent as they were on the fluctuating fortunes of the great empires, these imperial cities often declined as quickly as they grew. The population of Rome fell drastically as the empire in the west collapsed and the focus of the empire shifted to the east and the old Greek city of Byzantium, renamed Constantinople. By 600 CE Rome contained no more than 50,000 people and Constantinople contained over 500,000. As the Byzantine empire declined from the twelfth century that city too declined until it became the centre of the great Ottoman empire after 1453 – its population rose to 700,000 by 1700. Similarly Vijayanagar, the capital of the main Hindu empire in India in the thirteenth and fourteenth centuries, had a population of about 500,000 in 1500. Within a century, following the Mughal conquest, it was virtually deserted.

The other towns that developed had a variety of origins. Some grew up at important strategic positions – a river crossing or a fortified site. Others depended on their right to hold a market to meet the demands of the limited amount of trade. Some were deliberate acts of creation. From not long after 1000 BCE the Phoenicians and then the Greeks founded new cities across the Mediterranean as a way of removing surplus population. Many of these grew into major cities – the Phoenician colony of Carthage became the centre of its own empire and many of the Greek colonies such as Marseilles and Naples grew into important trading centres. Roman settlements were also established at key sites such as London, Paris and Cologne. Other empires founded cities too – Marakanda (modern Samarkand) was established as the Achaemenid empire pushed out from its homeland of Iran into Central Asia.

Pre-industrial cities had a number of characteristics that differentiate them from modern industrial and commercial cities. In nearly every case they were surrounded by walls. This not only provided defence but clearly demarcated the city (which often had a different legal system from the surrounding area). It also enabled city authorities to regulate and tax trade as it moved to and from the city through the small number of gates in the walls. Within these walls towns often retained a semi-rural appearance with many fields and orchards. Most of the inhabitants lived in narrow streets that were little better than alleys, suitable for pedestrians and pack animals but too small for carts which were confined to main streets that crossed the city. The centre of the city contained the public buildings and the houses of the rich, whilst the poor lived on the outskirts – the opposite of most modern cities. Residential areas and places of work were often carefully segregated – this was carried to an extreme in most medieval European cities which had ghettoes to which the Jewish population was confined.

Early Asian cities

Until the last two hundred years nearly all the world's great cities have been outside of Europe, reflecting the greater wealth and prosperity of the Islamic world and China. In 600 CE the largest city in Europe was Rome with a population of about 50,000. This was no bigger than the Mayan city of Tikal, it was half the size of Teotihuacan and a tenth of the size of the great cities of China – Ch'ang-an and Loyang. The first country to undergo incipient urbanisation was China under the Sung in the eleventh and twelfth centuries. At this time the Chinese economy was similar to that of Europe seven centuries later in the earliest stages of the 'industrial revolution'. (Iron production was twice that of England in the late eighteenth century, factories employing several thousand workers were common, mechanisation of production was beginning and there were sophisticated commercial systems – cheques, bills of exchange and promissory notes – and a paper currency was introduced in 1024.) The most important city until the twelfth century was the Sung capital K'ai-feng, which became a true commercial and trading city, not simply an administrative centre. By 1100 the population was probably about 500,000, making it the biggest city in the world following the decline of Baghdad after the fall of the Abbasid caliphate. The largest city in Europe was probably Venice with a population a tenth of K'ai-feng. Just one borough of the Sung capital ('Left Number Two' in the north-east corner of the city) had a population greater than the whole of Paris.

In 1100 the population of the great southern city of Hang-chou was probably only slightly smaller than that of K'ai-feng and its wealth was based on its trade networks into south-east Asia. In 1127 the Sung were driven to the south of the country by a Jürchen 'barbarian' invasion of the north and Hang-chou became the new capital. It rapidly became the largest city the world had ever seen. By 1200 its population was about two million – about ten times as big as any other city in the world (London had a population of about 40,000 at this time). This figure may seem extraordinarily high but the evidence from travellers from both the Christian and Islamic worlds (Marco Polo and Ibn Batuta) is remarkably consistent. They both state that the main street had ten market places evenly spaced at just over six kilometre intervals and that it took more than a day to walk the length of the main street. Probably about 800,000 people lived within the city walls but the suburbs spread for several kilometres beyond the walls so that it was probably about forty kilometres from one side of the urban area to the other. Hang-chou was undoubtedly the largest city in the world before nineteenth-century London. In the thirteenth century China was the one partially urban society in the world – possibly about 20 per cent of the population lived in cities. This phase was, however, short-lived. By the thirteenth century the Sung were under growing pressure from the Mongols, who already controlled the north of China. A long, highly destructive war destroyed much of the prosperity (and people) of China.

The earliest cities in Japan such as the imperial capital Kyoto were artificial cities constructed on a strict rectangular plan adopted from the Chinese. From the thirteenth century small trading cities and settlements around castles emerged. The long period of internal and external peace under the Tokugawa shogunate from the early seventeenth century laid the foundations for a highly prosperous economy. Rapid urbanisation followed. In 1600 there were only about thirty settlements in the country with a population of more than 5,000 but by 1800 there were over 160. The most spectacular growth occurred at Edo (modern Tokyo), which was a small fishing village when the Tokugawa made it their capital. Under the *sankin-kotai* system local landowners and rival aristocratic families had to reside in the capital for half the year and then leave family hostages behind when they left. The elite brought with them a large number of retainers and employees which swelled the population in a manner typical of many pre-industrial capitals. By 1800 Edo had a population of about one million, making it about the same size as London and one of the three largest cities in the world. However, it was more than a capital city and it had important commercial and industrial functions as well. Overall in the early nineteenth century about four million Japanese lived in cities – as high a proportion of the population as anywhere in Europe and far higher than many countries such as Spain and Italy.

Early European cities

In Europe, the Mediterranean was the centre for all the most developed societies and empires until at least the twelfth century and the size of the cities in the area reflected this fact. Constantinople was by far the largest (about 400,000 at its peak) but Cordoba, the capital of the Islamic world in the western Mediterranean, probably had over 100,000 people and other important cities such as Seville and Palermo were only slightly smaller. The north and west of Europe was very different. Under the Roman empire there were only a few towns, most linked to military settlements, and none had a population of more than a few hundred. After the end of Roman rule most of these cities declined, there was little trade and industry and the agricultural surplus was insufficient to support more than a very small urban population. In 1000 there were no more than about 100 towns or cities in the whole of Europe and half of them were in Italy which was the most developed part and had links to the prosperous Islamic world. In England in 1086 just 67,000 people lived in towns (10,000 of them in London) out of a total population of slightly less than two million.

As trade and industry slowly developed and the population increased between 1000 and 1300, towns developed away from the Mediterranean and others became larger. By 1300 there were

3–4,000 towns and cities in Europe but only nine of them had a population greater than 25,000 – Florence, Paris and Venice had about 100,000 people, London and Ghent about 50,000. Most towns were very small; even the flourishing Hanseatic port of Hamburg only had a population of 7,000, and the majority of towns in northern Europe had less than 2,000 people. Towns of this size were little more than large villages – most of the people worked in the adjoining fields and only a weekly market and the presence of a few craftsmen distinguished them from the rest of an overwhelmingly agricultural society.

Between 1300 and 1550 the pace of urbanisation slowed partly because of the population fall during the Black Death and because economic growth was stilted and patchy. (Between 1350 and 1550 the number of market towns in England fell by two-thirds.) Some cities such as Florence and Venice declined as their political and economic influence waned and Valladolid suffered a rapid decline when the Spanish capital moved to Madrid. Around the Mediterranean area, where economic development tended to be slower than in the north, many cities remained centres of elite residence and consumption. Naples, capital of the Kingdom of the Two Sicilies, with a population of over 400,000 in the 1770s, was a prime example but in Rome too the population depended upon the Papal court and administration. A few new cities were founded but they were capital or court cities in the same way as others thousands of years earlier. St Petersburg was the artificial construct of Peter the Great in the early eighteenth century. The growth of the Prussian monarchy in the eighteenth century transformed the small city of Berlin (which as late as 1700 had a population of 30,000 still living within the medieval walls) into a capital city with over 170,000 inhabitants a century later.

The pace of European urbanisation began to pick up from the mid-sixteenth century – the first area to be affected was Holland in the sixteenth and seventeenth centuries as the wealth created by its extensive trading empire filtered through society. In the early sixteenth century probably about a fifth of the population lived in towns with more than 10,000 people. Over the next century the urban population rose six-fold so that by 1622 about half the people of Holland lived in such cities. In the sixteenth century Paris was probably the largest city in Europe with a population of about 400,000 in 1550. The pace of growth in London was very fast. The population rose from 60,000 in the 1520s to 250,000 in the early seventeenth century and to 400,000 by 1650. By 1700 it was more populous than Paris and the biggest city in Europe. It was the centre for the court, government and law and therefore drew in merchants, traders, shopkeepers and craftsmen who catered for the elite. It also drew in a huge number of the rural poor in search of a job. In practice most finished up as beggars. The number of beggars in London rose twelve-fold between 1520 and 1600 even though the total population only quadrupled. Most were either expelled or consigned to 'workhouses'. These took the poor off the streets and fed them but they also provided cheap labour for employers and the conditions for the inmates were terrible and de-humanising. The death rate in London was so high that it needed an influx of 6,000 people a year simply to maintain its population and the London mortality rate was enough to cut in half the English popu-lation growth rate. By the eighteenth century about one in ten of the people of England lived in London.

The industrial city

In 1800 the world's population was still overwhelmingly rural – only about three in a hundred people lived in a city. Some of the cities were large – London and Edo had about one million people each, Beijing and Canton (Guangzhou) in China slightly less. Overall Europe had about 10 per cent of its population living in cities with Britain and Holland, the most heavily urbanised, having about twice this rate. In the nineteenth century Europe and North America witnessed a revolution in the environment in which people lived. The proportion of the population living in cities rose rapidly and these cities altered in a number of ways. The changes occurred first in Britain as the first country to begin industrialisation but the pace of change was still slow in the early stages. In 1851 six out of ten people still lived in the countryside. (Apart from Britain, only in Belgium did more than 20 per cent of the population live in towns in 1850 and only in France, Saxony and Prussia was the proportion higher than 10 per cent.) Yet by 1900 about three-quarters of the British people lived in cities and one in five of the population lived in London. The crucial statistic though is the number of people living in cities – it rose from two million in 1800 to almost thirty million a century later.

By 1900 most of Europe and North America had been transformed by urbanisation. Two-thirds of the world's urban population lived in these areas. Overall the urban population of Europe rose six-fold in the nineteenth century. For the first time in human history, cities, although necessarily still reliant on the countryside for food, ceased to be parasitic on the economy and made a major contribution through industrial output. This was only possible through the growing use of fossil fuels, the development of high-energy use societies and the consequent industrialisation. New factories and industries were set up and they attracted the rural poor in search of work. Cities eventually became self-sustaining in population as living conditions improved very slowly through the nineteenth century. In 1750 London was the only city in England with a population of over 50,000 – a century later there were twenty-nine such towns, nearly all of them dependent on manufacturing. These included the cotton towns of Lancashire, the woollen towns of Yorkshire and the mining and metal industry centres such as Sheffield, South Wales and the 'Black Country' of the Midlands. Manchester grew from 27,000 inhabitants in 1770 to 180,000 in 1830. Other new towns such as Swindon and Crewe were dominated by a single industry – in these cases railway works. Others were artificial constructs to house the workers in the new factories, for example Port Sunlight which was built to support the Lever Brothers plant. In London manufacturing also became important, especially the extensive sweated workshops of the clothing trade, but it remained the dominant commercial and financial centre.

In the rest of Europe and the United States the same pattern occurred later in the nineteenth century. New industrial towns developed across the continent of Europe, especially near the coalfields that provided the energy source for industrialisation – Belgium, the Ruhr and northern France. Industry here, as in Britain, provided the main source of employment – 80 per cent in Bochum in the Ruhr by the 1880s. Many towns were also dominated by single industries, for example Leverkusen and the Bayer chemical plants. Berlin continued to grow (from 170,000 in 1800 to one and a half million in 1900) not just as the capital of the newly unified country but as the hub of the railway system and the centre of the increasingly important electrical industry. Germany was more balanced in its urban growth than Britain, which was dominated by London – Berlin was offset by the great manufacturing cities of the Ruhr and the important ports such as Hamburg.

In North America the early colonial cities were small and little more than the centres for trade with Europe. In 1790 the United States had only five places with a population greater than 10,000, with New York the largest at 40,000 inhabitants. In total, probably no more than 200,000 Americans lived in cities in the early nineteenth century. In 1830 there were still only twenty-three towns with more than 10,000 people and only two big cities – New York (200,000) and Philadelphia (160,000). As immigration from Europe increased rapidly from the 1840s and industry developed, major industrial cities such as Pittsburgh grew. In the mid-nineteenth century the urban population of the United States doubled every decade to reach six million by 1860 with nine cities having a population of more than 100,000 (New York had reached 800,000). By 1910 there were fifty cities with a population of over 100,000. The flood of immigrants from Europe overwhelmed the social and economic systems of these cities. The infant mortality rate in New York doubled between 1810 and 1870. In the late nineteenth century the legal requirement for new tenement blocks (the old ones were much worse) was that they should have one lavatory for every twenty inhabitants and one water tap for each block.

Before 1800 most cities (apart from giants such as Hang-chou) were small in area – they were places which people could walk across to conduct their business. Rome in the second century CE was still contained within the Aurelian walls which enclosed an area of about thirteen square kilometres. Roman provincial cities were much smaller – London covered about 130 hectares and Bath less than ten hectares. The area of medieval London was a little under 300 hectares. Cities in the nineteenth century began to sprawl and take over the surrounding countryside. Market gardens and fields were destroyed, old villages (such as Hampstead and Highgate near London) were incorporated into the cities and large new estates were built of terraced houses or, in wealthier areas, villas with gardens. The wealthy and the new professional classes moved out of the centre of the cities into these new suburbs. They were replaced by an influx of the poor seeking work in the new factories and industries. The centres of cities were characterised by grossly overcrowded slums with people living in terrible squalor.

The new suburbs were largely unplanned – most were speculative developments often linked to or following on from new transport systems. It was these systems, increasingly dependent on fossil fuels, that revolutionised how people lived and worked. Suburbs became places where people lived but did not work – a fundamental departure from previous human history. Some of the first suburbs grew following the construction of bridges or ferries – for example Southwark in medieval London and Brooklyn in New York. However, it was public transport systems that transformed nineteenth-century cities. The first horse-drawn bus started in France but it was New York in 1829 that saw the first extensive system. However, it was the development of railways that was the key factor in changing urbanisation patterns. In London from the 1840s new residential suburbs such as Camberwell, Hornsey, Kilburn, Fulham and Ealing grew up along railway lines so that people could travel to and from work in the centre of the city. Residents of some of the first suburbs in the United States travelled on the horse-drawn street railway introduced in New York and adopted in eight other cities by 1860. From the 1870s this system acted as a feeder to the more ambitious system of elevated railways that enabled people to live even further from the city centre. The 1890s witnessed the end of horse-drawn urban railways and their replacement by electrified trolleys in the United States – in 1890 there were 9,000 kilometres of horse-drawn track and 2,000 miles electrified, but by the end of the decade only 400 kilometres were still horse-drawn and over 35,000 kilometres were electrified. The late nineteenth and early twentieth centuries saw the development of the underground railway – they became essential because of the slow pace of traffic in the streets jammed with horse-drawn vehicles. The first was built in London in 1863 but it was only with electrification that the system could develop on a large scale – Boston followed in 1897, Paris in 1900, Berlin two years later and New York in 1904. They were able to move huge numbers of people – in Boston there were fifty million passengers in the first year on a tiny network and the New York system was soon carrying over one million people a day on its one route. All of this infrastructure and increased energy consumption was needed to sustain the cities that had grown up in the nineteenth century.

Suburban development was a common phenomenon in every city in the world. The exact pattern and timing depended mainly on the construction of transport systems and their extent but other factors were also important. In the United States, where land was generally cheap, the density of settlement was far lower than in Europe and cities spread over far greater areas. In 1850 the edge of Boston was just over three kilometres from the city centre; fifty years later it was sixteen kilometres away. In New York cheap housing spread out along the developing subway lines and the population of Brooklyn, the Bronx and Queen's grew from 200,000 in 1890 to 2.7 million by 1940. In London the population grew from one million to four and a half million in the nineteenth century but another two million people lived in the adjacent areas of what came to be called 'Greater London'. In Berlin the suburbs grew faster than the centre from the 1870s and separate towns such as Charlottenburg and Spandau were absorbed into the city. The only major city that did not experience major suburban growth was Paris. The poor suburban railway system kept the city highly concentrated – the population density in the centre was about three times that of London. As the suburbs expanded the centre of the industrial city was transformed. The heart of most major cities became a hub of commercial and financial activity with only limited industrial work. As a result the population fell – that of the City of London began to decline in the 1850s and within a few decades went from 130,000 to almost none as offices took over. In New York, in 1905, over half the population lived within six kilometres of the city centre but twenty years later less than a third did so.

The twentieth-century city

During the twentieth century cities in the industrialized world have continued to expand, absorbing more and more of the countryside. In London the extension of the Metropolitan railway into the countryside north of the capital led to the development of a whole series of suburbs dubbed 'Metroland'. In Paris investment in suburban railways in the 1920s led to the development of cheap, often self-built housing in areas with minimal facilities and over one million people commuting into the city every day. The most significant development was the rise of the private car and therefore the United States was the first to experience this trend in the 1920s; Europe did not follow suit until the 1950s. Increasing car ownership made cities sprawl over ever bigger areas and made commuting distances rise still further. The urbanised area of New York in the 1980s covered about 5,700 square kilometres – five times the area it covered in 1920 yet the population had only doubled. Cities such as Los Angeles which were built around the car exemplified this trend on an even bigger scale. Two-thirds of central Los Angeles is now occupied by streets, freeways, parking facilities and garages. The overwhelming domination of the car in urban transportation has not made the system speedier or more efficient. In 1907 the average speed of traffic in New York was just over eighteen kilometres an hour but by the 1960s this had fallen to nine and a half kilometres an hour. In London traffic now flows as slowly as it did in the late nineteenth century, in Paris it is slower.

From the 1920s industry, particularly the new light industries such as radio, electronics, medical products and domestic appliances, moved out of the city to the newly built roads such as the A4 and A40 to the west of London and areas such as Enfield and Edmonton to the north. By the middle of the century industry in the United States had moved further out to the new peripheral motorways around cities such as Boston, New York and Washington. A similar pattern was found in London when the M25 orbital motorway was completed in the 1980s. As people moved out of city centres into ever more distant suburbs, major new shopping facilities were built on the edge of the built-up area, largely dependent on the ability of people to travel by car. These were common by the end of the twentieth century but the first was built in Kansas City as early as 1922. This trend reflected another phenomenon of the second half of the twentieth century – a small, but noticeable, decline in the importance of cities in the industrialised world. Nearly everywhere the urban population peaked at about three-quarters of the total population – Britain reached this level at the end of the nineteenth century, other countries a few decades later. The population of London, not just in the inner city but in the suburbs too, began to decline in the 1950s and the decline has continued, slowly but steadily, ever since. Paris experienced the same trend just a little later – it lost 10 per cent of its population between 1968 and 1975. In Germany, Canada and the Netherlands the proportion of the population living in cities has been falling since the 1960s.

The growth of the city was, for the most part, unplanned, or at best loosely organised, with the authorities struggling to provide or maintain services. Environmental improvements did come with the clearing of slums and redevelopment programmes – Haussmann's rebuilding of central Paris in the mid-nineteenth century was the first of these and Britain had a major slum clearance programme in the 1950s and 1960s. New towns, garden suburbs and the creation of new estates were all attempts at planning the urban environment. However, the forces behind urban expansion were difficult to control. A formal 'green belt' was established around London in 1937 in an attempt to preserve the surrounding countryside. The designated area has been kept largely free of housing but development has simply leapt over the belt and continued even further from the centre of the city. The area of the south-east of England that is socially and economically part of London is now greater than ever and commuting times and distances have also become ever greater. The relatively feeble planning controls in Britain did no more than avoid some of the worst excesses of urbanisation and in the United States controls have been even weaker and less effective.

The strength of the underlying forces behind urbanisation is demonstrated by the inability of even the planned economy of the Soviet Union to control the growth of Moscow. The urban population of the Soviet Union rose from 20 per cent in the 1920s to over 60 per cent by the 1980s. Some of this growth was contained by the construction of over 800 new cities. Nevertheless Moscow's population doubled between 1926 and 1939 when it reached four million, most of whom were crowded into the centre of the city because the suburban transport network was so poor (something that was not remedied until the late 1950s). In 1935 the government decided that the population of Moscow should be capped at five million (a rise of about 20 per cent). Even with draconian state controls and an internal passport system the government could not enforce that decision. In 1971 they decided the population should not exceed seven and a half million (the then current level) and that it should contract to six and a half million by the end of the twentieth century. The current population is over ten million.

The urbanisation of Japan followed the same course as Europe and North America but it occurred later because of the later industrialisation of the country. Although Edo was still one of the largest cities in the world in the mid-nineteenth century and other new cities such as Yokohama (founded as a fishing port in 1858) were growing rapidly, Japan remained a predominantly rural country. As late as 1920 four out of five people still lived in the countryside and a majority of the population was not urbanised until 1955. In 1884 there were only nineteen cities with a population greater than 50,000 – by the early 1970s there were over 600. The main expansion occurred around Tokyo (as Edo was renamed in 1868), especially as suburbs grew with the development of the electric tram system in 1903 and an underground railway after 1923. The population of the suburbs tripled to three million in the 1920s. By 1945 Tokyo covered an area twice as large as in 1923. The 'green belt' disappeared in the 1960s and weakly regulated development produced an urban area eighty kilometres wide.

Cities in the developing world

A century ago two-thirds of the world's urban population lived in Europe, North America and Australasia; now less than a third do so. The twentieth century saw the urbanisation of the developing world at a far faster rate than the industrialised world in the nineteenth century despite the slow, and in some cases non-existent, pace of industrialisation. In the nineteenth century the fastest pace of urban growth was about 2.5 per cent a year. In the developing world it was, on average, 3 per cent before 1940 and 4 per cent thereafter. (Growth at 4 per cent a year is enough to double the urban population every eighteen years.) In many cases rates were far greater than this. In Nigeria the population of Lagos rose from 126,000 in 1931 to over thirteen million in seventy years, a 103-fold increase. Nairobi in Kenya grew from 11,500 in 1906 to one million in 1982. In Mauretania the population of Nouakchott increased forty-fold in just twenty years after 1965. Ankara, the new capital of Turkey, was a small town with a population of just 30,000 in 1920. In 2000 its population was four and a half million – a 150-fold increase. The most rapidly urbanising area in the world in the twentieth century was Latin America. In 1920 only 14 per cent of the people of Latin America lived in cities with a population of more than 20,000. By the 1980s two-thirds did so and some countries such as Argentina and Uruguay were as urbanised as Europe. (Only Haiti, Honduras and Bolivia remained predominantly rural countries.) Despite high rates of urbanisation in Africa and Asia only about a third of the population live in cities. However, this comparatively low figure disguises very high rates of urban growth because of the very fast increase in the overall population.

One of the main reasons for this rapid growth in urban numbers is that major facilities tend to be concentrated in urban areas. For example, Abidjan in Ivory Coast accounts for nearly three-quarters of all the economic activity in the country although only a sixth of the population live there. Food supplies tend to be better in the cities for political reasons (like ancient Rome but without the circuses). To some extent other facilities are better in the cities although this is only in comparison with the countryside. About three-quarters of the urban population have access to some form of water supply (though this is defined as a tap no more than ninety metres away) – only about 40 per cent of the rural population have even this minimal access. The sewage system though is far worse. Only about a half of the urban population has any waste-water treatment and in some cities the situation is far worse. Over 80 per cent of the population of Manila, Dhaka and Karachi are not connected to a sewer system.

Overall the cities of the developing world are very similar to those of pre-industrial and early industrial Europe. Rapid growth rates have placed immense strains on the infrastructure. Unemployment is probably at least 20 per cent and in many cases higher than this. Housing is the most acute problem. In India about one and a half million people live on the streets because, although many have jobs, they are so badly paid they cannot afford housing. Probably more than a third of the urban population of the developing world live in illegal squatter, shanty settlements thrown up by the squatters themselves. They are usually built on land designated as unsuitable for housing by the local government and they have few, if any, facilities such as sewage, water supply, refuse collection and electricity. There are few schools and these areas tend to be riddled with crime and violence. The UN Habitat report of 2006 estimated that a third of the world's urban population live in slums – in sub-Saharan Africa the proportion is three-quarters of the total.

The rise of the metropolis

Concentrated industrialisation in the nineteenth century, based on the exploitation of coal and mineral deposits, brought about the formation of the first conurbations – large, formless, urban masses caused by the expansion and joining up of existing settlements. In Britain this was first noticeable in the 'Black Country' of the Midlands and the 'Five Towns' of the Potteries by the mid-nineteenth century. One of the most extreme examples of this urban form came in the Ruhr after 1850 following the start of coal mining and the building of a railway network. A mass of immigrants formed the labour force and the villages in the area grew in an unplanned fashion until they joined up into an urban sprawl. The population of the area was 900,000 in 1871, three and a half million in 1910 and four and a half million in 1939. By the 1980s the Ruhr was an agglomeration of eleven cities and four districts with a population of five and a half million.

In the Netherlands by the mid-twentieth century the *Randstad* or 'ring city' had emerged. This was brought about not by industrial concentration but by the acute shortage of land in the country. The chief cities were all close together and eventually spread until they joined together to form an almost continuous urban unit though with clearly defined functions. The *Randstad* comprises the cities of Dordrecht, Rotterdam (port/industrial), The Hague (government),

Delft, Leiden, Haarlem, Amsterdam (finance and culture) and Utrecht. This urban complex contains a third of the Dutch population living on just 5 per cent of the country's land area. Similarly in Japan there is now an almost continuous urbanised zone between Tokyo and Kobe. In the United States there is a string of cities linking Boston and Washington DC which contain over fifty million people (about a sixth of the population) in just 1.5 per cent of the area of the country

Number of cities with a population of more than one million

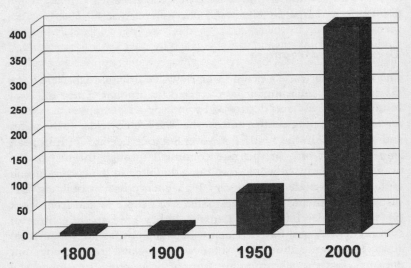

The twentieth century witnessed the growth of the large metropolis – a huge city, normally only one in each country, concentrating in most cases on tertiary (service) industries rather than manufacturing, and with as many links with the other major cities of the world as with the rest of the country. In 1800 there were just six cities in the world with a population of more than one million people. A century later this number had doubled to twelve. In 1950 there were eighty-three and by 2000 there were over 400 cities with a population above one million. However, a city with a population of a million is no longer a large city in world terms. In 1900 the largest city in the world was London with a population of about four and a half million followed by New York (2.7 million) and Paris (2.4 million). All of the large cities of the world were still in Europe and North America. By 1950 there were eight cities with a population over five million, of which the largest was New York (12.3 million) followed by London

(8.7 million) and Tokyo (6.9 million). The growth of the giant metrop-
olises in the second half of the twentieth century was phenomenal.
By 2000 there were twenty cities in the world with a population
greater than ten million. The largest was Tokyo (26.4 million) followed
by Mexico City (18.4 million) and Mumbai (18 million). There was
also a significant geographical change – of the top ten largest cities
none was in Europe, only two (New York and Los Angeles) were in
North America, four were in Asia (Calcutta and Shanghai in addition
to Tokyo and Mumbai), three were in Latin America (Mexico City,
São Paulo and Buenos Aires) and one was in Africa (Lagos).

The urban environment

The half of the world's people who now live in cities experience the
most artificial environment ever created by humans. Large areas of
countryside have been destroyed by the spread of houses, factories,
roads and shopping centres across what were once fields, open spaces
and woodland. In the 1990s alone over 800,000 hectares of European
land was built on – if this rate continued through the rest of the
twenty-first century it would result in a doubling of the current urban
area. Cities also depend on very high energy use in building and
sustaining them and in moving millions of people to and from work
every day. Cities have many benefits – they are usually centres of
cultural activity and have a much wider range of facilities than rural
areas. However, although in some working-class areas strong, but
informal, systems of community support developed, in general the
flood of people into cities destroyed existing social bonds and institutions without creating new ones capable of helping and sustaining
the inhabitants. Cities, as the nineteenth-century American writer
Henry David Thoreau wrote, tend to be places characterised by
'millions of people being lonely together'.

During the late nineteenth and twentieth centuries some of the
worst aspects of urban life were eliminated by rising prosperity and
government social programmes. Most cities in the developing world
still suffer from these problems. However, in the twentieth century
new problems emerged in the cities of the industrialised world. Public
transport declined under the pressure of increasing car use. As the
rich and middle class left the city to live in the suburbs the economic
decline of the inner city gathered pace. Jobs were lost (500,000 in
British inner cities in the 1960s alone), and a vicious downward spiral
of poverty, economic decline, crime and a deteriorating environment

proved very difficult to halt. By the late twentieth century a quarter of the population of New York (about four million people) had an income below the official poverty line, nearly 100,000 people slept on the streets, there were 500,000 drug addicts and only one in five of the babies born in Harlem was legitimate. In many inner cities about two-thirds of the population do not own a car and therefore have to rely on the decaying public transport system. In the black ghetto of Watts in Los Angeles the number of doctors per head is a fifth of the national average. In Britain the picture was little different. In Islington in central London nearly half the households in the late 1970s lacked their own hot water, bath or inside lavatory. In Tokyo the average family lived in a block of flats with only a tenth of the space of a European flat, many with shared lavatories and kitchens. Nearly a third of these flats were designated as sub-standard even by Japanese standards. Across Russia the old blocks of flats built in the Soviet era are no better and often worse than their Japanese equivalents.

The rise of cities on the scale seen in the contemporary world is a phenomenon directly linked to the exploitation of fossil fuels and the emergence of high-energy-use societies with complex commercial and financial linkages. Not only are they responsible for most of the energy consumption within society, they are also a focus for a range of environmental problems. In addition to crowded and often poor living conditions they are characterised by high levels of air pollution and noise, long commuting distances, either on over-crowded public transport systems or by car on congested roads, and a multitude of social problems.

14
CREATING THE AFFLUENT SOCIETY

Since the rise of settled societies some ten thousand years ago the overwhelming majority of the world's population have lived in conditions of grinding poverty. They have had few possessions, suffered from appalling living conditions and have been forced to spend most of their very limited resources on finding enough food to stay alive. In all societies the elite have lived at a higher standard of living but they too had access to only a limited range of goods. In the last two centuries a sizeable minority of the world's population has achieved a standard of living that would have been unimaginable to earlier generations. For those who have benefited from these changes it has brought a huge improvement not simply in their standard of living but also in the range of experiences available to them. However, this improvement has been obtained at a significant price – a vast increase in the consumption of energy and raw materials, widespread pollution and a variety of social problems. In addition, it has raised major questions of equity about the distribution of wealth within individual countries and, even more important, between the rich and the poor of the world.

Pre-industrial societies

Gathering and hunting groups keep possessions to a minimum since these are a hindrance to their mobile way of life. They may wear a few pieces of simple jewellery but most of their household goods are made from easily obtained materials, to be thrown away if necessary and quickly replaced when needed. The rise of agriculture and settled communities changed these priorities. First, more goods were needed to process and store food. Second, a sedentary way of life made it easier to acquire household and personal effects – they were no longer a hindrance and the benefits that they gave were obvious.

In the pre-industrial societies that characterised the world until about two hundred years ago agriculture dominated the economy. Average income was very low and nearly all of the limited spare resources were devoted to elite consumption or spent on big public projects such as the temples, palaces and pyramids of ancient societies or the cathedrals of medieval Europe. The low level of trade and poor transport meant that most regions had to be self-sufficient. The majority of the population had little contact with a money economy – barter and exchange were of far greater importance, usually farm produce for local crafts. People tried to be as self-sufficient as possible – even people in the towns kept a few animals and grew their own crops. Most goods that were produced were consumables and linked to agricultural production – food, drink, clothes, candles and a few raw materials. Industry was almost entirely limited to refining agricultural products – tanning, spinning and weaving, brewing and milling – or making items for the farming community – a few simple tools, pots and barrels. As late as 1600 80 per cent of England's exports were cloth and its four main imports were textiles, foodstuffs, timber and wine. Trade was at a low level and most states did not even have a system of national weights and measures. In the twelfth and thirteenth centuries most of Europe's north–south trade was done in a few days at the fairs of Champagne. Long-distance trade was restricted to a few luxury items. The range of goods available, even to the rich, was, therefore, very limited.

In medieval and early modern Europe (and conditions were little different elsewhere in the world) about 80 per cent of the spending of the mass of the population went on food but their diet was still poor. About half of this expenditure went on buying bread, which was often only baked every month or two (in the Tyrol only two to three times a year and it was often cut using an axe). A rapid rise in food prices could quickly drive spending on food to all of a family's income and even that might not be enough to stay alive. In these conditions it is not surprising that occasional bouts of feasting at major events such as at harvest and weddings, when meat might be eaten, played a vitally important social role. Even in relatively prosperous times and when harvests were good people might have no more than 10 per cent of their income to spend on clothing. Clothes were important items to be handed on to the next generation – the hospital regulations at Perugia in 1582 provided that the clothes of the deceased should not be stolen 'but should be given to lawful inheritors'.

Once food and clothing were provided there was very little money left for housing. The average peasant hut in Europe was made out of wattle and daub, with an earth floor, no windows or chimney. Cooking was done on a spit or a pot over an open fire and the smoke just accumulated in the hut. Few people had ovens, and bread (together with meat if it was available) was baked communally. For bedding there might be heaps of straw or bracken on the floor. The hut was usually shared with the few animals the peasant might own – at least they provided some warmth at night and during the winter. A detailed survey of the village of Navalmoral just south of Toledo in the 1580s gives a clear picture of rural Europe. It had a population of 243 families and about 1,000 people. Just twenty-two families owned half the land, there were sixty families of peasants, twenty-eight families had no land but grazed some livestock, ninety-five families were land-less labourers, there were twenty-one widows living alone with no visible means of support and seventeen families did not even have a place to live. As late as early nineteenth-century England there was dense overcrowding in the countryside and the conditions in Ireland were even worse. Conditions in the cities were usually even worse than in the countryside. In Florence in the 1630s houses in the poorest quarter contained eight to ten people to a room, ten to twelve families in each house, making a total of about a hundred people all living without water and sanitation. In Pescara in 1560 a quarter of the population was living in a shanty-town consisting of holes in the earth.

The rich might avoid the worst of these conditions but the range of goods available to them was still very limited. Most of their wealth was spent on housing (which would be more spacious and built out of stone or timber but still lacking any sanitary facilities), better clothing (certain items were often restricted legally to the aristocracy) and more elaborate food. They could also afford to employ servants in large numbers. In 1393 the monastery of Meaux had twenty-six monks and forty domestic servants. The employment of a large number of servants remained a mark of social distinction until well into the twentieth century. In most European cities before 1900 about one in seven of the population was a domestic servant.

Most people lived either in a state of destitution or on the edge of it. They had no savings and so the slightest problem such as illness or unemployment would quickly reduce them to starvation and begging. Little help could be expected from charity – the best estimate is that about 1 per cent of the limited wealth of pre-industrial societies was devoted to charity and monasteries gave about 1–3 per cent of their considerable wealth to charitable causes. The figures for the number of beggars in European society are startling. In Florence in 1457, at a time when the so-called 'Renaissance' was flourishing, official statistics give a very different picture of the reality of European society in one of the wealthiest areas of the continent. They show that just over 80 per cent of the population of the city were classified as either poor or destitute. At the end of the seventeenth century at Vauban in France beggars made up 10 per cent of the population, near beggars 30 per cent and the very poor 50 per cent. That left just 10 per cent of people not either destitute or on the edge of destitution. In early eighteenth-century Cologne there were 20,000 beggars out of a population of 50,000. In England at this time a quarter of the population was described as 'cottagers and paupers' in a permanent state of poverty and unemployment – 'miserable people and (harvest time excepted) without any subsistence'. At times of poor harvests and high unemployment about half the population of the country would be reduced to this condition. As late as 1815 in Sweden it was estimated that half the population were either landless peasants or beggars.

In these circumstances people often could not afford to keep their children. In the sixteenth and seventeenth centuries about 10 per cent of all babies born were simply abandoned on the streets of Italian cities. In Paris in the 1780s there were about 8,000 abandoned children every year – about a quarter of all births. A third of the population was aged under fourteen and so child labour was essential, particularly in the fields in the summer and at harvest time. The overwhelming majority of the population was illiterate and most children remained uneducated. Apprenticeships could start at seven years old though most began at ten or twelve. Children, orphans in particular, were very vulnerable to exploitation – the cloth manufacturers of Leiden imported 8,000 orphans from Aachen, Julich and Liège in the thirty years after 1638 in order to keep down adult wages. Adults were put out of work and most of the children soon died in the grim working conditions. Women worked too – about 80 per cent of the wool weavers of Florence in the early seventeenth century were female. In rural areas women worked in the fields, especially at harvest time, and also did piecework at home to supplement the family's meagre resources. Women were also employed in brickyards, coal mines, arsenals and metal works. Overall work in industry for both sexes was only available in a few areas and then only intermittently. Wages were usually just enough for subsistence – in the eighteenth century lace-making in the Massif Central (an intricate operation that almost always caused blindness) paid about two sous a day and that was enough to buy a loaf of bread if the harvest had been adequate.

There was no steady improvement in the human condition in pre-industrial societies. The standard of living tended to fluctuate according to the balance between population and food supplies. In 1300 when Europe was severely overpopulated, living conditions were very poor – food was often in short supply and the excess of labour made finding work very difficult. A few decades later conditions were better for those who survived the Black Death – the smaller population put less strain on food supplies and there was a shortage of labour that improved conditions for peasants and workers. By the mid-sixteenth century numbers had increased to near the medieval peak and the next century was a time of great hardship for most people in Europe as the sharply deteriorating climate reduced food supplies. After they had paid rent to their landlords and tithes to the church, most peasants had no more than a third of their crop to live on. Increasing state tax demands (to pay for the endless wars in Europe) could not be afforded. The only solution seemed to be revolt and a refusal to pay taxes – in Aquitaine there were over 250 revolts in the twenty-five years after 1635 and in Provence in the seventeenth century there were 374 revolts. From the mid-seventeenth century onwards some areas of Europe did begin to see a rise in living standards as agricultural productivity slowly increased. More trade and manufacturing and the expropriation of non-European resources also increased wealth. The increase was small and mainly confined to the Netherlands, England and France. In the rest of Europe conditions continued to stagnate or even deteriorate.

The cost of industrialisation

By the late eighteenth century Britain and some other parts of Europe had developed significant commercial and industrial sectors and were in the early stages of the transition from agricultural to industrial societies. There can be no doubt that industrialisation did, eventually, significantly raise the standard of living for the whole of society (though far more for some than the majority). However, the process of indus-trialisation required substantial capital investment and a redeployment of workers from the countryside into industry. The transition involved high social costs and there is little doubt that in the early stages of industrialisation the living standards of the majority of the population worsened. The process can be illustrated by comparing the experience of two very different societies in the first stages of industrialisation – the free-market capitalism of nineteenth-century England and the state-directed industrial expansion of the Soviet Union in the 1930s. (The next chapter will also show how both societies suffered from the same environmental problems during industrialisation.)

In the late eighteenth and early nineteenth centuries the rate of economic growth in Britain was low (apart from one or two short spurts, mainly during the railway boom) and industrialisation and factories took a long time to become predominant. Nevertheless the new manufacturing cities grew and an industrial workforce developed. However, there is no evidence of any improvement in the living standards of the bulk of the population until the late 1840s at the earliest. Throughout this period, because of the surplus of labour available, about half of the industrial working class lived at or below subsistence level, surviving on casual, part-time work, and only a small minority of skilled craftsmen were able to improve their position. Many, such as the handloom weavers (about 500,000 people), saw their jobs disappear with the rise of factory production. Like the peasants, few workers had reserves of money or possessions to pawn in times of recession when about three-quarters of the workers in some trades could be out of work. In the 1840s about 10 per cent of the population of England were still paupers and in some towns it was much higher. In Nottingham it averaged one in five of the population and at Clitheroe in Lancashire during the recession of 1842 there were 2,300 paupers out of a total population of 6,700. Living conditions in the industrial cities were squalid as workers flooded in to find jobs and swamped the inadequate housing that was available. In 1790 in Liverpool one in eight of the population lived in cellars, and in 1833 in Manchester there were 20,000 people living in cellars (about ten per cent of the population). Not surprisingly, mortality rates rose between 1810 and 1850 and only then began a slow decline. In 1840 in Manchester about six out of ten children died before the age of five – this was far worse than in the rural areas which had a rate about half this level.

In the second half of the nineteenth century living conditions for the bulk of the population slowly improved. There was more food available (much of it imported), city water supplies and sanitation were improved and housing became less crowded. Wages, especially for the skilled worker, rose from 1850 (much later in the century for the unskilled). Nevertheless many still lived in almost permanent want and sub-standard housing. In 1889 about a third of the population lived below the poverty line and were 'at all times more or less in want'. The diet of the poor was still seriously deficient and scurvy, rickets and anaemia were common. In 1800 boys recruited into the Royal Navy from the slums were, on average, almost twenty centimetres shorter than their upper-class contemporaries. In 1940 working-class children were still almost ten centimetres shorter than those who went to private schools. In the late nineteenth century the poorest sections of the community had about half the calorie intake of the richest. In 1899 when 10,000 men from Manchester tried to enlist in the army at the time of the Boer War only 1,000 were fit enough to be accepted. Housing remained a severe problem. In the 1901 British census conditions qualified as 'overcrowded' if there was a household of at least two adults and four children living in just two rooms without their own water supply and sanitation. Even on this restricted definition one in twelve of the population was 'overcrowded' and in some cities rates were far higher. In some of the inner city boroughs in London more than one in three of the population was 'overcrowded', in Glasgow the figure was more than half and in Dundee it was almost two out of three residents in the city.

Conditions in the industrialising countries of western Europe and North America were little different from those of nineteenth-century Britain. In the Soviet Union it was the government that embarked on a deliberate policy of massive industrialisation in the Five Year Plan of 1928. Although Russia had begun early industrialisation before the First World War many of these gains had been lost during the chaos of the revolution and subsequent civil war. With almost no outside help, industrialisation in the Soviet Union required the population to bear the cost – 80 per cent of the people were still peasants and the industrial working class, in whose name the revolution had been made, were only 3 per cent of the population. These problems were exacerbated by the rapid timescale over which the Soviet leadership (dominated by Stalin) decided industrialisation should take place. The 1928 plan called for a 236 per cent increase in industrial output and a 110 per cent increase in labour productivity.

The first intense period of industrialisation in the 1930s saw some of the fastest growth rates achieved anywhere in the world. Between 1927 and 1937 iron output quadrupled, coal production rose three-and-a-half-fold, electric power generation increased seven-fold and machine tool production rose seventeen-fold. The industrial labour force doubled in the five years after 1928. The consequences of this achievement were terrible for those involved. Millions of peasants died

during the collectivisation drive and the seizure of food to feed the
cities. The drive for greater productivity meant longer working hours,
a continuous working week, increased output quotas, 'shock brigades',
'socialist competition', the adulation of 'Stakhanovite' workers, the
'deprivileging' of skilled workers and, ultimately, the slave labour of
the Gulags. In parallel with worsening working conditions came a
falling standard of living. Food rationing was introduced, by 1932
meat consumption was at a third of 1928 levels and that year even
potato consumption fell. The living standards of industrial workers
fell drastically – in the four years after 1928 they fell by about half
and did not recover to the levels of the late 1920s until the mid-1950s.

The stages of industrialisation

Industrialisation has involved a succession of new technologies that
have transformed the number and range of goods available to people
and their production has affected the environment in a number of
ways. The first phase of industrialisation lasted until the last third of
the nineteenth century and was based on the greater output of textiles
through mechanisation and factory production, steam engineering,
iron production and the construction of railways. Textile production
remained central throughout the nineteenth century – as late as 1900
food processing and textile production still accounted for over 40 per
cent of industrial production in both western Europe and the United
States. Improvements in steam engineering to provide factory power,
iron production and the development of railways were all linked and
the latter was a vital factor in industrial growth. There were 72,000
kilometres of railways in the world in 1840 but 360,000 kilometres
forty years later. In the last third of the nineteenth century much of
the impetus provided by these new technologies had been lost and
they were replaced by a second wave of industrialisation based upon
electrical engineering and new processes in the chemical industry.

In the early twentieth century it was the rise of the vehicle industry
(with all its linked technologies) that provided the main thrust behind
continued industrial expansion. Artificial fibres also began to replace
natural fibres – rayon was manufactured before 1914 and output
increased greatly in the 1920s when it could be made for a quarter
of the price of silk. Nylon showed similar rapid expansion after 1945.
From the 1930s the increasing use of plastics derived from oil was a
major industrial growth area – world production of plastics doubled
every twelve years after 1945. By the 1970s it exceeded the combined

production of aluminium, copper, lead and zinc and per capita production had increased by over 1,000 per cent. In the main industrial states this technological phase was largely over by the 1960s and was replaced by the latest phase of communications and information technologies – computers, satellite communications, industrial robots, mobile phones and products based on the use of lasers.

In parallel with these technological changes there was an enormous increase in industrial productivity. In the industrialised world this process, combined with the similar increases in agricultural productivity which reduced the farm labour force, meant that far more people could be employed in the 'tertiary' sectors of the economy such as finance, advertising, tourism, education and health care. In the industrialised countries in 1900 about a third of overall output came from agricultural output and slightly more from industry. By the end of the twentieth century, despite the large increases in agricultural output and the even bigger increases in industrial production, agriculture contributed just 3 per cent of total output, industry about 35 per cent and over 60 per cent came from the service sector.

The industrial, economic and social revolution of the last two hundred years has had a fundamental impact on the world's environment through the consumption of energy, resources and the pollution this has produced. (The latter is the subject of the next two chapters.) Since 1750 world industrial output has increased more than 100-fold but the bulk of that increase came in the twentieth century and most of that in the second half of the century. In the early nineteenth century the world's *total* industrial output was roughly at the level of Brazil today. In 1900 Britain alone produced more than the total world industrial output in 1750 yet by 1980 global industrial output was about 100 times that of Britain in 1900. Much of this increase has been achieved because population growth increased the labour force and the demand for goods but industrial productivity has risen spectacularly through the various phases of industrialisation, mechanisation and automation. Industrial productivity is now about 200 times the level of 1750 and a modern American worker produces in one hour what it took a British worker two weeks of labour working a twelve-hour day to produce in 1800.

Mineral and metal production

In parallel with the massive increase in energy consumption in the last two centuries there has been a huge increase in mineral and metal production. The first metal to be used on any scale was lead (around 6000 BCE), followed by copper (around 3700 BCE) and then iron production (from about 1200 BCE). Iron became the main metal used for weapons and agricultural tools but output was low because of the limited demand and the production difficulties associated with small-scale furnaces and the need to use high-grade ore because of the limitations of the early smelting processes. As late as 1400 European production of iron was about 30,000 tonnes; China, the other major producer in the world, made slightly more but it unlikely that total world production was much over 100,000 tonnes. European production rose about six-fold between 1400 and 1700 but at the beginning of the eighteenth century world production is unlikely to have been much more than about 300,000 tonnes.

World production of iron and steel, 1700–2000

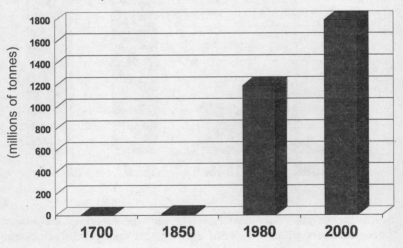

Since 1700 a multitude of new uses for iron (and then steel) formed the basis for many of the key products of modern industrial society – machines, railways, vehicles, buildings, ships, domestic appliances and many other consumer goods. The rise in iron and steel production is therefore a good indicator of the scale of industrial production. The first stages of industrialisation in Britain were marked by a steep rise in iron output – production increased ten-fold between 1788 and 1830 from 68,000 tonnes to 700,000 tonnes (twice as much as the whole world produced only a century earlier). By the mid-nineteenth century world production was about twelve million tonnes a year (forty times higher than in 1700). Huge though this increase was, it pales into insignificance compared with the increase by the late twentieth century. World production of iron and steel is now just under two billion tonnes a year. The overall increase in annual production since the first rough estimates of world output are available in 1400 is a staggering 20,000-fold.

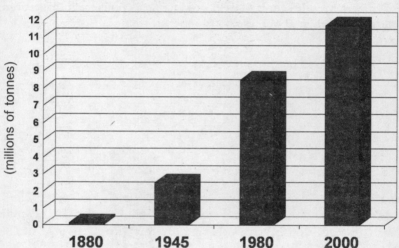

World production of copper, 1880–2000

Production of other minerals has also increased spectacularly. Three-quarters of all the gold ever mined was produced in the twentieth century. Copper is a key metal for the electrical industry and its production rose from about 120,000 tonnes in the 1880s to 500,000 tonnes in 1900. The twentieth century saw an increase to just over 11.6 million tonnes a year – a 100-fold increase since the 1880s. The production of nickel, mainly used to harden steel and in armaments production, rose over eighty-fold in the twentieth century. Aluminium was not produced until the late nineteenth century when the electrolytic process to refine alumina from bauxite was developed. Production uses huge quantities of electricity but aluminium was one of the key industries of the twentieth century

World production of aluminium, 1928–2000

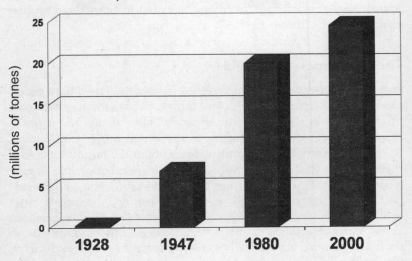

and its many uses ranged from aeroplanes to cars to household utensils and drinks cans. World production was just 223 tonnes in 1895 – a century later it was just over twenty-four million tonnes, a 10,760-fold increase. Recycling of metals is important in some areas, particularly copper, aluminium and steel, where cost pressures give industry an incentive to re-use materials, but in most cases metals are mined, processed, used and thrown away.

The environmental impact of metal production on this scale has been profound. About 70 per cent of the world's metal ore (95 per cent in the United States) is obtained by the most environmentally damaging of all methods – opencast mining. This keeps costs down but it involves the digging of vast pits, the removal of whole mountain tops, the destruction of topsoil and the creation of large amounts of waste. The waste leaves unsightly slag heaps, causes rivers to silt up, is often toxic and therefore creates uncultivatable land or leaches into water courses and poisons them. New technologies have enabled ever lower grades of ore to be exploited. In the case of copper the minimal workable ore had to contain 3 per cent of copper in 1900. By the late 1970s this had fallen to 0.35 per cent copper. However, this meant that in order to obtain a tonne of copper about 350 tonnes of rock had to be mined, transported and milled and eventually almost the same of amount of material disposed of as waste.

In the second half of the twentieth century over a fifth of the United States' huge consumption of energy was used to extract and process minerals.

The rise of mass consumption

Increased energy consumption and industrial production have generated steadily increasing wealth and transformed the living standards of everybody in the industrialised world. One of the first signs of this change was the development of shops and other retail outlets. In medieval and early modern Europe only in the largest cities was there enough wealth to support a few shops – most people bought what goods they needed either from local craftsmen or itinerant pedlars. In the sixteenth and seventeenth century the largest cities in Europe – Paris and London – saw the development of specialised shops but these dealt mainly with jewellery and clothing. Items such as furniture were usually made by craftsmen and sold direct. Gradually the range of shops grew larger and developed in the smaller country towns too. However, in the late nineteenth century most food was still sold in markets by the farmer rather than in shops. At this time in Britain the first national chains of shops began to emerge, at first in the food sector with companies such as Lipton's and Home and Colonial. The department store selling a wide range of products was a French invention, starting with Bon Marché in Paris in the 1860s. The idea spread rapidly across the world with stores such as Hermansky in Vienna and Tietz in Berlin. Many existing small retailers such as Harrods (originally a grocer's shop) and Debenhams (originally a draper's) diversified and new stores such as Selfridges and Army and Navy were set up. These new stores depended on a number of technical improvements – gas and electric lighting and lifts – and catered for the increasing number of relatively wealthy middle-class families who could afford the wide range of products on offer. The twentieth century saw the development of large supermarkets retailing food which soon dominated the market (at the expense of the small, local retailer) and then the new superstore stocking a huge product range. The latter began in the United States (Wal-Mart) and France (Mamouth and Carrefour) and they relied on increasing car ownership and the development of the out-of-town shopping centre. At the same time the products sold in these shops have become increasingly homogenised. At first brands were national but soon they became international as transnational corporations took over national companies and created international brand names.

By the early twentieth century the industrialised countries had almost met the basic needs of the population in terms of food, housing and clothing. Many industrialists were worried that workers would then decide to give greater leisure time a higher priority than work. In these circumstances would the economic growth that had characterised the nineteenth century come to an end? In practice this did not happen. Once workers had obtained a basic eight-hour day and forty-hour working week they were content to work more overtime to earn money to pay for the increasing range of consumer goods that was becoming available. Most of these goods and new industries developed first in the United States where greater wealth produced a mass market relatively early in the twentieth century. The 1920s marked the start of the consumer-durable boom – refrigerators, freezers and washing machines became common in prosperous American homes. A similar boom did not occur in western Europe until the 1950s and a decade later in Japan. In Japan in the 1960s the proportion of households owning refrigerators shot up from 5 per cent to over 90 per cent, for washing machines the figures were 29 per cent to 96 per cent and for televisions 16 per cent to 75 per cent. Saturation point for these new products was soon reached – by the 1980s in West Germany 90 per cent of households had a telephone, 84 per cent a car, 83 per cent a refrigerator, 80 per cent an automatic washing machine and 70 per cent a colour television. This boom was followed by another in the last two decades of the twentieth century based on advanced electronics – video recorders, portable cassette players, video cameras, personal computers, mobile phones, compact discs, digital cameras, DVDs, games consoles and MP3 players.

Apart from new products other mechanisms have ensured continuing high levels of production and consumption. First, firms have designed products with in-built obsolescence. Electrical companies are reluctant to make long-life light bulbs (which also use far less energy) because it would reduce their turnover and profits from selling short-life bulbs. When long-life bulbs are made they are deliberately made very expensive in order to keep demand low. In many areas products are not built to be highly reliable and repair is deliberately made difficult and expensive so that it is easier to buy a replacement. Second, obsolescence can be induced by frequent design changes. In clothing a whole industry now depends on the deliberate promotion of frequent change in what was once a basic necessity. In

1927 for the first time the number of replacement cars bought in the United States exceeded the number bought by new owners. With the market facing a period of slower growth, American car manufacturers started to introduce annual styling changes so as to increase pressure on customers to trade in old cars and buy new models for social and status reasons. This trend, combined with a deliberate fall in reliability, reduced the lifetime of cars and increased manufacturers' turnover and profits. In 1955, 80 per cent of the cars made by the three big American manufacturers were still on the road after nine years; by 1967 the figure had dropped to 55 per cent. Profits were also increased by other design changes that added 'value' as automation, robots and increases in productivity reduced the cost of making the basic car. At first these items were heaters, radios and cigarette lighters; later came air-conditioning, cassette decks, compact disc players, games consoles and navigation equipment. Production and consumption were also maintained by the increasing use of advertising to stimulate demand and create markets where none had existed before. This was linked to the development of credit mechanisms (at first hire-purchase and then credit cards and loans) so that people could buy more goods than their immediate income allowed. As early as 1926 three-quarters of all the cars sold in the United States were bought on credit terms.

The continuing pressure to consume reflected a force common in all societies – conspicuous consumption. One of the chief purposes of the medieval and early modern European laws about the wearing of luxury clothing such as furs, together with the habit of giving large banquets and maintaining a vast array of servants (at a time when these were almost the only forms of conspicuous consumption available) was to confine display to a small group of people so that they could flaunt their wealth. Over time the opportunities for display widened but the principle remained the same, as Adam Smith noted in the late eighteenth century:

> with the greater part of rich people, the chief enjoyment of riches consists in the parade of riches, which in their eye is never so complete as when they appear to possess those decisive marks of opulence which nobody can possess but themselves.

In 1899, the American economist Thorstein Veblen, in his book *The Theory of the Leisure Class*, analysed the way in which rich people continued to accumulate wealth well beyond the point of rational wants and he too concluded that display was an important part of expenditure. This phenomenon became ever more widespread from the early twentieth century. As increasing wealth permeated through society, goods which were once only available to the few became available to the many. Once they were readily accessible their symbolic value was destroyed so that the rich needed to possess new goods and signs of status in order to demonstrate their wealth. World tourism is a prime example of this trend. When large numbers of people are able to visit places that were once the prerogative of the rich then the wealthy have to travel further afield to ever more expensive, exotic and exclusive locations in order to maintain their status and distinction.

Cars

The new product that had the greatest industrial and social impact in the twentieth century was the car. The technologies that made the car possible were developed in the late nineteenth century but at the beginning of the twentieth century there were only a few thousand motor vehicles in the world. In 1900 the United States had 8,000 cars and there were still only sixty-two in Japan as late as 1909. They were built in small quantities by a large number of small-scale firms – in 1908 there were over 250 American car manufacturers. The devel-

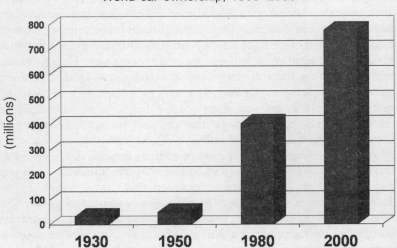

World car ownership, 1930–2000

opment of assembly-line techniques, epitomised by Ford's Highland Park factory in Detroit, increased output and reduced prices – a Model T Ford cost $825 in 1908 but only $345 eight years later. The result was a boom in demand – American car ownership leapt from 79,000 in 1905 to ten million by 1921. By the 1920s US car ownership was already at levels not approached in Europe until the 1950s. By 1930 there were twenty-six million cars in the United States but only just over five million in the whole of Europe. Car ownership in the United States continued to double every twenty years, reaching over 120 million by the mid-1970s. In Europe the great boom was between 1950 and 1970 – in Britain the number of cars rose ten-fold between 1939 and the 1980s from two and a half million to twenty-three million. Overall the number of cars in the world rose from thirty-two million in 1930 to 775 million at the end of the twentieth century – a twenty-four-fold rise. Car production now consumes more resources than any other industry. It uses about 20 per cent of world steel production, 35 per cent of the zinc, 50 per cent of the lead, 60 per cent of all natural rubber and 10 per cent of world aluminium production. In addition over a third of the world's oil consumption is accounted for by vehicles. In parallel, the rise of the car has brought into being a whole range of subsidiary industries – road construction, petrol stations, car sales dealerships and service garages.

The car is a striking example of the way in which growing affluence has brought not only many benefits but also major new problems and costs. For the individual, possession of a car has many attractions such as greater mobility, independence and personal convenience. At low levels of car ownership these benefits are clearly apparent and are one of the main driving forces behind the increase in car numbers. However, as numbers rise the costs of car ownership become more apparent – to the individual in terms of congestion and longer journey times and to society as pollution increases, cities have to be redesigned and roads, motorways and parking areas built; for those without a car life becomes increasingly difficult as public transport declines and shops move to the fringes of cities.

At first cars were hailed as environmentally beneficial. They would, it was hoped, be both cleaner and quicker than the horses that clogged city streets in the early twentieth century. The United States, as the first society to experience high levels of car ownership, was the first to experience the problems it brought in its train. The first widespread complaints about urban car parking were made in 1916 and seven years later the first proposals were made to ban them from city centres. By the mid-1920s about 24,000 people a year (10,000 of them children) were being killed and 700,000 injured in accidents involving cars. By the end of the twentieth century, despite improvements in car safety, there were 42,000 deaths a year and it was the major cause of death for those aged between five and twenty-seven. Overall in the twentieth century about three million Americans died in car accidents – about six times the number killed in war. Other societies too suffered the same high death toll as car ownership increased. By the end of the twentieth century car accidents killed 1,100,000 people every year across the world and nearly thirty-nine million were injured.

As car numbers increased it soon became apparent that existing roads in cities were inadequate. As early as 1903 a Royal Commission on traffic in London recommended parking restrictions and two new double-deck roads across the city to intersect near the British Museum. During the twentieth century cities had to adopt a number of methods to try to force an ever increasing volume of traffic through streets not designed for car use – parking meters, multi-storey parking, one way streets, underpasses, flyovers, traffic-free areas, selective bans and congestion charging. The attempt to deal with the symptoms of increasing car use have usually produced no more than temporary alleviation. Some cities gave up the struggle and tried to redesign the city round the car (as in Los Angeles), if necessary demolishing whole areas in order to build urban motorways.

As the number of cars in cities increased, the provision of public transport (one of the main features of the rise of the city in the nineteenth century) declined, caught in a vicious spiral of declining usage, rising prices and slower speeds because of the congestion caused by cars. It was perhaps fitting that it was the city of Detroit, the home of the American car industry, that decided in the 1920s to widen roads rather than build an underground railway to provide better urban transport. The pattern was repeated in other American cities – public transport fell by two-thirds in the twenty years after 1945 as the car took over. In the rest of the world the same pattern was repeated with trams and trolley bus systems being removed and public transport systems falling into decay – in Australia every city apart from Melbourne lost its trams and trolley buses between 1950 and 1970. In the United States, however, the car industry decided not to leave the decay of public transport to the vagaries of the market system and instead took action to close down public transport systems and force people to use cars. In 1936 three corporations connected to the car industry – General Motors, Standard Oil of California and the tyre company Firestone – formed a company called National City Lines whose sole purpose was to buy up alternative transport systems and then close them. Within twenty years over a hundred urban electric surface-rail systems in forty-five cities had been bought and closed. The company's biggest operation was the purchase in 1940 of part of the Pacific Electric system, which carried over 110 million passengers a year and served fifty-six communities. Over 1,100 miles of track were ripped up and by 1961 the whole network was closed.

The main response of governments to the rise of the car has been to spend more money on road construction. It was the easy option – it satisfied the powerful road construction lobby and it was popular with the majority of voters who were also motorists. All governments have spent far more on roads than public transport. The United States introduced federal aid for state road construction in 1916 and five years later began building interstate highways. The huge building programme of the 1950s was even justified as part of civil defence! In the United States the area devoted to roads and parking amounts to nearly seven million hectares, almost the same area as that planted for wheat. In Germany the autobahn network was started in the 1930s and since the 1950s every government in western Europe has built a large system of motorways as well as undertaking major programmes of improvements to existing roads. The problems soon emerged – road-building simply increased the amount of traffic because it made journeys easier and this then led to more congestion. Elsewhere improvements such as roundabouts, flyovers and by-passes did little more than shift traffic congestion from one place to another.

Building roads may be popular with motorists, the car industry and the construction lobby but the social and environmental costs involved are enormous, especially when compared to railways. The energy input in terms of steel and cement for road-building is three and a half times greater than that for an equivalent amount of railway construction and four times as much land is used. Overall railways are six times more efficient in energy terms than roads. Despite this in every industrialised country the capacity of the rail system has been severely reduced since the 1950s. In the United States railways account for 1 per cent of all intercity traffic, cars for 85 per cent. In Britain the movement of freight by road has risen by nearly 90 per cent since 1970 while that on the railway system has fallen by a quarter.

These problems are increased by the way in which cars are used and their energy efficiency. For the majority of journeys the driver is the only person in the car and the number of miles travelled per car per year has risen steadily – in Britain it has quadrupled in the last forty years. This trend is exacerbated by the energy inefficiency and high fuel consumption of most car engines. The average fuel consumption of American cars fell from sixteen miles to the gallon in the 1930s to thirteen miles to the gallon in 1973. The oil price rises of 1973–74 and 1979–80 improved this performance somewhat but the fuel economy standard for new cars has remained unchanged at 27.5 miles to the gallon since 1985. The highly popular SUVs (4x4) are exempt from these standards because, following extensive lobbying from the car industry, they are classified as light trucks. They now have half the US new car market and only achieve an average 17.7 miles to the gallon. This means that the average American car still travels only 22.3 miles to the gallon, far below European and Japanese standards. The amount of fuel used by each car every year in the United States is now four times the level of the 1930s. By the end of the twentieth century transportation in the United States was responsible for 7 cent of the world's energy use and was 25 per cent higher than *total* Japanese energy consumption. The same pattern of marginal improvements in fuel efficiency being offset by much greater car use are found elsewhere in the world. In Britain since 1960 fuel consumption per mile has fallen by 7 per cent but this has been more than offset by the rise in the number of cars and the greater distance each are travelled every year – the amount of motor fuel consumed in Britain has therefore tripled since 1960.

Leisure and tourism

As shorter working hours and greater affluence spread through the industrialised countries there was an increase in the amount spent on leisure activities. One of the earliest developments was the rise of mass spectator sports which began in Britain, then wealthiest country in the world, in the 1880s. The first professional football league in the world was formed in 1885 (mainly in the industrial towns of the North and the Midlands). Within twenty years seasonal attendances were over six million a year with an average of 300,000 a week. In the United States similar crowds watched baseball games and 130,000 people watched the Jack Dempsey and Gene Tunney heavyweight fight at Philadelphia in 1926. Participation was also high – in early twentieth century Britain about 300,000 men played football every week, there were over 20,000 registered anglers in Sheffield alone and tens of thousands cycled regularly. By 1914 the daily sports newspaper

L'Equipe sold over forty million copies a year in France and at times over 500,000 copies a day during the Tour de France cycle race.

The first decade of the twentieth century was marked by the huge boom in the new medium of cinema. There were two cinemas in Germany in 1900 but 2,500 by 1914. In Britain the weekly audience was about eight million by 1914. The addition of sound in the late 1920s and the production of newsreels made the 1930s the boom period for the cinema – by the end of the decade the weekly audience in Britain was about twenty million (equivalent to about half the total population). After the First World War there was a boom in radio stations and the ownership of sets. In the United States the number of households owning a radio set rose from 100,000 in 1922 to twelve million a decade later. In Britain it rose from 36,000 in 1922 to nine million by 1939. Very limited television broadcasts began in the mid-1930s (the first was in Germany in 1935) but it was not until the 1950s that the real boom began. Within a couple of decades almost every household in the industrialised world had a television set, by the 1970s it was a colour set and by the end of the century many households had two or three sets. Watching television soon became the main leisure 'activity'.

Increasing wealth and paid holidays led to a new industry in the twentieth century – tourism. In the eighteenth century only the European elite could afford to spend a few years travelling around on the 'Grand Tour' admiring the classical sites of Italy. Slightly more people could take the waters at spas such as Bath, Harrogate and Marienbad, or enjoy the sea air, but all travel was difficult, slow and expensive and foreign travel was confined to a very small minority. The construction of railways began the boom in travel. Thomas Cook organised the first special excursion train, from Leicester to Loughborough, on 5 July 1841 with 570 passengers each paying one shilling. As railways were built, new fast steamships were introduced and people had more money and more leisure time in which to spend it, travel could become tourism. At first the wealthier working class and the less-well-off middle class were confined to holidays at home at the developing seaside resorts such as Blackpool and Margate. Only the prosperous middle class could afford overseas travel. Gradually the 'package holiday' emerged. Firms such as Thomas Cook provided organised travel with prepaid vouchers for hotels and facilities for currency exchange (the American Express traveller's cheque started in 1891). Much of the tourism was confined to Europe, especially Switzerland, but as early as 1869 Thomas Cook organised the first tour to Egypt and Palestine and by the end of the nineteenth century such tours from both Europe and the United States were common. As tourism and travel developed so did the number of hotels (many were owned by the new railway lines) and the number of guides for travellers – the Michelin guide to the hotels and restaurants of France began in 1900.

The rise of the car further increased the range of holidays available – by 1926 the United States had over 5,000 motor camps across the country. (The same boom took place in western Europe in the 1950s.) Greater wealth and paid holidays made it possible for many of the working class across Europe to take holidays – in Britain Billy Butlin opened his first 'holiday camp' at Skegness in 1937. The major change came with the growth of air travel from the 1950s. The first commercial jet aircraft flew in the mid-1950s and the 'jumbo jet' entered service in 1970. In the last half of the twentieth century the number of passengers carried by scheduled airlines and the distance they flew increased seventy-five fold. The development of international tourism depended on the growth of cheap charter flights and then cheap airlines. The resulting growth was spectacular. In 1950 there were twenty-five million international tourists; fifty years later there were 760 million – a thirty-fold increase. Tourists now spend about £350 billion a year and 7 per cent of the world's workforce is employed in the industry. Tourism is the world's biggest single export earner and for some countries it is the mainstay of the economy. For four out of five countries in the world it is one of their top five export earners and in two out of ten countries it is the main source of foreign exchange. In Malta, for example, there are 380,000 permanent residents (many of them foreigners) and 1,200,000 tourists a year. A quarter of Malta's GDP comes directly from tourism and taking indirect earnings into account the proportion rises to 40 per cent. In many countries it is the main source of employment – 30 per cent in Gambia and an extraordinary 83 per cent in the Maldives.

Although the rise of modern tourism has brought many benefits it has also produced acute problems. Mass tourism has resulted in the construction of huge hotels, overcrowded beaches, the destruction of the local way of life and the development of synthetic 'local customs' to be shown to tourists in resorts that are almost indistinguishable from each other. Beaches and wildlife areas have been ruined and in areas such as the Mediterranean, which is already short of water, there have been major problems caused by the extra demand for water and the need for increased sewage disposal. Venice has almost been destroyed as a living city by tourism. The resident population has fallen from 175,000 in 1950 to the current 64,000 and the city has become little more than a museum (albeit a wonderful one). At times the number of visitors is so high that pedestrian movements have to be regulated along one way streets and entry restrictions imposed on the main attractions. These developments are seen in their most extreme form in much of tourism in the developing world. Luxury hotels and golf courses (also environmentally highly destructive) are isolated from the local community which derives few benefits apart from providing a source of cheap labour. The development of 'all-inclusive' resorts ensures that even less of the money reaches the local economy. About 80 per cent of the money spent on these holidays never leaves the home country of the tourists – it goes to airlines, tour operators, hotels and for the import of food and drink. The locals find that the prices they have to pay rise (usually by about 10 per cent) and the poorly paid employment they do gain is usually only seasonal.

The unequal world

The rise of societies dependent on high energy use, industrial production, advanced technology and affluence has led to a grossly unequal world because the benefits of this process have been limited to a minority of the world's population. The first settled societies and early empires were all agricultural and had roughly similar structures, were dependent on much the same technology, and their relative wealth only varied through the temporary gains from success in war and the looting of the wealth of the defeated. Until about 1500, when European expansion began, there is little doubt that western Europe was significantly poorer than the rest of Eurasia, in particular the Islamic world, China and India, although it was richer than the Americas and sub-Saharan Africa. In the next 250 years the establishment of the first European colonial empires, the looting of the wealth of the Americas, control over silver production in Latin America (which enabled Europeans to buy their way into the long-established Asian trading systems), and the development of plantation agriculture in the Americas (which was reliant on the import of twelve million African slaves), enabled Europe to catch up with the rest of the world. In 1750 western Europe was about as wealthy as China and probably slightly wealthier than India.

The exploitation of fossil fuel reserves in the form of coal and the growth of industrialisation in the nineteenth century meant that western Europe and North America became the wealthiest parts of the world. In the early twentieth century the richest countries in the world were about ten times wealthier than the poorest. That inequality was magnified on a colossal scale in the twentieth century. By the early twenty-first century the richest countries in the world were seventy-one times wealthier than the poorest. The disparities in wealth were enormous. The average wealth per head in a country such as Luxembourg was 113 times greater than that in Burundi. Even in the same region differences can be huge. The average wealth per head in Israel is thirty-seven times higher than in the Gaza Strip. In many countries the situation was deteriorating rapidly in the late twentieth century. By the mid-1990s, people living in eighty-nine countries of the world were poorer than they were in 1980 and in forty-three countries they were poorer than in 1970 – this was an absolute not a relative decline.

In the early twenty-first century the poorest 20 per cent of the world's population (about 1.3 billion people) receive about 1 per cent of the world's income. They are classified by the WHO as living in 'extreme poverty' and lack adequate food, housing and drinking water. Over three billion people (about half the world's population) have to live on less than $2 a day. A third of the world's children are classified by the WHO as 'undernourished' – indeed 60 per cent of the children born in India would be placed in intensive care if they were born in California because they are so underweight. Infant mortality rates in the forty-eight poorest countries in the world are seventeen times higher than in the industrialised world. A sixth of the world's population do not have access to safe drinking water and twelve million children a year die as a direct result of drinking polluted water and because a simple oral rehydration therapy costing thirteen pence is not available. (This is at a time when sales of bottled drinking water are worth $100 billion a year even though it costs 2,000 times more than high quality tap water and the profit margins for companies are around 25 per cent.)

The quality of life for people, however, cannot be measured simply by dividing national income by the number of people. In the 1990s the GDP per head in the United States was 40 per cent higher than in Italy but life expectancy was lower by almost two years because of the poor health system. Indeed the life expectancy of African-Americans is lower than the average in China and the infant mortality rate in cities such as Washington DC, Baltimore and St Louis is higher than in cities such as Bangkok and Cairo. Sri Lanka in the 1990s had a GDP per head just one-fifth of that in Malaysia, but it had a similar mortality rate, food intake and number of doctors per head and its literacy rate was substantially better.

The problems of poverty tend to be nastier and more obvious than those of affluence. Their environmental impact is different too – the search for land to grow food produces deforestation and soil erosion and the failure to provide clean water and adequate sanitation leaves appalling pollution. The desire for affluence though is global (especially at a time of global communication) and many countries have tried to move from poverty to affluence (and the environmental problems this produces). In a world economy dominated by the industrialised world and its financial and commercial interests, that process has been extremely difficult. In the early twentieth century 90 per cent of the world's industrial output came from the countries of western Europe and North America. That proportion was reduced by the industrialisation of the Soviet Union and Japan but at the end of the twentieth century half the world's industrial output took place in three countries – the United States, Japan and Russia – and three-quarters came from seven countries – those three plus China, Germany, France and Britain. Of the newly industrialising countries Brazil was the most important but it contributed less than 2 per cent of world industrial output. For the so-called 'Asian tigers' the figure was lower still – for both South Korea and Taiwan it was less than 1 per cent. Most countries, especially those in Latin America and Africa, saw their share of world industrial output fall in the twentieth century.

Since the 1950s the industrialised world has been committed to a policy of providing aid to promote economic development so as to narrow the gap between rich and poor. The policy is founded on the assumption that the correct course to follow is one of free-market capitalism and increasing integration into the world economy dominated by the industrialised countries. That policy has failed utterly as the figures for comparative wealth demonstrate. The current UN target is that 0.7 per cent of national income should be given as aid. That is a significant reduction from the original target of 1 per cent but even so only five countries meet it (Denmark, Sweden, Norway, the Netherlands and Luxembourg). In Britain the proportion fell from 0.52 per cent to 0.31 per cent in the last two decades of the twentieth century. The lowest percentage is given by the United States (0.13 per cent), equivalent to $40 a head, whereas in the Netherlands it is $385 a head. In practice the overwhelming majority of the aid budget never leaves the country that 'gives' it – it is tied to the purchase of equipment in that country or used to subsidise the prices of firms competing for contracts in the developing world. Many of the deals that are struck have little to do with aid. In the mid-1980s £65 million of British aid went to India to buy helicopters from an ailing aircraft company. Aid from the United States goes to countries judged to be strategically important such as Israel. Britain's aid programme paid for a £7 million hospital in the Falkland Islands – equal to £5,500 per person at a time when aid to India was 15p per person.

In addition substantial aid has come from the multilateral lending agencies – the World Bank accounts for over a fifth of all world aid. However, much of this money has been given for large-scale construction programmes which have provided work for the large corporations of the industrialised world. Many of these projects, in particular the building of large dams, have been highly damaging environmentally. Until the late 1980s the World Bank took no account of environmental concerns in its policy-making and since then consideration has usually been little more than cosmetic. The World Bank also supported Indonesia's transmigration programme, designed to relocate three and a half million people from Java to the outlying islands. In theory this was designed to reduce over-crowding in Java. Its environmental impact was disastrous as deforestation increased dramatically. In practice the project was a 'national security' programme intended to swamp ethnic minorities and tribal peoples in the outlying islands with people of Indonesian origin.

The industrialised world has also funded aid on strictly commercial terms. In the late 1970s when interest rates were low this seemed to be a reasonable policy but the policy soon turned sour. The oil price rise of 1979–80 badly affected developing countries dependent on imported oil and as interest rates rose rapidly. Most countries were also dependent on a small number of commodities for their export earnings. Zambia for example relied on copper for 90 per cent of its export earnings. Overall by the late 1980s primary commodity prices were, in real terms, at two-thirds of their level in 1957. The overall result was catastrophic. In 1970 the world's sixty poorest countries owed $25 billion, by 2002 they owed $523 billion. However, during those thirty-two years they had repaid $550 billion in both interest and capital. By the early twenty-first century for every $1 these sixty countries received in aid they spent $13 on debt repayment. However, very different rules apply to the rich and the poor countries of the world. The poorest countries have only 1 per cent of the world's total

Debt of 60 poorest countries, 1970–2002

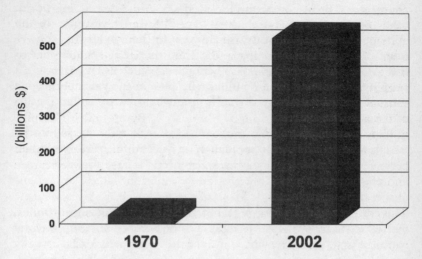

debt (the public debt of the United States is over $5,000 billion and rising rapidly). They are expected by the World Bank and the International Monetary Fund to put about 20–25 per cent of their export earnings into debt repayment whereas no country in the European Union repays its national debt at more than 4 per cent a year.

Developing countries have therefore had to turn to the IMF for assistance. (Promises of debt cancellation by the industrialised world have been highly conditional and have either involved acceptance of IMF policies, or have not been kept.) IMF 'adjustment policies' are designed to maximise export earnings and so keep up debt repayments. The policies countries are forced to accept usually consist of currency devaluation, drastic cuts in government spending, elimination of food and consumption subsidies, abolition of price controls and privatisation of state industries and 'liberalisation' of internal markets so that corporations from the industrialised world can move in. Because these policies are usually linked to high levels of corruption and high spending on arms they have been catastrophic for a majority of the world's people. Their impact can be seen clearly in the case of Morocco. In the 1950s it adopted a World Bank/IMF programme to concentrate agricultural production on fruit and tomato exports to Europe rather than wheat for the home market. This involved dam construction and irrigation (all on the land of large landowners). The debt accumulated to pay for this programme was $16 billion by 1984 – more than the country's GDP. Per capita wheat production was lower than in the 1930s and food imports, which took up most of the limited foreign exchange reserves, more than tripled between 1970 and 1983. Massive food subsidies kept prices low and also meant wages could be low so that exports were cheap. However, an IMF rescue package was required to enable the country to go on paying the interest on its debts. Food subsidies were cut (but not producer subsidies) so that the burden of the adjustment was borne by the poorest. By the mid-1980s nine million out of twenty-one million Moroccans were judged to be 'absolutely poor'. Half the population was aged under fifteen but their prospects were poor because two-thirds of those aged between fifteen and twenty had no job and received no education.

The emergence of affluence has not changed the persistent historical fact of internal inequality in society. Everybody in the industrialised world now is far richer than anybody two hundred years ago but massive differences in wealth and income remain. However, the development of high-energy-use societies and industrialisation has brought about a huge shift in the pattern of wealth distribution worldwide. Domination of the international economic system has enabled the industrialised countries to utilise the vast majority of the world's resources and develop unprecedented levels of consumption. One part of the world can now be dubbed 'affluent', while the great majority of the world's population still live, as they always have done in the past, in conditions of absolute poverty, want and disease. The changes that opened the way to these high levels of consumption for a minority of the world's people involved a substantial increase in the sources and amount of pollution.

15
POLLUTING THE WORLD

Pollution has been one of the distinguishing features of every human society. For much of human history the chief struggle was over sanitation and obtaining unpolluted water supplies. As population rose and more people lived in cities this problem became ever more acute. The development of industrial societies introduced new pollutants on a major scale and brought about new risks to human health and greater damage to the environment. Pollution was at first mainly localised – it was normally confined to a city, river, mine or waste dump. As industrialisation increased the areas affected widened to whole regions, continents and oceans. By the late twentieth century pollution was on such a scale that there was a clear threat to the global regulatory mechanisms that make life on earth possible.

One of the basic ecological rules is that the earth is a closed system and that everything must go somewhere – 'disposal' of waste merely means placing it somewhere on the planet. However, human understanding of the consequences of pollution has always lagged well behind the creation and release of pollutants. Attempts at controlling pollution are as old as the problem itself but the response has usually been belated and inadequate.

For archaeologists the wastes produced by societies are a major source of information. Many of the earliest cities in the Near East are built on huge piles of rubbish, known as tells, accumulated over the centuries. As archaeologists dig through the layers of discarded household refuse they find old pottery and tools, the charred remains of food and coprolites (fossilised faeces) which provide information about the different groups occupying the area, their activities and the food they ate. Gathering and hunting groups also accumulated large piles of rubbish – waste pieces of rock from tool making, animal bones and the tools thrown away when they became blunt. The late Upper Paleolithic site of Gare de Couze in France covers an area about 270 metres by fifty metres and it contains between one and two million discarded stone tools. However, the amount of pollution produced by gathering and hunting groups was small – their numbers were low and they had few possessions.

Water supply

The emergence of settled societies created a problem that has still not been solved across the world – disposing of human excrement and urine whilst obtaining decent drinking water. Failure in this area has been a major contributing factor to poor human health and high death rates for most of the last ten thousand years. The difficulty of separating the two functions is illustrated by the history of a small village in Cambridgeshire called Foxton, where a stream ran through the centre of the settlement and acted as both a sewer and a source of drinking water. The problems involved in regulating use of the stream can be seen from the fact that bye-laws for cleaning it out had to be made on eight occasions between 1541 and 1698, which suggests that they were not being obeyed. The main problem was that gutters, cesspits and what were described as 'other noysome sinks' discharged into the stream. More laws to stop wastes being put into the stream except after 8 p.m. on pain of a fine of twelve pence were issued in 1562, 1594, 1598, 1600, 1611, 1643, 1665 and 1698, which again suggests that the regulations were ineffective. In addition doing washing in the stream or keeping ducks and geese on it were banned. Even if all these various rules had been followed it would not have solved the problem for villages downstream, which had to cope with the contents of the 'noysome sinks' of Foxton even if they were only discharged at night.

The early agricultural villages needed a reliable source of water and many grew up around a stream, spring or well. For most rural settlements around the world these are still the only source of supply. The growth of cities meant that more complex arrangements were needed. The cities in the Indus valley such as Harappa and Mohenjo-daro had fresh water tanks to store supplies, central baths and drains and lavatories to try to keep water and sewage separate. Soon even more elaborate systems were required. Some of the first systems to bring water from outside the city were constructed in Greece as underground tunnels at Samos and Athens. Raised aqueducts were soon a familiar sight across the Mediterranean from Spain and southern France to Carthage and Alexandria. In 312 BCE the growing city of Rome decided that the river Tiber was too polluted to use as drinking water and constructed the first aqueduct. Within three hundred years Rome depended upon nine aqueducts, which in total extended over 420 kilometres and supplied all the city's drinking water and 1,000 public baths. Many of the Roman aqueducts, which were impressive and enduring feats of civil engineering, continued to be used long after the fall of the empire. The building of these centralised water supplies established a pattern that was to persist in virtually every city until the late nineteenth century – water was not normally provided for individual houses (unless they had their own wells) but was obtained from fountains and taps in public places.

The rise of cities in north and west Europe from the eleventh century led to exactly the same problems as those experienced in the Near East and Mediterranean thousands of years earlier. By the early thirteenth century the Thames was already polluted and in 1236 the first water was brought to the city of London from Tyburn spring (near what is now Marble Arch) in lead pipes. Other cities such as Zittau (1374) and Breslau (1479) also laid water pipes but often these were no more than hollowed-out logs, a method still employed on Manhattan Island in the nineteenth century. In 1610 the New River Company was established to bring clean drinking water from the New river to London. Other companies supplying the capital drew their supplies from the increasingly polluted Thames leading to regular outbreaks of disease. In 1852 Paris had to abandon taking water from the polluted Seine and switched to using a large artesian well at Grenelle. As the new industrial cities grew maintaining an adequate and safe water supply became a severe problem. As early as the 1820s English industrial towns were building artificial reservoirs. Soon these grew into large lakes as valleys were flooded, often many miles from the city – much of Birmingham's water now comes from Wales, and Manchester gets much of its supplies from the Lake District.

Although water was transported to cities its use was limited by the lack of supply to individual houses. In eighteenth-century Paris water was taken round the city by 20,000 water carriers using buckets. In the mid-nineteenth century 17,000 houses in the centre of London relied on their own wells and the other 53,000 depended on taps in the street, about one for every twenty or thirty houses. They normally only supplied water for about an hour a day for three days a week. Coventry had water supplies in just 350 of its 7,000 houses. Few houses were built with bathrooms; indeed, when Queen Victoria moved into Buckingham Palace in 1837 she found that there were none in the building. As late as 1908 there were no bathrooms at 10 Downing Street. Even public baths were not always available – in 1842 Liverpool was the only town in Lancashire that had them. In the less developed and prosperous parts of Europe this situation persisted well into the twentieth century. In Moscow in 1914 only 9,000 houses in the centre of the city had piped water, the rest relied on 140 wells. By 1926 only half the city's population had piped water and 250,000 still depended on ponds and rivers of dubious purity. In 1962 almost one in seven dwellings in Paris still had no running water and in the Soviet Union the situation was worse – over 60 per cent of urban housing had no water supply of its own.

Although in the industrialised world large volumes of water are supplied to the urban population its purity is being called into question again. The main problem in the second half of the twentieth century was not human wastes but contamination by industrial effluent and the run-off of fertilisers, pesticides and herbicides from agricultural land. Not all of these pollutants can be removed by modern treatment and filtration methods. The overall scale of the problem is unknown but alarming situations have been discovered in many areas. Twenty per cent of the wells in California had pollution levels above official safety limits in the 1980s, in Florida 1,000 wells had to be closed because of contamination, in Hungary 773 towns and villages had water unfit for consumption and in parts of Britain water in some places could not be given to new-born babies because of high nitrate levels. Overall, though, there have been significant success stories. Many of these are the result of the decline in industrial activity in the rich world as economies have become dominated by the service sector. In 1957 the Thames was declared biologically dead. Over the next few decades industry in the area collapsed, sea traffic into the Docklands area ended and better water treatment plants were built. By 2005 there were over 100 species of fish in the river and animals such as seals, dolphins and porpoises have been seen.

In much of the developing world the problems of the last ten thousand years have not been solved. Over a billion people (a sixth of the world's population) have no access to safe drinking water. In India official estimates suggest that 70 per cent of water supplies are polluted. The problems are well illustrated by the Ganges, which drains a quarter of India's land mass. In 1900 the basin had a population of about 100 million and ten million people dumped waste into the river. By 1990, 450 million people lived in the Ganges basin and seventy million discharged waste into the river, nearly all of it untreated. In addition thousands of cremated and partly cremated bodies are dumped into the river at Varanasi together with numerous animal carcasses. The Ganges may be a sacred river but it is now a heavily polluted and disease-breeding nightmare.

Sanitation

Until the development of water treatment facilities in the late nineteenth century there was hardly a city in the world that had succeeded in keeping its water supply clean and uncontaminated by human wastes and other rubbish. Disposing of human wastes posed huge organisational problems and normally it was simply put into streams and rivers in the hope that it would be carried away. There is no doubt that if someone living in the rich world in the early twenty-first century were transported back to any city at any period before the mid-nineteenth century they would be horrified and overwhelmed by the smell. People rarely bathed (it was thought to be bad for health by draining away bodily strength) and an occasional cursory wash was the most they indulged in. In the streets and rivers of cities there would be piles of rotting rubbish and human and animal excrement mixed with pools of urine and polluted water. The overall state of cities was well summed up by the mother of the Regent of France, who described Paris in the mid-seventeenth century as 'a horrible place and ill smelling . . . one cannot linger there because of the stench of rotting meat and fish and because of a crowd of people who urinate in the streets'.

Most houses had no lavatory facilities and even in the richest palaces and castles of medieval and early modern Europe the lavatory was no more than a hole in the floor that discharged into the street, the moat or simply outside the wall of the building. The palace of Versailles was only provided with portable commodes while visitors to the Louvre or Palais de Justice relieved themselves in corners, and the outside walls of buildings were stained by chambermaids throwing the contents of commodes out of the windows. Town houses might be provided with a cesspit but not normally with a sewer (there were some but they were little better than elongated cesspits because they were not flushed out with water). In town houses the cesspits were usually located in cellars and not well sealed. In October 1660 Samuel Pepys went to his cellar where: 'I put my foot in a great heap of turds, by which I find that Mr Turner's house of office is full and comes into my cellar, which doth trouble me.' Three years later the problem was worse and Mr Turner had to extend his vault into Pepys' cellar to try to contain his cesspit.

Since people did not have their own lavatories they used any available open space. In eighteenth-century Paris a row of yew trees in the Tuileries functioned as an open air toilet and when the authorities drove people away they used the Seine instead. City streets were full of animal wastes, dead animals, refuse and the left-overs of the butchering trade. In fourteenth-century Paris about 300,000 animals a year were slaughtered and the offal and carcasses were left to rot in the streets and streams. In September 1366 the butchers of Paris were forced to move out of the city and then they dumped their rubbish in a rural stream. Street cleaning, if there was any, was primitive and small-scale. Around 1300 the Piazza del Campo in Siena (the centre of the town) was kept more or less clean by five pigs which ate the refuse. At this time there were twelve rubbish carts in the whole of London and they tipped their loads into the Thames. In the American city of Boston in 1652 the town council issued an ordinance prohibiting the throwing of the 'entrails of beasts or fowls or garbage or dead dogs or cattle or any other dead beast or stinking thing' into the streets. It was obviously ineffective because the city had to maintain a scavenging system to clean the streets but that was abandoned in 1720 because it cost too much.

Allahabad did have such a system in the nineteenth century – it removed 100 tonnes of rubbish a day but it employed 638 sweepers and drivers and took up over half the city's budget. Few cities could afford to maintain such a system. A visitor to Madrid in 1697 described the streets of the town as:

> always very dirty because it is the custom to throw all the rubbish out of the window. One suffers even more in winter because carts carry several barrels of water which are emptied into the streets in order to carry away rubbish and let the filth run off; it often happens that one encounters torrents of this evil water which blocks one's way and poisons by its stench.

The cities of the Near East were little different as a French traveller in Isfahan discovered in 1694 when he commented on the streets full of mud and noted that:

> This great filthiness is still further increased by the custom of throwing dead animals, together with the blood of those killed

by butchers on to the squares and of publicly relieving oneself wherever one happens to be.

Even when rubbish was removed from the streets it was often only thrown over the walls of the town where it decomposed in a stinking pile. In 1512 when an English attack on Paris seemed possible, the garbage was so high against the city walls that it had to be dug away to make them defensible. When the French writer Jacques Caille visited Rabat in the early nineteenth century he reported that the city's waste was removed by just five or six donkey carts, which did not operate on holidays, in the winter or when it rained. The result was that:

> the streets of the city often show a layer of liquid mire more than ten centimetres deep. When waste matter has been removed it is thrown into the sea; or it is simply heaped up at the gates of the city, where it forms a veritable cess pool.

In some cities outside Europe and the Near East a more effective system operated. An English diplomat who visited Beijing at the end of the eighteenth century was surprised to find that people did not throw rubbish into the streets. Most Chinese cities had well organised systems where people left waste in tubs outside their houses where they were emptied by teams of cleaners and used to manure paddy fields. In Tenochtitlan, the Aztec capital, refuse was collected by boats on the canals and taken to manure the chinampas (floating fields). The problem with using human faeces as fertiliser or using irrigation canals as lavatories (as was the custom in Egypt and Mesopotamia) is that it spreads intestinal diseases, worms and flukes very easily. In China, where it was normal to use human waste as fertiliser, about 90 per cent of the people in the early twentieth century suffered from worm infestation and in the mid-twentieth century a quarter of all deaths were due to faecal-borne infections. Cities in Britain used much the same system in the second half of the nineteenth century. In Manchester the city council sold 100,000 cartloads of human excrement every year to local farmers. On the other side of the Pennines in Halifax the council developed a dry sewage system where the excrement was dried in open pits and ground up before being sold to farmers.

The rapid growth of cities in the industrial world in the early nineteenth century with their grossly overcrowded housing produced appalling sanitation problems. The almost complete lack of sanitation produced unimaginable conditions. One of the worst places was the 'Potteries' of northern Kensington in London, an area of just over three hectares originally dug out to provide brick clay for the surrounding suburbs. It filled up with the sewage from the neighbourhood and was full of open sewers and stagnant lakes. In the early 1850s over a thousand people lived there together with 3,000 pigs that ate some of the refuse. But almost everywhere in London was in a disgusting state. In 1847 John Phillips, the engineer to the Metropolitan Commission of Sewers, reported that:

> There are . . . thousands of houses in the metropolis which have no drainage whatever, and the greater part of them have stinking overflowing cesspools. And there are hundreds of streets, courts and alleys that have no sewers . . . I have visited very many places where filth was lying scattered about the rooms, vaults, cellar areas and yards so thick and so deep that it was hardly possible to move for it.

Frederick Engels in his investigation into the condition of the working class in Manchester in the 1840s decribed one area of the town near the river Irk where 200 people shared one lavatory:

> In one of these courts, right at the entrance where the covered passage ends is a privvy without a door. This privvy is so dirty that the inhabitants can only enter or leave the court by wading through puddles of stale urine and excrement.

Improvements in the water supply and the invention of the water closet merely transferred the pollution problem. Sewers could now be flushed through with water – it was legal to connect water closets to surface streams in England after 1815 and compulsory after 1847. However, this turned the rivers into open sewers. During his tours of Manchester, Engels stood on Ducie Bridge over the river Irk and described the scene below:

> At the bottom the Irk flows or rather stagnates . . . [it] receives as well the contents of the adjacent sewers and privvies. Below the Ducie Bridge, on the left, one looks into piles of rubbish, the refuse, filth and decaying matter of the courts on the steep left bank of the river. [It] . . . is a narrow, coal-black stinking river full of filth and garbage which it deposits on the lower-

lying right bank. In dry weather, an extended series of the most revolting blackish green pools of slime remain standing on this bank, out of whose depths bubbles of miasmatic gases constantly rise and give forth a stench that is unbearable even on the bridge forty or fifty feet above the level of the water.

In London sewers emptied into the Fleet river which flowed into the Thames, where the sewage was left to float up and down on the tide in the centre of the city. During hot weather the smell caused by this decaying filth spread over much of the city. Near the Thames it was unbearable. During one particularly bad period in 1858, known as 'The Great Stink', the smell was so powerful that sittings of the House of Commons had to be abandoned. Many water companies were still taking their drinking water from the Thames and, not surprisingly, this was a major source of intestinal illnesses and outbreaks of cholera. In 1853 when the Lambeth Water Company finally moved its source of supply further upstream away from the most polluted area, the death rate in the area it supplied rapidly fell from 130 per thousand to thirty-seven per thousand.

Improvements, in the form of sewage and water treatment plants, were slow to take effect even in a relatively prosperous country such as Britain. The provision of sanitation in houses took even longer. As late as 1911 less than half the houses in Manchester had their own lavatory and in one part of the city 700 people shared thirty-three lavatories. At this time in Dundee only three hotels and two private houses had water closets and they only worked with buckets of water. The rest of the town had to use about 1,000 privately owned dry privies and fourteen public conveniences. Elsewhere improvements took even longer. The first sewer in Moscow was not built until 1898 and in 1905 only 6,000 houses were connected to the system. Before 1917 only eighteen cities in Russia had even a rudimentary sewage system. In the 1980s nearly all of Moscow's sewage was still dumped, untreated, into the Moscow river. The sewage input was twice the volume of the river flow and so the river was little more than an open sewer. In Paris in 1925 half the houses were not connected to the sewage system and until the 1960s half of the sewage from the system was dumped, untreated, into the Seine. Tokyo, as late as 1912, still lacked even a primitive sewage system and after one was built half the population in the mid-1970s still lacked mains drainage. Generally it was only with the slum clearance programmes of the mid-twentieth century that the overwhelming majority of people in even the prosperous countries of Europe and North America had a decent sanitation system. Even then raw sewage was often still dumped a hundred metres offshore from where it was washed up on to beaches.

In most of the developing world the current situation closely resembles that of Europe and North America in the middle of the nineteenth century. In the European empires the colonial rulers did not think that the 'natives' wanted or needed sanitation. Governor Lugard of the Gold Coast (modern Ghana) wrote that 'such a community has no desire for municipal improvement. It neither appreciates nor desires clean water, sanitation or good roads or streets.' Street cleaning, piped water and sanitation were therefore limited to the areas where the Europeans lived (always carefully segregated from the 'natives'). The newly independent countries of the mid-twentieth century had few resources to start building water and sanitation systems. Descriptions of their cities have changed little over the centuries. A report on Hyderabad in the 1950s commented that:

> A high majority of the citizens commit nuisance promiscuously in open spaces . . . Public latrines are few and far between . . . [and] are not kept clean by the scavengers, and it is an annoying sight to see many a scavenger emptying his bucket full at some street corner.

In Manila, where nine out of ten homes are not connected to a sewer system, untreated sewage makes up 70 per cent of the volume of the Pasig river. An overwhelming majority of the world's citizens still lack decent sanitation and still suffer from the disease and squalor this causes.

One problem that has largely disappeared from the streets of the industrialised world is the huge quantities of waste left behind by animals on their way to market and from the large number of horse-drawn vehicles. (Dogs are the main exception – in Britain they put 1,000 tonnes of excrement and three million gallons of urine on to the streets every day.) For centuries the streets of every city in the world were polluted with horse droppings and urine. Roads would be turned into rivers of liquid manure when it rained or in hot, dry weather large quantities of dried manure would blow about as dust. The flies that were attracted were a major nuisance. So too was the noise from the sound of horses' hooves and metal-wheeled vehicles on cobbled streets – hence the habit of laying straw on the streets in front of hospitals or houses where people were ill. In most cities there was an army of crossing sweepers who charged a small fee so that people could cross the road without fouling their shoes and clothes. In 1830 animals left about three million tonnes of excrement on the streets of Britain – most of it was simply piled up into rotting, stinking heaps. The great increase in horse-drawn traffic in the nineteenth century made this situation even worse. By 1900 about ten million tonnes of excrement were deposited on Britain's streets every year. Most of the horses were grossly over-worked and under-fed. Few lasted more than a couple of years and many died in the streets. In 1900 New York had to clear 15,000 dead horses from its streets every year. The problem was never solved – it disappeared when horses were replaced by cars, buses and lorries, which brought their own form of pollution.

Smoke

Apart from the appalling smell, the piles of rubbish, the rotting carcasses and the human and animal excrement in the streets, one of the other distinguishing characteristics of cities, even before major industrialisation, was the clouds of smoke that hung over them. Wood burnt on open fires, using poorly constructed chimneys or in rooms without any chimneys (as was the case in most peasant huts), caused large amounts of smoke (and produced eye diseases). However, it was the shift to coal-burning that produced the main problems. Not surprisingly, it was in England, the first country to use coal on a major scale, that the first complaints about the smoke it produced were heard. In 1257 Queen Eleanor was driven from Nottingham castle by smoke from the numerous coal fires in the town. Thirty years later a commission was set up to investigate complaints about smoke levels in London. In 1307 the burning of coal in London was banned – the edict was ignored because people often had no other fuel available.

It was the great increase in coal-burning in the sixteenth and sevententh centuries that produced the first large-scale pollution problems in London. The city could be seen at a distance by the huge pall of smoke that hung over it. At this time the western parts of the city, especially Westminster, became the most fashionable quarter because the prevailing westerly wind normally kept the smoke away. In 1661 the diarist John Evelyn wrote his *Fumifugium* – an attack on the state of London – and described the capital as more like:

the face of Mount Aetna, the Court of Vulcan ... than an Assembly of Rational Creatures and the Imperial Seat of our incomparable Monarch ... For there is under Heaven such Coughing and Snuffing to be heard, as in London churches and Assemblies of People, where the Barking and Spitting is uncessant and most importunate ... It is this horrid Smoake which obscures our Churches, and makes our Palaces look old, which fouls our Clothes and corrupts the Waters, so that the very Rain, and refreshing Dews which fall in the several Seasons, precipitate this impure vapour, which, with its black and tenacious quality, spots and contaminates whatever is exposed to it.

Provincial cities were in the same state. In 1608 visitors to Sheffield were warned that they would be 'half choked with town smoke' and in 1725 when William Stukeley was in Newcastle he found that 'the perpetual clouds of smoke hovering in the air make everything look as black as London'. Even in Oxford the air was so bad that marbles brought back from Italy were damaged very quickly.

These problems became far worse in the nineteenth century as industrialisation took off, cities grew in size and coal became almost the sole form of domestic heating and cooking. London, the largest city in the world at this time, produced some of the worst conditions. In 1880 there were 600,000 homes in the central parts of the city with three and a half million fireplaces. London fogs (or smogs as they became known) became common and a major health hazard. In the middle decades of the century the number of foggy days every year tripled and the death rates from lung complaints rose sharply every time one occurred. In December 1873 there were about 500 deaths from a severe fog and in February 1880 over 2,000 people died in just three weeks. The pattern was repeated year after year for decades as smoke pollution worsened. The best estimate is that between 1840 and 1900 about 1.4 million people in Britain died as a direct result of urban air pollution. Between 1920 and 1950 the average number of hours of sunshine in central London was about 20 per cent less than in the outer parts of the city less affected by coal smoke. In the United States it was steel towns such as Pittsburgh which were badly affected. The first attempts to control smoke were taken in St Louis in 1940 – Pittsburgh followed in the next year but the measures were not enforced during the war. Controls were introduced on industry in 1946 and on domestic use the next year. Slowly the air began to clear during the 1950s. In London it was the terrible smog of December 1952 when there was no daylight for almost a week and over 4,000 people died (some walked into the Thames because they could not see the river) that finally brought about action. The Clean Air Act of 1956 introduced controls on the types of fuel that could be burned in city centres and by 1970 the amount of smoke in the air over London had fallen by 80 per cent and the amount of December sunshine had increased by 70 per cent. The enforced use of smokeless fuels, the decline in domestic coal consumption and the greater use of electricity, gas and oil for heating and cooking in cities across the industrialised world drastically reduced smoke concentrations.

Early industrial pollution

All industrial processes produce waste products, many of which are harmful to life in any significant concentration. The impact of pollution from the earliest mining and industrial production was limited and its effects usually fairly localised while industrial output remained on a small scale. Significant pollution came from the mining and refining of metal ores, in particular lead and gold (which normally requires the use of highly poisonous mercury). Lead production increased significantly in the Roman empire as it was easy to work and the main material of water pipes. The Roman writer Strabo commented on the need for lead smelting works to be equipped with chimneys 'so that the gas from the ore may be carried high into the air, for it is heavy and deadly'. The impact of Roman lead-smelting can be detected thousands of miles away in cores taken from the Greenland ice-sheet. They show lead deposition was ten times natural levels during the period of the Roman empire. These cores also show heavy concentrations of copper deposition in two phases: first, around 500 BCE when copper coinage became common in the Mediterranean area, and then around 1000 CE during the early industrialisation of Sung China. These early production processes were so inefficient that about 15 per cent of the copper was released into the air. The levels found in the Greenland ice are the highest known before the early nineteenth century and are at about 10 per cent of current levels even though overall production was probably far less than 1 per cent of current output.

In 1556 Georgius Agricola published his *De Re Metallica*, a treatise on mining in Germany, and described a scene that could have been found anywhere in the world where early mining took place:

the fields are devastated by mining operations . . . The woods and groves are cut down . . . then are exterminated the beasts and the birds . . . when the areas are washed, the water which has been used poisons the brooks and streams and either destroys the fish or drives them away.

In Japan in 1610 the Ashio copper mine, which produced almost half the country's copper, was opened. Within a few years the area around the mine was ruined (exactly as Agricola described). The mine was closed in 1790 but reopened in the 1870s. The waste was, as before, dumped into the Watarse river. Within a year the fish were poisoned as was the adjoining farmland when the river flooded. People, animals and plants in the area died – within twenty years twenty-eight villages were badly affected and about 40,000 hectares of land contaminated.

Although the energy inputs into early industry – human, animal, water and wind – were largely non-polluting, the waste products were not. The tanning of ox, cow and calf hides and the tawing of sheep, deer and horse hides produced large quantities of acid, lime, alum and oil which, together with the remains of the hides, were usually dumped into the local river or stream. Other industries such as brewing needed clean water and complained regularly about pollution upstream whilst putting their own wastes into rivers. Cotton dyeing and sugar refining also polluted water supplies and in 1582 Dutch authorities had to order linen bleachers not to dump their wastes into the canals but to use separate disposal channels known as 'stinkherds'. Just over twenty years later, James I of England was issuing proclamations against the pollution caused by starch makers in London. In 1627 there were complaints about the fumes from the alum factory at St Katherines near the Tower of London. Even the aristocratic part of the city was not immune to pollution. In the early eighteenth century the Duke of Chandos was railing against the conditions at his new town house in Cavendish Square, which he claimed was 'poisoned with the brick kilns and other abominate smells which infect these parts'.

Industrialisation

The concentrated phase of industrialisation that began in the late eighteenth century brought about a revolution in the scale, intensity and variety of pollutants released into the environment. For many decades there were no controls over this pollution. The result was that the 'industrial revolution' created areas of concentrated pollution and environmental degradation – ruined landscapes of chimneys belching smoke and poisonous gases, huge slag heaps of waste materials, rivers and canals full of a cocktail of industrial wastes and where vegetation was destroyed for miles around. People lived, worked and died in these conditions.

A major offender was the early chemical industry, which made large quantities of sodium carbonate for use in glass, soap and textile-making. An unwanted by-product was a huge amount of highly corrosive hydrogen chloride, which was released from chimneys without any controls. In the early 1860s the British government finally formed the Alkali Inspectorate in an attempt to control the problem. But, as with so many attempts to control pollution, the interests of the industrial lobby prevailed – controls were not implemented until new, less polluting processes were in use. In 1875 Queen Victoria complained that the ammonia fumes from the local cement works were making Osborne House, her residence on the Isle of Wight, uninhabitable. Government inspectors found they had no powers to close the works.

Largely unregulated industrialisation produced huge amounts of pollutants in every industrialised area from the Ruhr and Limburg to the Black Country of the English Midlands, to the Monongahela valley near Pittsburgh, which had 14,000 smokestacks belching fumes into the air. Many areas were turned into poisoned wastelands. A century of industrial production (mainly copper works) in the Lower Swansea valley led to the destruction of nearly all the vegetation and huge piles of toxic waste. A major regeneration project in the late 1960s still left the area unfit for housing. In January 1839 the Newcastle-upon-Tyne council made an official protest about conditions in the area:

> The gas from these manufactories is of such a deleterious nature as to blight everything within its influence, and it is alike baneful to health and property. The herbage of the fields in their vicinity is scorched, and gardens neither yield fruit nor vegetables; many flourishing trees have lately become rotten naked sticks. Cattle and poultry droop and pine away. It tarnishes the furniture in our houses, and when we are exposed to it, which is of frequent occurrence, we are afflicted with cough and pains in the head.

A visitor to the city confirmed this picture a few years later:

> The sturdy hawthorn makes an attempt to look gay every spring; but its leaves . . . dry up like tea leaves and soon drop off. Cattle will not fatten . . . and sheep throw their lambs. Cows too . . . cast their calves; and the human animals suffer from smarting eyes, disagreeable sensations in the throat, an irritating cough, and difficulties of breathing.

Industry put its wastes into rivers as well as the atmosphere. Factories were often built along river banks to make waste disposal easier and the mix of chemicals produced by the different industries was more than enough to kill off all life in the rivers and endanger human health. In 1866 the British Royal Commission on River Pollution found that some river water was so contaminated that it made good ink and they used it for part of their report. They reported that it was possible to set the Bradford Canal on fire – it was a popular pastime with local children. The River Irwell was described in 1869 as being 'caked over with a thick scum of dirty froth'. Nothing much had changed by 1950 when it was a vivid orange in the morning and jet black by midday. Even in the mid-twentieth century rivers still caught fire because of the load of chemicals they contained. Incidents were reported on the Iset near Sverdlovsk (1965), the Cuyahoga at Cleveland (1969) and the Volga (1970). In Tokyo in the late 1960s three-quarters of the city's rivers had no fish in them. An official report on the state of the Sumida river in 1972 bears an eerie similarity to descriptions of nineteenth-century England:

> As a result of pollution, the famous events which once took place on the river – swimming, regattas and firework displays – have vanished. The gases rising from the river corrode metals, blacken copper and silver ware and shorten the life of sewing machines and TV sets.

During the twentieth century most industrial countries introduced some regulation of water and air pollution by industry. But pollution has not been stopped. Governments usually gave greater weight to the demands of economic growth, industrial profitability and maintenance of employment than to demands for more stringent controls. In most cases industry has been able to obtain permission to release chemical waste leading to a situation of regulated pollution. Infringement of regulations and the occasional prosecution have not usually produced more than nominal fines that account for perhaps a few minutes' worth of profit-making. The main factor that has reduced industrial pollution in the industrialised world has been the decline of heavy industry, the decreasing importance of industry as a whole in the economy and the gradual shift into tertiary activities.

The great industrialisation drive in the Soviet Union from the late 1920s duplicated conditions found in western Europe and North America during the first wave of industrialisation. The overwhelming pressure on industry was to meet targets and maximise output. The impact on the environment was usually ignored. The results were predictable. Air quality in Moscow declined rapidly in the early 1930s – trees started dying and their growth rates fell by 90 per cent. By the early 1960s emissions of sulphur dioxide and nitrogen oxides were reaching near-lethal levels in some suburbs of the capital. Three-quarters of all the industrial sludge produced in the Soviet Union was simply dumped untreated into rivers. At the Karaganda Central Ore Concentration Mill carbon monoxide levels were twenty times and sulphur dioxide levels four times higher than the officially allowed maximum. At the Krasnoural'sk copper smelters pollution levels were ten times higher than the permitted level. By the late 1980s over 40 per cent of the milk produced in the Soviet Union contained chemicals dangerous to human health.

The similar expansion of heavy industry in eastern Europe after 1945, when the Soviet model of economic development was adopted, produced a major environmental catastrophe. The most heavily devastated area was the triangle formed between Dresden in East Germany, Prague in Czechoslovakia and Krakow in Poland. This region had a massive concentration of iron and steel plants, metal industries and chemical plants nearly all powered by the poor-quality lignite coal found in the area which produced large amounts of impurities and pollution. By the late 1980s even the Communist government in Czechoslovakia was prepared to admit that the area around Prague was a 'disaster zone'. In the town of Most, sulphur dioxide concentrations were twenty times the maximum levels recommended by the WHO and school children had to carry portable respirators. At this time East Germany had the highest ouput per head of sulphur dioxide of any country in the world. In Poland the Communist government described Upper Silesia as an 'environmental disaster area' where sulphur dioxide levels in the air were one hundred times official safety levels and 170 tonnes of lead, seven tonnes of cadmium, 470 tonnes of zinc and eighteen tonnes of iron were dumped from the atmosphere on to the historic city of Krakow every year. Smog occurred on a third of the days in the year, two-thirds of the food produced in the area was so contaminated that it was unfit for human consumption and 70 per cent of the water could not be drunk. A third of all the rivers in Poland were devoid of life and the Vistula was unfit even for industrial use over two-thirds of its length because it was so corrosive. The poisons brought down by the rivers had created an area of about 100,000 square kilometres in the Baltic that were biologically dead. Pollution levels improved dramatically in the 1990s but not because of regulation. It was the result of the collapse of heavy industry across the region once government subsidies and protection were removed by the first non-Communist governments.

Other countries in the early phases of industrialisation produced the same environmental damage. In Brazil much of the rapid expansion of industry from the mid-1960s was concentrated around Cubatão near São Paulo – it produced about 40 per cent of Brazil's steel and fertilisers without any form of pollution control. By the early 1980s it was one of the most polluted places on earth. Air pollution was at twice the level considered as lethal by the WHO, a third of all infants died before their first birthday, there were no fish in the rivers, no birds in the area, nearly all the insects had died and all the trees were dead. The fall of the military government in 1984 led to some action to control pollution. By the late 1990s the situation in the area was better but still bad by any normal standards.

Overall between 1850 and 1990 the amount of heavy metals put into the atmosphere by various industries across the world increased dramatically. Levels of cadmium emissions per year rose fifteen-fold, for copper the increase was twenty-six-fold and for zinc a massive one-hundred-and-thirty-seven-fold. From the 1960s air pollution in western Europe, North America and Japan fell as heavy industry went into decline. However, this was offset by increasing levels of pollution in eastern Europe, the Soviet Union, China, India and Latin America. The collapse of heavy industry in eastern Europe in the 1990s had only a marginal effect on total pollution because of rapid industrialisation elsewhere, particularly in China. It is difficult to measure the total amount of global air pollution from heavy metals but levels at least doubled over the course of the twentieth century and they may have risen about five-fold. Overall in the early twenty-first century about a fifth of the world's population (well over 1.2 billion people) breathe seriously unhealthy air and about 700,000 die as a direct result of air pollution every year. During the twentieth century about forty million people died as a result of breathing polluted air and about three-quarters of that death toll came in the second half of the century.

Acid rain

One of the most widespread and significant forms of air pollution is acid rain. The burning of fossil fuels, especially coal and lignite, and the smelting of sulphide iron ores, are the main sources of artificial sulphur dioxide in the atmosphere. The burning of fossil fuels also oxidises nitrogen to produce a variety of nitrogen oxides. When all these oxides escape into the atmosphere they are altered, through a complex series of processes, into sulphuric and nitric acids. The result is precipitation (in all its forms) that is unnaturally acid. The huge increase in fossil fuel use and developing industrialisation are the causes of a major increase in acid rain in the last two centuries. Global emissions of sulphur dioxide were probably about one million tonnes a year in 1850 but had risen to seventy-five million tonnes a year by the end of the twentieth century. Emissions of nitrogen oxides rose from about two million tonnes a year in 1900 to about twenty-seven million tonnes a century later.

The phenomenon of acid rain was first identified in Manchester,

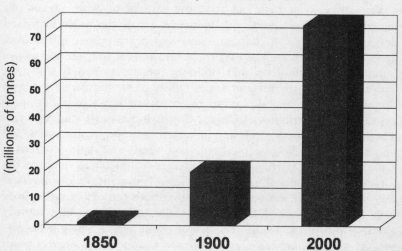

World emissions of sulphur dioxide, 1850–2000

one of the centres of British industrialisation, in the 1850s and explained in detail by one of the earliest British pollution inspectors, Robert Smith, in his book *Acid and Rain*, published in 1872. Acidity is measured on a pH scale where 6.5 is neutral; the scale is logarithmic so that a reading of 5.5 is ten times more acid than 6.5. Precipitation is normally slightly acid (5.7) because of carbonic acid produced from carbon dioxide, which is naturally present in the atmosphere. Increasing amounts of sulphur dioxide and nitrogen oxides in the atmosphere have produced abnormally acid rain. At times readings as low as 2.1 have been recorded (more acid than vinegar, which is about 2.4) and once at Wheeling, West Virginia, in the heart of one of the most polluted areas in the United States, a pH of 1.5 was noted (battery acid is 1). Acid rain leads to the gradual acidification of streams, rivers and lakes with devastating effects on ecosystems. A pH of 6.0 is the critical level – below this animal life begins to be affected. The processes involved are complex but one of the most serious is the higher concentrations of heavy toxic metals as the pH level falls. Fish are badly affected by the precipitation of aluminium as aluminium hydroxide on their gills, which reduces oxygen intake and causes major imbalances in internal salt levels. Once water reaches a pH of 5.5, salmon are affected and molluscs are rare. Between 5.5 and 5.0 there is severe damage to eggs and larvae and snails cannot survive below a pH of 5.2. Fish cannot live with a pH level much below 5.0 and at a level of 4.5 the flora is badly affected. A major problem is short bursts of very acidic conditions caused by heavy rainfall or the spring melt of snow and ice. Another consequence of acid rain is the slow destruction of buildings as the acid eats away at the stone. Krakow suffered severely until the mid-1990s but the phenomenon was found in most industrial cities and in the great cathedrals of northern France such as Rheims, Beauvais, Tours and Orléans.

At first, with low chimneys on most factories and power stations, acid rain was a localised phenomenon concentrated in the main industrial cities. However, in an attempt to reduce the high levels of pollution found in these areas, a policy of building ever higher chimneys was adopted in the hope that the pollutants would be dispersed. The policy worked and pollution levels dropped in the immediate area but the consequence was that acid rain increased over a wide area downwind of the main industrial zones. Some of the greatest damage to rivers and lakes therefore occurred in eastern Canada and the north-east of the United States, to the east of the main industrial areas, and in

Sweden and Norway, which received most of their acid rain on the prevailing westerly wind from Britain. In upstate New York in the Adirondack mountains the proportion of lakes with a pH below 5.0 rose from 4 per cent in 1930 to over half forty years later. Levels of acid rain in Norway and Sweden by the 1980s were as high as in any of the heavily polluted industrial districts even though their own output levels were low. The thin soils and underlying granite rocks provided little capacity to absorb acid rain and this was combined with the fact that half of the year's precipitation fell as snow, giving a high burst of acidity in the spring and early summer. The result was that while most lakes in Sweden had a pH of about 6.0 in 1950, thirty years later many were below 5.0 – in the west of the country the lakes were a hundred times more acid than they had been in the 1930s. In the south of Norway fish stocks fell by half in the forty years after 1940 and the Atlantic salmon disappeared after 1978 because of the acidification of its spawning grounds. By the late 1980s 20,000 lakes in Scandinavia were badly acidified and about a quarter of all the lakes in the region no longer contained any life.

Norway first complained about acid rain coming from Britain in the 1860s. However, it was not until the 1980s that the first international action was taken to control this form of pollution. The general principle that countries should stop transnational pollution was accepted at the UN Environment Conference in Stockholm in 1972. However, nothing was done about acid rain until 1984 when its effects were becoming very clear (probably a quarter of all the trees in Europe were affected). Even then only some countries were prepared to take action – the so-called 'Thirty per cent Club' who agreed to make a 30 per cent cut in their sulphur dioxide emissions by 1993 (measured against 1980 output). Some countries such as Austria, Switzerland and France did better than this and made cuts of over 50 per cent by the late 1980s. These cuts were finally extended on to a European scale in 1985 through a largely moribund body – the United Nations Economic Commission for Europe (UNECE), which was the only suitable forum that brought together the countries of both eastern and western Europe at a time when the Cold War was still a defining factor in international relations. All countries did make the 30 per cent reduction by 1993. In parallel the European Union brought in a directive on pollution from large power stations in 1988 which drove a switch to using natural gas, which has a much lower sulphur content than coal, and the introduction of better technology in the older power stations. In 1994 the UNECE agreed a further set of reductions (70 to 80 per cent in western Europe and about 50 per cent in the east) by 2000 (again over 1980 levels). Most of these targets were met. At the same time the Sofia Protocol covering nitrogen oxides was agreed in 1988 with a complex series of targets for reductions over 1987 levels. Overall the effects of these agreements were noticeable – by 2000 European sulphur dioxide output was 40 per cent lower than in 1980. In 1999 the agreements were taken further with the UNECE Gothenburg Protocol, which set targets for individual countries for reductions in sulphur dioxide, nitrogen oxides, ammonia and volatile organic compound emissions by 2010. All of these agreements have significantly improved air quality in Europe and reduced acid rain levels. Acid rain levels in Norway are now half the level of 1980 but recovery in rivers and lakes has been slow. Although improved technology played a part, the cost of these reductions, particularly in Britain, was borne by the coal-mining industry, which was deliberately run down with the switch to natural gas, leaving tens of thousands of miners unemployed.

The United States, which was responsible for half of the acid deposition in Canada, finally accepted it needed to take action in 1991 when an agreement between the two countries was reached. By the late 1990s the sulphur dioxide emissions of the US and Canada were about a third less than in 1980. However, the rapid industrialisation of China since 1980 has seen coal consumption (which provides four-fifths of China's energy), rising at 15 per cent a year. Much of this coal is poor-quality lignite which has a high sulphur content. As a result Chinese sulphur dioxide output rose by over 70 per cent in the 1990s and is now increasing at over 10 per cent a year. China is now the largest emitter of sulphur dioxide in the world and the situation was officially described as 'out of control' in 2003. The area of China affected by acid rain grew by 60 per cent in the 1990s and it is now estimated to affect a third of all the land in the country. Large amounts are exported too – as far as Japan, Taiwan, Korea and the Philippines. There are no signs of any regional agreement on controls, let alone reductions, as occurred in Europe and North America.

Pollution and health

The pollution caused by industrialisation in western Europe and North America, and later in Japan, the Soviet Union and eastern Europe, had a severe impact on both workers and the people who lived in these areas. Early industrial processes had few, if any, safety measures and as a result many workers suffered from disabilities and early death. A number of industrial diseases were first identified by Bernardino Ramazzini, a professor of practical medicine at the universities of Modena and Padua between 1682 and 1714. He pointed out that potters suffered from trembling, paralysis and loss of teeth from the large amounts of lead in the glazes they used. Glass-makers had ulcerated lungs and sores in the mouth from using borax and antimony to colour the glass. Gilders and hatters were subject to mercury poisoning – hence the expression 'mad as a hatter' or 'the Danbury shakes' (named after the American centre of the hat industry). In addition dust was a hazard everywhere but particularly affected grinders, woodworkers and miners. The great expansion of coal consumption in the nineteenth century was built on the labour of millions of coalminers, who suffered high rates of pneumoconiosis through inhaling large amounts of coal dust. Other industries had their own diseases. Lead production works were the most dangerous of all places to work – only the most desperate and starving would accept jobs because they were equivalent to a death sentence. Nevertheless employers could find enough workers. Cotton mills were unhealthy because of the large amount of lint in the air, which produced the lung disease byssinosis. Exposure to coal and oil products increased the risk of contracting cancer. This was first noticed in 1775 among boy chimney sweeps (who climbed up chimneys) and again in the 1870s when 'mule spinner's cancer' became common in the textile industry following a switch from natural lubricants (mainly whale oil) to mineral oils. In one coal-tar dye plant in the United States over a quarter of the workforce employed in the fifty years after 1912 contracted cancer.

Industrial pollution also affects the people living in the surrounding area. Pollution in the early industrial cities was very high because of the extensive coal-burning, heavy metals and a wide range of other pollutants in the air and water. It had a severe effect on human health. The statistics available from parts of eastern Europe in the 1980s show in detail the effects of heavy industry pollution and are comparable to conditions in the mid-nineteenth century in western Europe. In the 1980s Upper Silesia had the highest infant mortality rate in Europe (forty-four out of every thousand children died before their first birthday), three-quarters of the children aged ten needed constant medical treatment and the childhood leukemia rate doubled in the 1980s. Over a third of the children in Katowice had symptoms of lead poisoning. At most in Czechoslovakia only a third of children were free of illness, the respiratory disease rate was twice the national average and life expectancy was four years less than the national average.

Heavy metals are particularly dangerous to human health. By 1980 a tenth of Japanese rice paddies were unsuitable for human food production because they were contaminated with cadmium. As early as the 1930s the first cases of *'itai-itai'* – a bone disease caused by cadmium poisoning – were found in the Jinzu river valley. What happened at Minimata, a small fishing village, was far worse. The Nippon Chisso chemical factory was built there in 1910 and after 1932 it manufactured acetaldehyde using mercury as a catalyst. The waste was simply dumped into the bay where bacteria converted the mercury into methyl mercury, an organic compound that slowly accumulated, in ever increasing concentrations, up the food chain. In the late 1940s fish in the bay began to die and then, following a big increase in output in the early 1950s, cats in the village went mad – they had been eating the fish. By 1956 extensive brain damage was found in many children in the area. The company doctor knew the cause was mercury poisoning but was pressurised to keep quiet. The local authority did nothing and sided with the company because of the jobs and money it brought to the town. By 1990 over 1,000 people in the area had died as a result of mercury poisoning and the rest found it impossible to marry outsiders because of fears about inheritance of brain damage. Eventually the company was sued and in 1977 paid $100 million to the victims. Nets were put across the bay to stop fish entering and the government spent $400 million dredging out the toxic waste. In 1997 the bay was declared free of mercury and the nets were removed. It is very unlikely that all the mercury has in fact been removed.

Modern industrial pollution

Since the late 1940s there has been a major change in industrial processes and the types of pollution produced. In the second half of the twentieth century pollution levels rose far faster than the increase in industrial production. Until the mid-twentieth century the overwhelming bulk of industrial pollution came from the burning of fossil fuels, iron and steel production, other metal manufacture and the chemical industry. The key change in the last half century has been the manufacture of a wide range of synthetic chemicals, many of which are highly toxic in even minute quantities and which cannot

World organic chemical production, 1930–2000

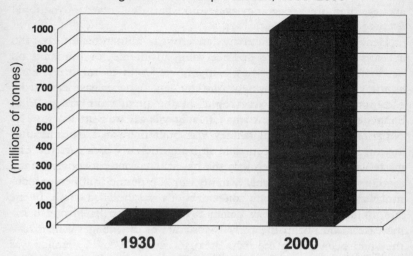

be broken down in the environment. Since 1950 about 75,000 new chemicals have been produced and about 2,000 new ones are now added every year. The overwhelming majority have not been subject to safety tests. World organic chemical production rose from one million tonnes a year in 1930 to 1,000 million tonnes at the end of the twentieth century. Traces of about 300 of these chemicals can now be found in people who live in the industrial world.

Modern industrial production has shifted towards more polluting products – plastics, detergents, synthetic fibres, fertilisers and pesticides – in place of natural, less polluting products such as soap, natural fibres and organic fertilisers. The switch from soap to detergent-manufacturing has increased phosphate production more than twenty-fold (and used far more energy). Companies prefer detergents because profit levels are far higher than on soap even though cleaning power is often little different. The increasing use of detergents, which finish up in streams, rivers, lakes and oceans, has substantially increased phosphate contamination of water supplies. Phosphate levels in American water courses rose two-and-a-half-fold between 1910 and 1940 but another seven-fold in the next thirty years. Excess nitrate and phosphate levels produce a rapid growth of bacteria and blue–green algae which, when they die, remove oxygen from water, and produce lakes and water courses unable to sustain much life. This is now common across the world.

The impact these chemicals can have on the environment can be illustrated in two cases – pesticides and polychlorinated biphenyls (PCBs). Before the mid-twentieth century farmers relied on natural products such as pyrethrum or chemicals with no long-term damaging effects such as 'Bordeaux' and 'Burgundy' mixtures for pest control. Since the mid-1940s pesticide production has become a major chemical industry with use increasing on average at over 12 per cent a year. The first highly toxic pesticides were the organochlorines such as DDT, followed from the early 1950s by the far more toxic, though less persistent, organophosphates. Because only about 1 per cent of the amount of pesticide sprayed on the crop or weeds actually hits the pests they have to be highly toxic in even minute doses and many of the compounds are highly carcinogenic. This toxicity affects wildlife and plants in the vicinity, workers applying the chemicals (often with poor or non-existent safety equipment), local residents subject to spray drift and water supplies when the pesticide excess runs off into streams.

The effect of using DDT can be seen by what happened when the Clear Lake area of California was sprayed in 1949, 1954 and 1957 in an unsuccessful attempt to eliminate gnats. The pesticide was sprayed at a concentration of one-fiftieth of a part per million. The organisms at the bottom of the food chain, the plankton, were found to have DDT residues 250 times greater than the water of the lake. In frogs it was 2,000 times higher, in fish it was 12,000 times more and in the grebes that fed on the fish the DDT concentration was 80,000 times greater than the water. The result was that eggshells of the grebes became so thin that they cracked under the weight of the bird. Of the 1,000 pairs of grebes in the area not one hatched a chick between 1950 and 1962. It was the implications of this disaster (repeated elsewhere with other chemicals) that led Rachel Carson to write *Silent Spring*. The controversy surrounding the book led, despite strong opposition from the chemical industry lobby, to the eventual banning of DDT in the industrialised world. However, the United States continued to export DDT to the developing world into the 1990s.

The impact these chemicals can have is shown by how far they can travel in the environment. In 1983–84 East Germany sprayed DDT in many areas. It was easy to detect elsewhere because its use was banned in western Europe. The East German DDT increased West German DDT levels four-fold and in Poland contamination increased ten-fold. The DDT was detected over a range of 1,500 kilometres from north of Stockholm to the south of France. A few years earlier islands on Lake Superior showed traces of a pesticide that had only been sprayed on to cotton crops in the southern United States over 1,500 kilometres away. The widespread use of these chemicals – apple crops in southern England are now sprayed over twenty times during the growing season – leaves residues in food and water yet they have not reduced crop losses. Despite the massive increase in pesticide use in the United States between the 1940s and the 1980s crop losses increased by a sixth.

The production and use of PCBs illustrates how the effects of using synthetic chemicals are often not recognised for decades, by which time it may be too late to avoid major environmental damage. PCBs are chlorinated hydrocarbons, closely related to DDT, and are now recognised as some of the most carcinogenic compounds known to science. From the late 1930s they were used as insulators in electrical products, especially transformers, additives in paint and in 'carbonless' carbon paper. In total about two million tonnes of PCBs were made before production was banned in the United States and Japan in the 1970s and the European Union a decade later, under pressure from environmental groups. As early as 1936 chloracne (a very serious skin complaint) was identified among people who handled PCBs and special controls were introduced over its production but not disposal. Disposal of about half the world's production of PCBs has caused major environmental and health problems. Most were dumped in the oceans or left to rot in waste dumps. High-temperature incineration seemed the only safe route for disposal but even here residues can escape into the atmosphere. PCB contamination has been found in human milk across the industrialised world and even small traces have resulted in birth defects. In 1968 over 12,000 people in Japan suffered from terrible illnesses after eating cooking oil contaminated with PCBs. The impact on animals was also severe. By the 1980s half the seals in the Wadden Sea off the Netherlands were sterile because of PCB poisoning. Whales in the St Lawrence river had residues in their bodies sixteen times higher than that required to have them classified as toxic waste anywhere in the industrialised world. PCB residues have also been detected in Antarctic penguins (thousands of kilometres from any PCB use) and in dolphins and seals in the Indian Ocean. After the ban on PCB production a new class of chemical was introduced – PBDEs, polybrominated diphenyls. These too have recently been found to cause major problems. Residues have been detected in whales in the Arctic and significant numbers of female polar bears at the top of the Arctic food chain are now hermaphrodites and unable to reproduce as a result of the accumulation of PBDEs in their bodies.

Industrial accidents

Industrial accidents are nothing new but a number of factors in the last fifty years have greatly increased their likelihood. The increasing dependence on oil, especially the exploitation of the vast fields in the Near East, has meant that the volume of petroleum carried at sea has almost quadrupled and the distance over which this extra oil has been carried has also quadrupled. Oil tankers are now more than thirty times bigger than they were in 1945 and so not only has the possibility of an accident increased enormously but so have the consequences. The first major accident was in 1967 when the *Torrey Canyon* ran aground off the coast of Cornwall, spilling 100,000 tonnes of crude oil. Other major spills occurred in 1978 when the *Amoco Cadiz* broke up off the coast of Brittany and in 1989 when the *Exxon Valdez* ran aground in Prince William Sound in Alaska. In the last thirty years there have been almost 750 major oil spillage incidents across the world. In every case beaches and coastlines were polluted and large numbers of seabirds and animals were killed. The local environment can take years to recover from even a minor spill. Offshore oil production can also be environmentally damaging – in 1979 the Ixtoc 1 oilrig in the Gulf of Mexico released 600,000 tonnes of oil into the sea. In Nigeria the delta region where oil production has been concentrated for the last forty years is now an environmental disaster area.

Accidents to synthetic chemical plants can be particularly damaging because of the highly toxic nature of the products they manufacture. In 1976 a broken valve in a plant at Seveso in northern Italy released about 3,000 kilograms of the highly toxic chemical dioxin. No action was taken at the time and it took several days before the impact became apparent. Over 37,000 people were exposed to the chemical and over 700 had to be evacuated from the most contaminated area. Many suffered from disfiguring skin diseases and there was a significant increase in the number of babies born with defects. The top twenty centimetres of soil over an area of eighteen square kilometres had to be removed and buried as highly toxic waste. Much of the area was turned into a park because it was unfit for human habitation. Ten years later a fire at the Sandoz chemical plant at Basel in Switzerland resulted in thirty tonnes of pesticides, fungicides and dyes being washed into the river Rhine. No life survived in the river for 200 kilometres downstream and even lower down it took several years for life in the river to recover.

The worst chemical disaster in the world occurred at Bhopal in central India on the night of 2–3 December 1984. About thirty tonnes of methyl isocyanate gas escaped from the plant into an adjacent slum area. About 500,000 people were exposed to the gas, about 20,000 people died and another 120,000 were severely affected, many of whom were blinded. Despite the fact that none of the plant's six safety systems was operational on the night of the accident no prosecutions have been brought. Union Carbide, the owners of the plant, undertook no clean-up in the area and after a complex series of deals with the Indian government (which seemed to be mainly interested in keeping foreign investment flowing into the country) only minimal compensation was paid. In July 2002 Dow-Carbide (following a take-over of Union Carbide) offered about $500 to each family affected. When asked how this compared with a $10 million out-of-court settlement the company had recently made with the family of a brain-injured American child, the official spokesperson for the company replied: '$500 is plenty good for an Indian'.

Toxic waste

Industrial accidents are regarded as no more than an incidental hazard that has to be accepted if modern materials are to be produced. Equally the disposal of the waste produced after the use of these materials is regarded as a fact of life. It is part of a wider problem of disposing of the waste produced by modern societies with their high levels of consumption. In the second half of the twentieth century the trend was towards increased packaging for all goods (partly because supply chains became longer) and for containers and other products to be 'non-returnable'. Milk bottles that were re-used have been replaced by milk in cartons. In the United States beer consumption rose 37 per cent between 1950 and 1967 but the number of 'non-returnable' beer bottles increased by almost 600 per cent. These bottles are, of course, returnable – it is the companies who do not want to organise the return and re-use system. In Britain people now throw away six billion 'disposable' nappies, seventeen billion plastic bags from supermarkets and 972 million plastic bottles every year. Waste disposal or, more euphemistically, 'waste management' has become a major growth industry. Recycling is still at low levels – in Britain on average the

United States hazardous waste production, 1970–2000

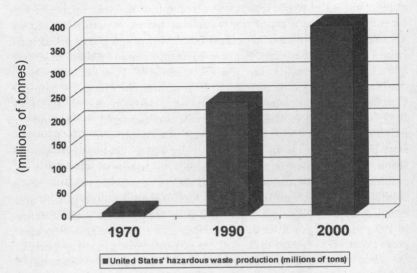

rate is 13 per cent (half that in the United States and a quarter of the German level) – apart from some metal industries (especially copper, steel and aluminium). The only options are seen to be incineration (expensive and it still releases pollutants into the air), dumping in the oceans or putting the waste into landfill sites. Every year New York dumps about ten million tonnes of waste in the Atlantic off the mouth of the Hudson river, creating an area of toxic black sludge stretching over 100 square kilometres. Many countries are running out of landfill sites and those that do operate have problems dealing with non-degradable items and the leaching of chemicals into the adjoining soil and water.

Disposal of toxic waste poses even greater problems. As the production of toxic chemicals has increased so has the amount of toxic waste. In 1970 the US produced about nine million tonnes of hazardous waste every year – thirty years later the figure had risen to 400 million tonnes a year. Until the 1970s there were few controls over toxic waste dumping in the industrialised world and even the location of many of these early dumps is still unknown. Much of the waste was put into landfill sites without adequate containment (often no more than a lining of clay) or even mixed in with ordinary household waste in order to 'dilute' it. Toxic waste dumps became major environmental and health hazards. One of the worst affected areas is near Niagara

Falls where, in the 1980s, there were over 200 dumps containing eight million tonnes of waste which was seeping into the water supplies for six million people. Part of the waste was Mirex, a pesticide used to kill fire ants. Over 1,000 kilograms leached through the soil and into Lake Ontario in the 1970s – it is so toxic and carcinogenic that under both US and Canadian law the allowable level in water supplies is nil. Nearby, at the sadly misnamed Love Canal, the Hooker Chemical Company buried a potent cocktail of toxic waste between 1942 and 1953. Then they sold the land to the local community for $1 and an indemnity against any future liability. A school and about 250 homes were built on the site. In 1978, after acute suffering among the population the federal government declared it a national disaster area, evacuated the residents and fenced off the entire zone. In three separate incidents the Dutch authorities also had to demolish 500 homes and evacuate 1,500 people who were living on top of toxic waste dumps. In 1984 the Georgswerder dump in Hamburg exploded and for several years released over 100 million litres of poisonous gas every year. In the early twenty-first century there were still 55,000 contaminated sites across Europe and 50,000 in the United States with little or no possibility of an effective clean-up being undertaken.

Toxic waste trade

Following these scandals and increasing public concern about the dumping of toxic waste, governments across the industrialised world began to introduce tighter controls on dumping. Dumping at sea was effectively banned and controls on putting toxic waste in landfill sites raised the cost of disposal in the United States from an average of $15 a tonne in 1980 to $250 a tonne by the end of the decade. With incineration costing about $1,000 a tonne the producers of toxic waste began to look elsewhere for disposal sites. Developing countries became the main target. Philadelphia (the 'city of love') ran two incinerators and in 1987 simply dumped 4,000 tonnes of toxic waste on a beach in Haiti. A year later they dumped 15,000 tonnes in an old quarry in Guinea. Poor states could be bribed (as could their politicians) to take the waste. In 1988 Guinea-Bissau agreed to take fifteen million tonnes of toxic waste for a fee of $600 million – this was worth four times its GDP. In nearly every case the toxic waste was just left to rot and seep out from rusting drums near homes and schools.

Growing concern about this trade and its consequences for the poorest people in the world led to the Basel Convention in 1989 designed to limit the trade. However, the convention did not ban the trade in toxic waste – it allowed a 'notification and consent' process so that as long as the importing country knew the contents of the waste it was accepting and agreed to take it the trade could continue. That convention was ratified and came into effect. The amendments made in 1995 – which banned the trade for 'disposal' but allowed trade for 'recycling' until 1998 had, in 2005, still not been ratified and so the trade continued. Most important of all the largest producer of toxic waste, the United States, did not ratify the original 1989 convention and refused to be bound by any restrictions on the trade in toxic waste. Indeed the United States has no controls on either the import or export of toxic waste and does not even keep records of what is traded.

From the early 1990s increasing amounts of toxic waste were designated for 'recycling' so as to avoid criticism of dumping. But this 'recycling' was highly dangerous – it usually involved the hand-sorting of solvents, work on highly poisonous lead-acid batteries, asbestos-impregnated materials and PCB-contaminated electronics. Safety measures were usually non-existent and the work was under-taken by the poorest in society and, in many countries, the prison population. The fastest growing trade is now electronic waste. The United States discards about 100 million computers a year and Lagos in Nigeria currently receives about 500 containers of used computers from the US and Europe every month. Companies claim that this provides cheap computers for developing countries but more than three-quarters of the computers are unusable. Mobile phones are classified as toxic waste even when they do not contain the highly poisonous batteries. The US has thrown away 130 million phones and 80 per cent of them have been sent to Pakistan, India and China for final disposal.

The disposal of ships has become an increasing problem. The industry argues that they are not covered by the Basel Convention (even though they clearly are) and instead are creating their own very lax series of rules under the International Maritime Organisation. The shipbreaking industry has moved to the beaches of South Asia where vessels are broken up by hand. At Alang in India 40,000 people are employed, without safety equipment, to break up ships containing asbestos, PCBs and toxic paints. In early 2006 when the French government tried to use the beaches to dispose of an old warship full of toxic materials they were forced, by public pressure, to bring the ship back to Europe for safe (but expensive) disposal. The United

States, which has to dispose of 225 obsolete warships, has decided, facing costs of $1.9 billion for safe disposal, to go for a mixture of dumping at sea ($500 million) and export to Asia for disposal ($170 million).

Nuclear pollution

Since 1945 one of the most hazardous of all new technologies has been the production of energy from nuclear power. All forms of radiation are inherently dangerous – at high levels it is lethal within a short period and at lower levels there is a significantly increased risk of cancers and genetic defects. There is no known 'safe dose' – instead the nuclear industry works to exposure levels (both for its workers and for the public) which it hopes will not produce an 'unacceptable' level of danger. Humans are exposed naturally to background radiation from materials in the earth's crust and from outer space. Between 1945 and the early 1960s this was significantly increased by the large-scale atmospheric testing of nuclear weapons. In addition the production of nuclear weapons led to major pollution. The major US nuclear weapons site at Hanford in Washington state leaked 422,000 gallons of radioactive material from storage tanks between 1945 and 1973. Over half a tonne of plutonium (lethal in doses of a few milligrams) has been buried around the site. A full clean-up is impossible but the latest estimates of the cost of even a partial clean-up amount to somewhere between $100 billion and $1,000 billion over a seventy-five-year period. The USSR programme produced an even greater disaster around the Mayak nuclear fuel reprocessing plant in the Ob river basin in Siberia. Liquid nuclear waste (including twenty-six tonnes of plutonium – fifty times the amount dumped at Hanford) was released into the Techna river, a tributary of the Ob. By the 1950s the highly radioactive material had reached Lake Karachai where it slowly accumulated. The heat generated by the nuclear waste, accompanied by a drought in the region, evaporated the lake and exposed the highly radioactive lake bed. It released radioactivity equivalent to 3,000 Hiroshima-sized nuclear bombs over 500,000 people. The lake bed was covered in concrete to stop the wind blowing more away. The lake is now the most radioactive place on earth – standing on the lake shore for half an hour is enough to give a lethal dose of radiation.

Although theoretical calculations show that the chances of a nuclear accident are highly remote, in practice nuclear power generation has, throughout its history, suffered from serious accidents. In 1957 the core of one of the reactors at Windscale in north-west England caught fire and there was a major release of radioactivity across much of Britain. Over two million litres of contaminated milk had to be destroyed but the number who died or contracted cancer as a result is unknown because no studies were ever carried out. At Three Mile Island in Pennsylvania in 1979 one of the reactors suffered a partial core meltdown. A catastrophic accident was narrowly avoided but the reactor had to be left permanently entombed in concrete. The worst nuclear accident occurred at Chernobyl in the Ukraine in 1986 when an explosion in one of the reactors released a gigantic cloud of radioactive debris that spread across Scandinavia and western Europe. Vegetable and animal sales had to be stopped across a wide area and many of the reindeer of Lapland were slaughtered because they had eaten large quantities of contaminated lichen. Twenty years after the accident sheep in parts of north-west England and Wales were still too radioactive to be eaten. The effects in the immediate area of the explosion were, of course, far worse. About thirty people died trying to deal with the accident but contamination of a large area of the Ukraine was serious. 220 villages had to be abandoned and over 600 towns and villages decontaminated. An area covering 640,000 hectares was declared too dangerous for human habitation. Over 135,000 people had to be evacuated (the numbers were kept down when the Soviet government increased the 'safe dose' of radiation for the public to 50 per cent more than the lifetime dose allowed for workers in the nuclear industry). The death toll is unknown but by the early twenty-first century 32,500 people had died as a direct result of the explosion and the number was continuing to rise as more people contracted fatal cancers. There were more accidents after Chernobyl. In 1999 an uncontrolled chain reaction at the uranium reprocessing plant at Tokaimura in Japan released radioactive gas into the air – two workers were killed and thirty-five badly contaminated. Five years later four workers were killed at the Mihama plant in Japan.

Lower-level exposure to radiation, though less spectacular than major accidents, can be highly dangerous. The whole process of providing nuclear fuel and disposing of waste is highly dangerous. The mining of uranium exposes the miners to highly radioactive dust. The very high lung-disease rates experienced by the metal-ore miners in the Erzegebirge region (in the Sudetenland) in the mid-sixteenth century was almost certainly caused by uranium in the rocks. In the twentieth century half of all uranium miners died of cancer – a rate five times higher than the population as a whole. The next stage, the milling of urnaium ore, causes 4,000 deaths a year from lung cancer in the United States alone. Large amounts of waste are produced from civil nuclear uses, from the clothes of workers in nuclear installations to the isotopes used in medical machines. (In 1987 at Goiânia in Brazil 244 people were contaminated from a radiotherapy machine that had been sold as scrap – four died.)

The major problem with nuclear pollution is that created by civil nuclear power plants when they cease operation. In the mid-twentieth century, governments in the industrialised world embarked on programmes to build nuclear power stations irresponsibly – they knew there was no solution for dealing with the nuclear waste it would create. They hoped that a solution might be found in the next few decades but they also knew they would be out of office when the problem had to be faced. Until the mid-1980s the industrialised countries dumped some of the waste (mainly intermediate level) at sea – the total rose from 20,000 tonnes a year in 1967 to 100,000 tonnes by 1983. The radioactivity in the main dumping ground off the Spanish Atlantic coast rose seven-fold in this period. Growing concern about the dangers involved, especially corrosion of supposedly safe containers, led to an international agreement to stop dumping at sea. With domestic public opinion strongly opposed to the building of nuclear waste dumps this left governments with no realistic option for dealing with nuclear waste, especially as the older nuclear plants began to stop generating power and the earliest nuclear-powered warships went out of service.

By the early twenty-first century the British government was faced with the problem of disposing of two million cubic metres of low-level waste, 241,000 cubic metres of intermediate-level waste and 1,340 cubic metres of high-level waste. The latter contained materials that would need to be protected for 100,000 years and which were lethal in minute doses. (How this is to be done is unknown – most suggestions look for a deep – about 500 metres – mine shaft in what is hoped to be a geologically stable area but protection and political stability for an unimaginable length of time would also be required.) Only 8 per cent of this waste was securely packaged and the search for a deep site to bury the waste was effectively abandoned in the 1990s – it is currently held in thirty-seven 'temporary' surface sites. In 2006 the British government asked local authorities in suitable areas to volunteer to take this nuclear waste. This incentive was a government 'bribe' of extra investment in the area if they did so.

The cost of decomissioning the existing nuclear power plants is estimated at a minimum of £70 billion.

Vehicle pollution

At first cars, buses and lorries were welcomed as less polluting than the horses that fouled city streets in the early twentieth century. As vehicles spread across the world the problems they produced rapidly became apparent. All internal combustion engines produce a range of pollutants – carbon dioxide, carbon monoxide, nitrogen oxides, a range of volatile and toxic organic compounds and smoke. Once released into the air they produce other forms of pollution including peroxides (which irritate the eyes) and ozone (which at ground level reduces photosynthesis in plants and makes breathing more difficult). In combination all these different pollutants react with sunlight to produce 'photochemical smog' – a poisonous brown haze now found over most modern cities. Vehicle exhausts were also made more polluting by the addition of lead to fuel. By the 1920s the higher compression engines then coming into production required higher octane fuel to operate correctly. This could have been achieved through improved refining techniques. However, Thomas Midgely, a scientist we will meet again in the next chapter, suggested that it was easier and cheaper to add tetraethyl lead to petrol to achieve the same result. The corporation which manufactured the tetraethyl lead was half owned by General Motors and so there was an incentive to keep increasing the compression ratio in car engines. Between 1946 and 1968 the amount of lead used per vehicle mile in the United States rose by 80 per cent. As every society has discovered lead is poisonous and its effects include brain damage, especially in children.

The effects of the pollution produced by vehicles were first noticed in the United States, the first country with high levels of car ownership. The Los Angeles valley has a natural inversion layer, which results in air being trapped in the valley on many days in the year. Concentrations of exhaust pollution can, therefore, rise very rapidly. In 1943 the first photochemical fog was noticed and by the late 1950s the problem was acute. Eye irritation was experienced on half the days in the year and by August 1969 doctors had to warn residents not to play golf, jog or do anything involving deep breathing during a particularly bad period of pollution. By the late 1980s photochemical smog affected over 100 American cities, eighty million Americans lived in areas where ozone was above permitted levels and San Franciso suffered from dangerous air quality on over 200 days a year. In the early twenty-first century the air quality in Los Angeles was officially defined as 'unhealthy' on three-quarters of the days in the year. As car ownership increased elsewhere other countries experienced similar problems. Smog became serious in Tokyo in the late 1960s and over 50,000 people were disabled by it in 1970–72. In Mexico City, which, like Los Angeles, has a natural inversion layer in the valley, there were 312 days of severe smog in 1988 and early in the next year school children had to be kept at home for a month because of high pollution levels. In Athens (which contains a third of the Greek population) problems became acute as the number of cars rose from 100,000 in 1965 to one million by 1983. Combined with a lack of public transport (there was only one metro line until 1997), four rush-hours a day because of the siesta and large-scale industry on the outskirts of the city, vehicle pollution produced a disastrous situation. Photochemical smog (known as 'nephos') became common, severely damaging the ancient monuments in the city. Ozone levels in 1980 were double those of 1940 and by the 1990s air quality was about six times worse than that in Los Angeles.

Controls on vehicle emissions have only been introduced belatedly and no government has been prepared to take action to limit car ownership. A few, such as Venezuela and Greece, have experimented with restrictions on vehicles in their capitals by limiting use to alternate days according to whether the number plate has an odd or even number. These restrictions have either been ignored, not enforced or avoided by the rich who own two cars. Controls have therefore relied on technological fixes aimed at the symptoms of vehicle pollution rather than their cause – increasing vehicle ownership and increasing use of every vehicle on the road. Most of these controls have only been introduced under considerable pressure from environmental groups and against stiff resistance from car makers and petrol companies.

The first ban on lead in petrol came in the major cities of the Soviet Union in 1967 (there were few cars anyway). This was followed by the US Clean Air Act of 1970 which started the phasing out of lead additives. Lead concentrations in American air fell ninety-five per cent between 1977 and 1994 as a result. The European Union did not start on this process until the mid-1980s. This could be achieved at relatively low cost – petrol companies had to produce a slightly modified product (usually with tax incentives as part of the package) and car manufacturers had to make a small adjustment to engines. The reduction in exhaust gases was technically more difficult. After 1975 all new cars in the United States had to be fitted with a catalytic converter which removed some of the most harmful chemicals in exhausts. The European Union did not introduce a similar requirement until 1993. However, catalytic converters do not remove what is now recognised to be the most potent pollution produced by vehicles – carbon dioxide.

16
The Threat to Global Systems

The pollution described in the previous chapter, even acid rain, toxic waste, nuclear and vehicle pollution, had its impact at the local and regional levels. Although the effects of acid rain and vehicle pollution might be felt in most areas of the world solutions did not have to be found at a global level. Acid rain in Canada and Scandinavia could be brought under control through action taken by their neighbours. Pollution in rivers like the Rhine and seas such as the Mediterranean could be dealt with through regional agreements. However, in the second half of the twentieth century two forms of pollution affected the global systems that make life on earth possible. The scale of the problems was such that action by a single country could not solve the problem. Complex international agreements were required but this process involved the extremely difficult task of reconciling very different national interests. In one case – the ozone layer – they succeeded, in the other – global warming – they have so far failed.

The destruction of the ozone layer

At ground level ozone is a pollutant that reduces plant photosynthesis by about 20 per cent and makes breathing more difficult. However, in the stratosphere (about twenty to thirty kilometres above the earth) the ozone layer absorbs most of the ultraviolet radiation from outer space and stops it from reaching the earth's surface where it would be harmful to nearly all forms of life. In humans excess ultraviolet radiation causes cataracts and skin cancer; in plants it slows down photosynthesis and it severely damages phytoplankton, the foundation of all oceanic food chains. During the second half of the twentieth century the protective ozone layer was partially destroyed by artificial chemicals called chlorofluorocarbons, or CFCs. These gases are very stable and can survive in the atmosphere for up to a century. As they rose into the stratosphere they were broken up by ultraviolet radiation causing a chlorine atom to split away. In a complex series of chemical reactions, this resulted in the destruction of ozone. A single chlorine atom is capable of destroying up to 100,000 ozone molecules.

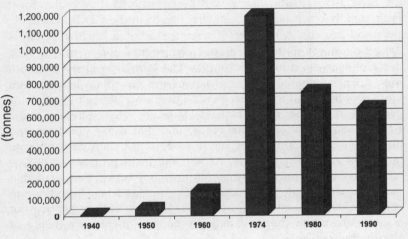

World CFC production, 1940–1990

CFCs were invented in the late 1920s by Thomas Midgely – the same scientist who first suggested tetraethyl lead as an additive in petrol (a notable double in environmental destruction). CFCs were adopted in many industrial processes because they seemed to be inert – they did not burn, did not react with other substances and they were not poisonous. They were first used as refrigerants, then in car air-conditioning systems, as propellant gases in spray cans, to make expanded foam containers and to clean electronic circuit boards. CFCs were cheap to produce and this low cost combined with their apparent safety meant that there was no incentive to minimise usage and there appeared to be no reason to take any special precautions in their use or when they were thrown away. About a third of CFCs in car air-conditioners were lost through 'routine' leakage and another half during servicing. When old fridges were scrapped the gases were left to escape into the atmosphere. About three-quarters of the CFCs that entered the atmosphere did so through the use of spray cans – by the early 1970s Americans were using 1,500,000 spray cans every year. Production of CFCs rose rapidly from about 100 tonnes in 1931 to 40,000 tonnes in 1950. Usage nearly quadrupled in the 1950s so that output reached 150,000 tonnes in 1960. The rate rose exponentially in the 1960s as use of spray cans took off – by 1974 output was eight times higher at 1,200,000 tonnes. Nearly all of this CFC usage took place in the industrialised world.

The accumulation of CFCs in the atmosphere was first noticed in the early 1970s. Until then it had been assumed they broke down fairly rapidly but it became clear that nearly all the CFCs ever made were still present. A number of scientists pointed out the potential dangers and environmental groups campaigned for a ban on their use. However, the industrial lobby, especially firms like Dow Chemical, were strongly opposed to any restrictions and denied any link between CFC emissions and ozone depletion. There were plenty of politicians prepared to support them. Donald Hodel, Interior Secretary in the Reagan administration of the 1980s, said that using suncream and wearing hats and sunglasses made more sense than damaging an important American industry. At first the main pressure to reduce CFC use came from a growing consumer boycott of spray cans that used the gases. By the late 1970s the US, Canada and the Scandinavian countries had banned the use of CFCs in spray cans – this was not a problem for the chemical industry because there were plenty of alternatives available.

International action was much slower to get off the ground. In 1985 the Vienna Convention did no more than establish the framework around which an agreement might be reached. Despite its limitations it was at least a recognition that there was a problem. It was the sudden discovery of a large 'hole' (in fact a huge thinning) in the ozone over the Antarctic by the British scientist Jo Farman in 1985 that finally brought about action. (The 'hole' had been there for some years but the computer analysing the data had been programmed to ignore extreme measurements on the grounds that they must be false!) Conditions in the Antarctic winter were ideal for ozone destruction and the hole grew from about one million square kilometres in 1979 to over twenty million square kilometres by the mid/late 1980s. (It has reached this size in every year since 1988 and in 2000 was over thirty million square kilometres in extent.) In the early spring the hole starts to drift northwards (as far as 45°S) and by the late 1980s spring and early summer ozone levels over Australia and New Zealand were about 15–20 per cent below normal. In the northern hemisphere similar thinning became evident over the Arctic by the late 1980s (it was not as severe as in the Antarctic) and this thinner ozone layer also drifted southwards in the northern spring to cover much of Canada, Scandinavia and even western Europe.

International action was promoted by the UN Environment Programme and led to the Montreal Convention of 1987. For the first time targets for a reduction in CFC production were set although the agreement as a whole was weak. It allowed production to increase by 10 per cent above 1986 levels (the baseline for all subsequent reductions) before reductions of 20 per cent by 1994, rising to 30 per cent by 1999. The chemical industry was still opposed to these reductions and so therefore were key governments such as the US and Britain. They only agreed to the 1987 convention under strong public pressure. However, both Dow in the US and ICI in Britain were researching possible alternatives to CFCs and once these became available they, and their governments, quickly became keen advocates of CFC reductions – they stood to gain considerable commercial advantage through controlling the new replacement technology. It was, however, clear that reductions by the main industrialised countries (who had caused the ozone destruction) would not be sufficient to deal with the problem on a global scale. CFC use in the rest of the world was increasing rapidly – the number of people owning fridges in Beijing rose from 3 per cent of the population in 1975 to 60 per cent a decade later. Major questions of equity arose – it was unfair to ask developing countries to forgo the benefits of CFC use (particularly in their nascent electronics industries) to deal with a problem for which they were not responsible. Equally if the industrialised countries insisted on shifting to the new chemicals why should relatively poor countries have to pay considerable costs to access the new technology controlled by the chemical companies of the industrial world?

The next stage in the series of international agreements was to devise a mechanism for the transfer of technology to the developing countries so that they too could switch from CFCs. This was achieved at a series of conferences – London (1990), Copenhagen (1992) and Vienna (1995). These agreements meant that CFC production would be banned in the industrial world after 1996 and from 2010 in developing countries. However, it was soon clear that the substitutes developed by the chemical industry – hydrochlorofluorocarbons (HCFCs) – also destroyed the ozone layer. They were as damaging as CFCs but their only advantage was that they were not as long-lived and so the period of ozone destruction would be shorter and more concentrated. Nevertheless the agreements of the 1990s only called for their phase-out in a series of steps between 2004 and 2015 with a final end to production in 2020 in the industrialised world and 2040 in the developing world. This long timescale reflected the demands of the chemical companies that they be allowed a 'reasonable' period of production in order to recoup their initial investment. Over the next few decades HCFCs will slowly be replaced by HFCs – hydrofluorocarbons – which, because they contain no chlorine, will not damage the ozone layer. The power of the industrial and agricultural lobby is demonstrated by the lack of action in a number of areas where, it is argued, no alternative exists to ozone-damaging chemicals, for example in fire retardants. Exemptions apply in this case and also in the far more important case of methyl bromine, which is used as a pesticide. Bromine destroys forty-five times more ozone than an equivalent amount of chlorine yet its production continues. Indeed in 2005 the United States argued for an increase in its production instead of its phasing out.

Despite the slow timescale of some of the changes (and small loopholes) it is clear that the international community has been able to devise a set of agreements to limit the damage to the ozone layer. Nevertheless it will take many decades to recover from the damage caused by the indiscriminate use of CFCs. This is because of the very long life of CFCs and the shorter but more intense damage caused by HCFCs. Until 1997 the amount of ozone in the stratosphere was falling at about 8 per cent a decade. Since 1997 the rate of decline has been about 4 per cent a decade. In 2006 there were the first signs that the size of the Antarctic ozone 'hole' might be stabilizing. Current estimates are that the peak of ozone destruction will come in about 2010 and that some time shortly after that date there should be the start of a slow recovery in ozone levels. The latest estimate is that ozone levels will not reach those found in the 1970s until 2065 at the earliest. The earth will, therefore, have suffered at least a century of abnormally low ozone and higher than normal levels of ultraviolet radiation reaching the surface of the planet. As a result there will be a significant impact on human health. For every 1 per cent drop in ozone concentrations there is an equivalent rise in the number of malignant cancers and cataracts. In the last thirty years there have been between one and two million extra deaths from skin cancer because of damage to the ozone layer. Over the next five or six decades that figure will be much higher. The effects on sensitive ecosystems, particularly the oceans, are impossible to quantify.

Global warming: the basics

Without greenhouse gases in the atmosphere to trap outgoing terrestrial infrared radiation the average temperature on the earth would be about -18°C and too cold for life. These gases, primarily carbon dioxide and methane, maintain the average temperature of the earth at about 15°C. However, in the last two hundred years human activities have added extra quantities of these greenhouse gases – carbon dioxide, methane and nitrous oxide – and introduced new ones in the form of the various CFCs. The effect of these changes has been to turn a vital life-sustaining mechanism into the world's most threatening and potentially catastrophic environmental problem – global warming.

The major human contribution to global warming has been the addition of extra carbon dioxide to the atmosphere, most of which has come from the burning of fossil fuels. Every time coal, oil and natural gas are burned, whether in homes, factories, power stations or vehicles, carbon dioxide is produced and escapes into the atmosphere. The main secondary source is from deforestation and the clearing of land. This has an impact in two ways – first trees and plants are burned (producing carbon dioxide) but then there are fewer trees and plants to absorb carbon dioxide during photosynthesis. When carbon dioxide is released into the atmosphere about half of it is absorbed by various natural 'sinks', in particular the oceans. The second major source of additional greenhouse gas production is methane, which has escaped to the atmosphere from a number of sources. First, in the paddy fields of Asia methane is released from the decaying animal and vegetable matter (used as a fertiliser) at the bottom of the water. Second, domesticated animals, in particular cattle, have bacteria in their guts which release methane as a waste product. Third, termites, which feed on decaying wood, also produce methane as a by-product of digestion. The third source of greenhouse gases is nitrous oxide which is produced by vehicle engines and escapes through the exhaust and is also produced by nitrate fertilisers. CFC production in the twentieth century was the final source of extra greenhouse gases.

Global warming: the facts

The increase in the amount of carbon dioxide in the atmophere in the last 250 years reflects the second great transition in human history – the exploitation of fossil fuels and the development of societies dependent on high energy use. Coal production is now 350 times higher than in 1800 and oil production is 350 times higher than in 1900. The number of vehicles in the world rose from almost nil in 1900 to 775 million in 2000. In parallel, forests – particularly tropical forests – have been destroyed on an unprecedented scale in the last two hundred years. All of these forces have had an impact on the earth's atmosphere. Since 1750 about 300 billion tonnes of carbon dioxide has been added to the atmosphere through human actions – but half of that total has been added since 1975. The result has been a rise in the amount of carbon dioxide in the atmosphere. Concentrations are measured in parts per million (ppm) and in 1750, before the widespread use of fossil fuels, there were about 270 ppm of carbon dioxide in the atmosphere. This is the baseline against which additions are measured. The slow pace of industrialisation in the nineteenth century can be judged from the fact that concentrations had only risen to 280 ppm by 1850 and 295 ppm by 1900. By 1950 this had still only risen to 310 ppm – a roughly 15 per cent rise in 200 years. From 1959 there are very accurate measurements of the carbon dioxide

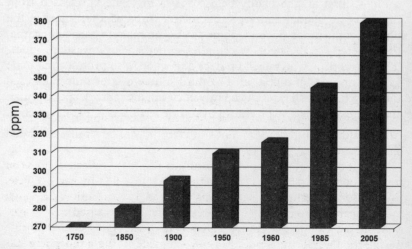

Atmospheric concentrations of carbon dioxide, 1750–2005

in the atmosphere taken from Mauna Loa on Hawaii. In 1959 the figure was 316 ppm, by 1985 it was 345 ppm and by 2005 it was 381 ppm – a 20 per cent rise in forty-five years, reflecting the huge increase in carbon dioxide output in the second half of the twentieth century. Overall carbon dioxide concentrations in the atmosphere are now about 40 per cent higher than they were when the second great transition was in its early stages. The rate of increase is also rising – it was about 1.5 ppm a year in the mid-twentieth century but reached a record 2.6 ppm in 2005. Carbon dioxide is not the most powerful of the greenhouse gases but because of the huge volumes that have been released it accounts for about two-thirds of the total effect of greenhouse gas emissions.

Extra methane in the atmosphere is largely the result of the expansion of agriculture in the last two centuries, reflecting the unprecedented and rapid rise in the world's population from under one billion in 1800 to over six billion in the early twenty-first century. The number of paddy fields in the world increased substantially in that period and was increasing by just under 1 per cent a year in the late twentieth century. The last two centuries have also seen a vast increase in the number of domesticated animals in the world. In the hundred years before 1990 the numbers of the five main domesticated mammals rose four-and-a-half-fold and they all produced methane as a by-product of digestion. Termites have flourished as the tropical forests have been destroyed – in the second half of the twentieth century the area covered by the world's tropical forests fell by 46 per cent. Each individual termite may not produce much methane but there are many billions of them steadily munching their way through decaying timber across the world. The consequence of all these activities has been to raise the amount of methane in the atmosphere two-and-a-half-fold since 1750 from 0.7 ppm to 1.72 ppm. The current rate of increase is about 1 per cent a year. Although methane is still only present in minute quantities in the atmosphere it is about twenty times more powerful than carbon dioxide at absorbing infrared radiation and so contributes almost a fifth of the overall additional greenhouse effect.

The increase in nitrous oxide in the atmosphere reflects the rise in vehicle numbers and the increasing distance travelled by each vehicle. In addition inorganic fertiliser use makes a major contribution. Although not all artificial fertilisers are nitrates overall use increased nearly 350-fold in the twentieth century and much of this consisted of nitrate fertilisers. Overall the amount of nitrous oxide in the atmosphere has risen by 17 per cent since 1750. It is only found in the lower levels of the atmosphere but it is 120 times more powerful than carbon dioxide as a greenhouse gas. The other problem is that it is extremely long-lived and the amounts so far put into the atmosphere will remain there for several centuries. As we have seen, CFC production increased 240-fold between 1940 and 1974 and production continued into the twenty-first century. Although CFCs are in the atmosphere only as trace gases (one of the reasons why their accumulation was so difficult to detect) they make up about 12 per cent of the total additional greenhouse effect because they are many thousands of times more potent than carbon dioxide as infrared absorbers.

Global warming: the impact

In 1896 a Swedish scientist, Svante Arrhenius, was the first to forecast a rise in global temperatures from the burning of fossil fuels – the science behind this prediction was, and is, fairly straightforward. His warning was largely ignored for decades. Environmentalists began to warn of the dangers and potential problems ahead in the 1960s but they too were ignored. It was not until the 1980s that the scientific community as a whole began to accept the reality of global warming. In the late 1980s the UN Intergovernmental Panel on Climate Change (IPCC) was established and the three reports that it has published so far have come to dominate the scientific debate about global warming.

The major problem the IPCC faced was to identify to what extent climatic changes can be attributed to artifically induced global warming through human activities rather than their being the result of natural variability. The Milhankovic effects (the long-term variability of the earth's orbit) are the primary determinants of the ice ages and the warm interglacials. In addition there are smaller, short-term fluctuations caused by factors such as sunspots, variability in the sun's output and dust in the atmosphere (particularly from volcanoes). Climatic research is not made any easier by the lack of reliable meteorological records before the twentieth century. Nevertheless a great deal of research in the last fifteen years has produced a clear and generally reliable picture of changes in the earth's climate in the recent past and the impact that humans have had on these changes.

Global temperatures have increased since 1861. Over the twentieth century the average increase was 0.6°C, although in Europe and other high-latitude areas the increase was 0.95°C. The warming in the twentieth century took place in two periods – 1910–1945 and 1976–2000 with a cooler period in between. Overall the rate of increase between 1976 and 2000 was three times the overall rate of increase since 1861. The key climatic facts are:

- The twentieth century was the warmest century in the last millennium.
- The 1990s were the warmest decade ever recorded.
- 1998 was the warmest year recorded in the last thousand years.
- Nineteen of the twenty warmest years ever recorded occurred after 1980.
- The four warmest years ever recorded were 1998, 2002, 2003 and 2004.
- In September 2006 NASA reported that the earth had warmed by 0.2°C in *each* of the last three decades.

The overwhelming scientific opinion is that these changes have been caused by human actions and that these changes cannot be explained by 'normal' variability in the climate. Indeed the natural changes in the second half of the twentieth century – variability in the sun's output and volcanic activity – will have had a cooling effect and without human actions the cool spell of the mid-century would have increased and been prolonged. In 1998 an American report from the Global Change Research Programme concluded that 'much of the warming that has been observed can indeed be attributed to human actions'. In 2001 the IPCC report, while still couched in the language of scientific caution, stated: 'The balance of evidence suggests a discernible human influence on global climate . . . The warming of the past 100 years is very unlikely to be due to internal variability alone.'

The other signs from across the world in the twentieth century are all consistent with the observed levels of global warming:

- Since the late 1960s global snow cover has decreased by 10 per cent.
- Air temperatures over the Antarctic have risen by 2°C since 1975.
- Lakes and rivers in the northern hemisphere are now frozen for two weeks a year fewer than they were in 1900. The western European spring is now six to eight days earlier than it was in 1975 and in places such as Spain it is two weeks earlier. Autumn occurs an average three days later.
- Glaciers across the world have retreated and lost volume in every

year since 1982. On average glaciers are now six metres thinner than they were in 1980. The area covered by glaciers in the Alps is now less than half that of 1850. Mount Kilimanjaro in east Africa has lost over 80 per cent of iuts snow cover since 1912.

- As ice has melted and the oceans have expanded as they have become warmer, sea levels have risen – by between ten and twenty centimetres in the twentieth century.
- Warmer air is able to hold more water vapour and rainfall has increased by just under 1 per cent a decade in the mid to high latitudes of the northern hemisphere. The number of heavy storms has also increased – by about 4 per cent over the twentieth century.
- More water vapour in the air has also increased cloud cover – it is now 2 per cent greater in the northern hemisphere than it was in 1900.
- Additionally, the extra water vapour in the atmosphere is acting as a greenhouse gas.

These changes have also had an impact on agriculture and natural ecosystems.

- Rice yields at the International Rice Research Institute in the Philippines have fallen by 10 per cent since 1980. This is because the temperature rise at night (when rice needs time to cool and respire) has been 50 per cent more than the average temperature rise.
- Over 70 of the 110 species of harlequin frogs have disappeared from the tropical forests of Central and South America since 1989. Global warming has produced more cloud cover leading to cooler days but warmer nights. In these conditions a fungus that kills the frogs has spread rapidly.
- In Britain 80 per cent of terrestrial species have either moved north by between 30 and 60 kilometres in the last 25 years or to a higher ground at a rate of 5 to 10 metres a decade.
- The temperature of the North Sea has risen by 1°C since 1980 and as a result cold-water species such cod and haddock have moved north by about 100 kilometres and been replaced by fish that prefer warmer water. Sand eels have also moved north leaving many seabirds without food. Around Britain small marine animals and seaweeds have, on average, moved 150 kilometres north in the last 50 years.
- In 2002 the majority of basking sharks in British waters were found off the coast of Cornwall. In 2005 the overwhelming majority were

found off the coast of Scotland as they followed plankton north wards as water warmed.

- A University of Texas report in 2003 showed that ninety-nine species of birds, butterflies and alpine herbs were moving northwards at an average rate of 600 metres a year as temperatures rose.

Global warming: the forecast

Forecasting future climate change is a difficult business and computer models are still relatively simplistic. Nevertheless certain broad assumptions can be made. The first is that climate change will be much faster in the twenty-first century than it was in the last few decades of the twentieth century. There is no doubt that the level of carbon dioxide in the atmosphere will pass 400 ppm within the next ten years. The IPCC has produced a range of forecasts depending on what action is taken to control future emissions of carbon dioxide. They suggest carbon dioxide levels of between 540 and 970 ppm by 2100 – an increase of between 100 and 250 per cent over 1750 levels. This would produce an average rise in temperatures in the twenty-first century of between 1.4°C and 6.4°C, although the mid–high latitudes will see rises about 40 per cent greater than this – between 1.9°C and 8.1°C.

The impact of even the lowest of these temperature rises will be enormous. Tropical and semi-tropical crop yields are likely to fall dramatically and there will be at least a tripling in the number of poor harvests in Europe and North America. The number of people living in water-scarce areas will rise to between a third and a half of the world's population as glacier meltwater from the Himalayas and the Andes declines. As temperatures rise mosquito-borne diseases, especially malaria, will spread in Africa and North America and are also likely to return to Europe. The number of deaths from climate-related diseases will probably double by 2030. Energy demands will rise as more air-conditioning systems are used (only partially offset by lower heating requirements in the winter). As ice across the globe melts and the oceans expand as they get warmer, sea levels will rise. The likely rise will be between ten and eighty centimetres. Some low-lying island states such as the Maldives are on average no more than a metre above current sea levels and they will effectively disappear long before the end of the twenty-first century. Tuvalu in the Pacific is likely to be the first state in the world to be submerged. In addition the number of people at risk from coastal flooding will rise from seventy-five million to 200 million and many ports and cities could become uninhabitable. The main problem for human societies is that the effects of global warming will be felt in the highly unequal world created in the last two centuries. The rich countries are far better able to cope with the effects of climate change. They have the resources to build coastal defences and to obtain the food they need – if necessary surpluses currently dumped on the world market can be kept for domestic consumption. In general they have enough water and they have the political power to keep out the flood of environmental refugees that is likely to arise in the next few decades. As the cautious words of the 2001 IPCC report put it: 'Those with the least resources have the least capacity to adapt and are the most vulnerable.'

The major problem for natural ecosystems is the speed of the temperature rise which is already far quicker than any natural rate of increase in the past. Some species may be able to migrate but many will become trapped in small enclaves and eventually will become extinct. Polar bears will not survive if winter ice in the Arctic melts earlier and earlier and summer ice disappers. It is unlikely that as temperatures rise and the great taiga – coniferous forests of Siberia – die out that they will be replaced in the same timescale by deciduous trees adapted to warmer climates. In January 2004 an article in *Nature* suggested that the minimum likely temperature rise ($1.7°C$ by 2050) could wipe out about a third of all the species on the planet and that a rise of more than $2°C$ would drive about half of all species to extinction.

It is now generally accepted that a $2°C$ rise in average temperatures is the threshold level beyond which the risks to human societies and ecosystems grow significantly and that a $2°C$ rise will become inevitable once carbon dioxide levels increase beyond 400 ppm as they will do in the next decade or so. The reason for this is that much of the carbon dioxide released over the last few years still has to have an effect on the climate. The stabilisation of the temperature increase at about $2°C$ will require drastic action and the longer action is delayed the deeper future reductions will have to be. The IPCC estimate is that to stabilise temperatures at about a $2°C$ rise would require global carbon dioxide levels to drop to *below* 1990 levels within a couple of decades at most and they would then have to 'continue to decline steadily thereafter'. Eventually carbon dioxide emissions 'would need to decline to a very small fraction of current emissions'. Even if this were achieved the changes produced by little more than two centuries of massively increased energy usage will be very long-lived. The thermal expansion of the oceans will continue for centuries and so will the rise in sea levels.

Global warming: the latest evidence

Since the third IPCC report was published in 2001 increasing doubts have been cast on its modelling and forecasting techniques. In order to reach a consensus the IPCC has, even in its 2007 report, been deliberately conservative in its estimates. It has assumed that the effects of rising temperatures will be steady and will not increase dramatically and unpredictably once an unknown threshold is passed. However, there is now a considerable and accumulating amount of evidence that this is not the case and that in the past climate change began slowly but suddenly accelerated through various 'positive feedback' mechanisms. There has also been a change in the assumptions used by climatologists about how the earth's climate has altered in the past. Until well into the second half of the twentieth century it was assumed that the earth's climate was roughly stable and self-regulating so that changes when they did happen – such as the shift from an ice-age to an interglacial and vice-versa – took place steadily over tens of thousands of years. Gradually more careful and detailed analysis from an increasing range of sources reduced the timescale over which dramatic changes occur to perhaps a millennium or even a century. These timescales were reduced still further when two deep ice cores taken from the Greenland ice cap in 1993 were carefully analysed. They showed that a temperature variation of 7°C could occur within fifty years and that major shifts in the climate of the North Atlantic could take place in as little as five years. All of this evidence has reinforced the view that the earth's climate does not change slowly and steadily as temperatures rise and fall. Instead there are thresholds beyond which change becomes very rapid and self-reinforcing.

Climatologists have, for some decades, been aware that these mechanisms exist. For example, as polar ice melts the surface of the earth becomes darker, causing it to absorb more heat which then speeds up the process of warming. In addition much of Siberia consists of frozen tundra under which huge quantities of methane are stored from decaying peat. If Siberia were to warm then it is likely that the methane would be released, thus producing very rapid temperature rises (because methane is such a powerful greenhouse gas) which would, in a positive feedback mechanism, release still more methane and so cause temperatures to rise still further. On the other hand a warmer earth should have more clouds as the water vapour in the atmosphere increases and this should reflect more heat away from the surface, thereby slowing up global warming. Striking the right balance between these various forces is a difficult task. Over the last few years climate scientists have become increasingly concerned that there are indeed thresholds beyond which global warming could speed up dramatically and that major changes in the earth's climate could occur very rapidly. In February 2005 there was a major conference at Exeter on the possibility of a fundamental change in climate systems. The report was published in January 2006 and endorsed by the British government. It stated: 'The emission of greenhouse gases . . . is causing global warming at a rate that is unsustainable.' Why did they reach this conclusion?

There is increasing evidence that climate change is now occurring on a far greater scale and at a far faster rate than at any time in the past. In November 2005 the results of the European Project for Ice Coring in the Antarctic were released. The core was taken down to a depth of 3,270 metres, enabling measurements of the climate back to 800,000 years ago to be made. This showed that carbon dioxide levels in the atmosphere are now 30 per cent higher than at any time in the last 800,000 years and that methane levels are also 130 per cent higher than at any time in that period. Even more important, carbon dioxide levels in the atmosphere are now increasing 200 times faster than at any time in the last 800,000 years. This confirms that not only is rapid change taking place but that it is on a scale and a rapidity unknown in the past even in the transition from a glacial to an interglacial period. Another study published in *Science* in November 2005 showed that sea levels are now rising twice as fast as in the past. Between 3000 BCE and 1800 CE the rise was about one millimetre a year. Since 1850 the rise has been two millimetres a year and the sea level at Brest is now 250 millimetres above the level in 1800. Nearly all of this rise can be attributed to the thermal expansion of the oceans as the world has warmed – the effect of melting glaciers and ice sheets is still to be felt. The rate at which sea levels are rising is accelerating and the latest estimate is that sea levels will rise in the twenty-first century by about 28–34 centimetres. A paper in *Science* in December 2006 suggested that the rise in the twenty-first century could be between 50 and 140 centimetres.

It has always been recognised that the rise in the average global temperature disguises very different patterns. The tropics are expected to warm only a little whereas the change in high latitudes, especially the polar regions, is likely to be much more dramatic. This could have a major impact because the removal of polar ice will cause the earth to warm more rapidly and the melting of polar ice will substantially affect sea levels. There is now an accumulating amount of evidence that climate change in the polar regions has been far more dramatic than expected given that overall global temperatures only increased by about 1°C in the twentieth century. This evidence reinforces the view that dramatic changes in the earth's climate may happen far sooner than had previously been assumed. The Antarctic peninsula has been warming rapidly in the last fifty years – average temperatures have risen by 2.5°C since 1950 – although the rest of the continent is getting colder. The result is that about nine out of ten of the glaciers on the peninsula are in retreat – on average by fifty metres a year. In addition about a third of the west Antarctic ice sheet is thinning – on average by about ten centimetres a year but in places by as much as three to four metres a year. In 1995 the small Larsen A ice-shelf broke off into the sea. The thinning of the ice continued and in the spring of 2002 the large Larsen B ice-shelf (it was the same size as Luxembourg) fell away in just thirty-five days.

Similar patterns are found in the Arctic. Alaska is now on average 2.2°C warmer than it was in the 1950s and the Columbia glacier is now fourteen kilometres shorter than it was only twenty-five years ago. In the Arctic region the temperature rise has been even more dramatic. Average temperatures are now 3°C higher than the average for the whole of the second half of the twentieth century. The summer ice-melt is seventeen days earlier than the twentieth-century average and in 2005 the extent of Arctic ice was the lowest ever and covered 500,000 square metres less than the average for the last two decades of the twentieth century. In the winters of 2004–05 and 2005–06 the Arctic ice failed to re-form fully, which will only speed up the summer melting of the ice. In the summer of 2006 the Arctic ice area was two million square kilometres less than usual. The latest forecast is that all summer ice in the Arctic will have gone by 2040 at the latest. Even more dramatic has been the speed-up in the retreat of glaciers on Greenland and the melting of the ice sheet. Previously it was assumed that the huge Greenland ice sheet would melt steadily over many centuries as temperatures rose. Recent evidence suggests that this will not be the case. In 2005 the Kangerdlugssuaq glacier on the east coast of Greenland retreated five kilometres in a single year. The flow of the glacier also became faster by up to 300 per cent and in one year it became 100 metres thinner. Overall the Greenland ice sheet lost 224 cubic kilometres of ice in 2005 – in 2003 it was still assumed the ice sheet was in balance. The old models that treated the ice sheet as a single block of ice are clearly wrong. What seems to be happening is that the ice sheet is breaking up, not melting. As more water flows out from the surface of the ice sheet under the glaciers the speed of the glaciers increases and they deposit more icebergs into the sea. In 1996 the glaciers on Greenland dumped about fifty cubic kilometres of ice into the sea but by 2005 the rate had tripled to 150 cubic kilometres. Most of the high levels of melting have occurred in the south of Greenland but already there are signs that the north of the ice sheet is warming rapidly too. The latest estimate suggests that the rate of loss of the Greenland ice sheet has almost tripled since 2004. Continued melting on this scale would raise sea levels very rapidly – possibly by five metres in a century.

The increase in cold water pouring into the North Atlantic from the melting Arctic ice could also have another very significant climatic impact. Parts of north-west Europe, in particular Ireland, Britain and Norway, are kept warmer than would be the norm for their latitude by the North Atlantic Drift – a mass of warm water that moves eastwards from the Caribbean. In the past large quantities of cold water from the north have stopped this ocean circulation. It happened about 8,000 years ago when, as the ice sheets in North America were in retreat, a large quantity of cold water escaped from a massive lake that had been dammed by the ice. The best estimate is that it took only five years to shut down the North Atlantic Drift. Scientists have been measuring ocean temperatures in the North Atlantic for several decades using remote sensing buoys. Until 1998 there was no significant change from levels found in the 1950s. However, there has been a 30 per cent drop in the strength of the ocean current in the last few years. This could be the first sign of a dramatic change in climate. If the North Atlantic Drfit were to be shut down by the large volumes of cold water from the Arctic then average temperatures in Britain and Ireland would fall by about 4–6°C producing a climate like that of Newfoundland, which lies in roughly the same latitude.

Another worrying sign of what may be the beginnings of dramatic climate change came in 2005 and 2006. It was discovered that the entire western Siberian sub-Arctic region was beginning to melt. The temperature rise here, 3°C in forty years, has also been far higher than the global average. A frozen peat bog the size of France and

Germany combined has begun to form a series of large lakes from the bottom of which the huge quantities of the stored methane are starting to escape into the atmosphere. Methane escape in Siberia is now five times higher than previously estimated. (The area contains a quarter of all the stored methane under the earth's surface.) It appears a threshold may have been crossed in the region and that a positive feedback mechanism may be beginning to take effect.

There are other possibilities which suggest that climate change may happen far faster than some climate models predict. The reduction of pollution in the industrialised countries, notably the decline of heavy industry and cleaner burning technologies elsewhere (especially the decline of lignite burning), are likely to remove significant amounts of smoke and dust from the atmosphere. This would cause temperatures to rise faster than expected. A study published in the spring of 2006 suggested that between 1950 and 1980 the amount of solar energy reaching the earth's surface fell by about 2 per cent every decade because of the increasing levels of dust and pollution in the atmosphere. That trend has been reversed since the 1980s and now increasing amounts of solar energy are reaching the earth's surface as the overall level of air pollution declines. This will speed up global warming. In 2006 evidence from Antarctic ice cores suggested that another positive feedback mechanism exists. As the earth warms, carbon dioxide held in ecosystems, in particular soils, escapes. Past evidence suggests this could increase predicted temperature rises in the twenty-first century by seventy-five per cent. Even more significant, most climate models assume that as carbon dioxide output continues to rise then the earth's natural sinks, in particular the oceans, will continue to absorb about half of the global output of carbon dioxide. However, there is an increasing amount of evidence from climatic change millions of years ago that this is not the case. At a certain point, which cannot be determined, the ability of the oceans to absorb carbon dioxide may be reduced significantly. If this happened then there would be a very much faster rise in atmospheric concentrations.

All of the possibilities – the rapid melting of the polar ice, the release of methane from the frozen tundra, the reduction in atmospheric pollution, extra carbon dioxide releases and the decreasing ability of 'sinks' to absorb carbon dioxide – suggest that the conservative estimates of climatic change may be far too low. The lowest estimate – a 1.4°C rise in the twenty-first century – already seems far too low and unattainable. The IPCC's worst-case estimate in 2007 was a rise of 6.4°C by 2100, which most observers agree would be catastrophic for the world, may also be too low. If any of the five possibilities above occurred then the best available estimate is that global temperatures would, on average, be about 10°C warmer than they are now.

Global warming: the heart of the problem

Global warming is the greatest threat that the world faces and finding a solution will be extremely difficult for reasons that lie deep within the way human societies have evolved in the last 10,000 years and, in particular, the transition to high-energy-use societies in parts of the world since 1800. The other major threat to global regulatory systems – ozone depletion – could be solved relatively easily. The chemicals involved, though important, were not central to economic activity, and substitutes were developed within a few years. This meant that people and industries did not have to give up using air-conditioners, fridges, freezers and spray cans, or stop cleaning electronic circuit boards. The chemical industry could continue to make profits from the successors of CFCs (HCFCs and HFCs). The problems of equity could also be solved. It was not necessary for developing countries to give up the use of these chemicals just as they were beginning to benefit from the increasing use of fridges. Though costly, mechanisms for the transfer of technology could be set up. Although the world was slow to act on ozone depletion and, as a result, will experience about a century of depleted ozone leading to a large number of early deaths from cancer, the problem was solved.

None of these factors applies in the case of global warming. The burning of fossil fuels only began on any scale after 1800 and most of the output of carbon dioxide took place in the twentieth century. The use of fossil fuels is the foundation upon which modern industrialised, urbanised, high-consumption societies are built. They are high-energy-use societies and could not survive in their current form without such use. These factors have been heightened by the rise of the car in the twentieth century – about a fifth of the world's carbon dioxide output now comes from transport of various kinds. Any attempt to reduce fossil fuel consumption would require fundamental adjustments in the nature of modern societies. In democratic societies attached to large-scale car ownership and use this is extremely difficult – voters mainly judge governments by their ability to keep up the pace of economic growth and the increased levels of consumption that this produces. Both of these factors require continued growth in energy consumption.

Issues of equity are also much more fundamental in dealing with global warming than in the case of ozone depletion. The industrialised countries became wealthy on the back of huge increases in energy consumption and the carbon dioxide that this produced. Responsibility for the current situation in the world's climate lies overwhelmingly with the rich countries of the world. It is therefore not unreasonable for the developing countries to claim that it is the industrialised countries that should bear the burden of any effort to reduce carbon dioxide levels. It is surely unfair to demand that the poorest countries and people in the world should forgo the benefits that increased consumption and affluence can bring merely because some other countries industrialised first and brought about the current situation. However, some of the developing economies, in particular China, India and Brazil, are already large users of fossil fuels. They plan to expand their economies rapidly in the next couple of decades and this will raise their carbon dioxide emissions dramatically. Unless they too were to make some effort to reduce their output then any cuts that might be made by the richer countries would be offset very quickly.

In addition there are large, powerful vested interests around the world that see their future as dependent on the continued use (and growth) in fossil fuels. A number of economies are almost entirely dependent on fossil fuel production, in particular the oil exporters of the Middle East. Australia too is a major fossil fuel exporter (in this case coal). The oil industry depends on continued use of fossil fuels and few corporations have begun to consider their future in a different type of world. Other corporations are equally dependent on current energy use and consumption patterns. These groups are powerful lobbyists and politicians of all types, reluctant to face the problems involved in dealing with global warming, and are more than willing to listen to these siren voices arguing that nothing major needs to be done and that a few technical adjustments are all that is needed.

The fundamental problem in dealing with global warming is that there is no foreseeable technology that would, in the next few decades, stop the release of carbon dioxide every time fossil fuels are burned. 'Carbon capture' might be possible in a few decades' time on large power stations but the internal combustion engine is not likely to be susceptible to this treatment in the foreseeable future. There are, of course, many technical improvements that could be made. If vehicles in the US were as fuel-efficient as those in Japan and western Europe then large amounts of carbon dioxide output would be saved. Houses, and the electrical goods in them, could all be more energy-efficient.

Greater use could be made of public transport. However, past experience suggests that extreme caution should be applied to any idea that technical improvements and increased energy efficiency will solve the problem of global warming. During the twentieth century all the industrialised countries became far more energy-efficient but this did not stop a huge rise in energy consumption. Indeed there is plenty of evidence that increasing energy efficiency, combined with greater wealth in society, simply increases demand for energy by far more than any efficiency savings.

The nuclear industry also sees itself as part of the solution to the problem of global warming. This is partly because of the large number of old nuclear power stations that will have to be scrapped by 2020 and the need, as the industry sees it, to secure its future. The generation of electricity from nuclear power does not produce carbon dioxide. However, this ignores the large amount of energy required in the construction of the power stations, the mining of uranium and its enrichment into nuclear fuel. Although nuclear power generates about a fifth of the world's electricity its contribution to overall energy use is much smaller – about 7 per cent of the total. This creates a major problem. Even if nuclear power were to do no more than double its share of world energy use then the existing 450 or so stations in the world would have to increase not to about 900 but to over 1,300 because of the inevitable rise in world energy use. Not only that but a large proportion of the existing stations would also have to be replaced as they finally cease to operate. Overall a doubling in the share of energy coming from nuclear power would require about 1,500 new nuclear power stations to be built. Each plant takes about ten years to construct and so more than one a week would have to be started from 2007 even to maintain the current share of world energy use currently provided by nuclear power.

The consequences of undertaking such a programme are substantial. No nuclear power station has ever generated electricity cheaper than other sources of power and so major government subsidies would be required. In a world increasingly dominated by the threat of various forms of terrorism the security threat posed by such large amounts of highly radioactive nuclear fuel would be very high. In addition there is the permanent possibility of an accident with potentially devastating consequences. Finally there is the problem of disposing of the nuclear waste produced by the power stations. In 1976 the UK Royal Commission on the Environment commented:

It is irresponsible and morally wrong to commit future generations to the consequences of fission power on a massive scale
unless it has been demonstrated beyond reasonable doubt that
at least one method exists for the safe isolation of these wastes
for the indefinite future.

A previous generation of politicians embarked on this course in exactly
these conditions but the problem remains unsolved. In 2003 a British
government briefing paper on nuclear power stated: 'For intermediate
and high-level [waste] there is as yet no long-term management option.'
To embark on a programme to triple the amount of nuclear waste in
the world in these circumstances would simply compound the mistakes
of the past when doing so will not solve the problem of global
warming.

Global warming: responsibility

The responsibility for most of the carbon dioxide put into the atmosphere in the last two centuries lies with the main industrialised
countries, in particular the United States, Britain, Germany, Japan
and the USSR/Russia. In the nineteenth century, when emissions were
still relatively low, the primary contributor was Britain, followed by
Germany and the United States. As industrialisation spread, other
countries, in particular Japan and the USSR, made significant contributions. However, it was the vast increase in industrial output, the
growing use of the car and expansion of electricity use in all the
industrialised countries in the twentieth century, especially after 1945,
that brought about the very rapid rise in carbon emissions in the last
third of the twentieth century. The United States, as the largest
economy in the world, made the biggest single contribution. It was
only very late in the twentieth century that the newly industrialising
countries, in particular China and India, made a significant contribution. Between 1940 and 1980 the United States emitted six times
more carbon dioxide than China and sixteen times more than India.
Even if the historical perspective is put to one side, the picture
for carbon emissions in the early twenty-first century is clear. The
United States makes up about 5 per cent of the world's population
but it is responsible for just under a quarter of the world's output of
carbon dioxide. China accounts for 14 per cent, Russia just over 6
per cent, Japan just under 5 per cent and India 4 per cent. (Britain
produces just over 2 per cent of the world's carbon dioxide.) However,

Percentage share of world carbon dioxide emissions, 2003

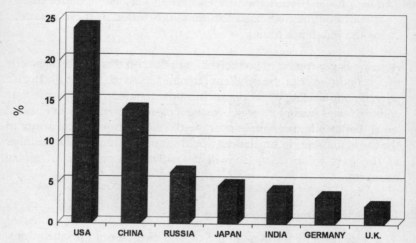

this picture is misleading because it does not take account of population size. Each person in the United States is responsible for the emission, on average, of just under twenty tonnes of carbon dioxide every year. This is seven times more than for each person in China and twenty times more than for each person in India. On this basis each person in the United States is responsible for emitting about six times the world's average emission per person. Even compared with other industrialised countries the United States is profligate in its use of energy and therefore in its output of carbon dioxide. Average emission levels are more than twice those of a citizen of the European Union and only slightly less than twice those of a highly energy-inefficient economy such as that of Russia. Compared with the poorest countries in the world the responsibility of the average American is staggering. Each person in the United States is responsible for 332 times more carbon dioxide emissions than a person living in Burundi and 997 times more than a citizen of Chad.

Forecasting future emission levels is problematic but a few predictions can be made. On the assumption that over the next twenty years Chinese and Indian carbon dioxide emissions rise at about 4 per cent a year, twice the rate in the United States, then by 2025 China would be the largest emitter of carbon dioxide in the world (its population is about four times larger than that of the United States). Even so, taking the period between 1980 and 2025, the United States would

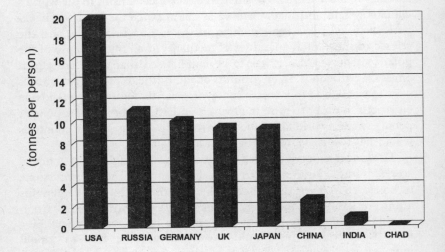

Per capita carbon dioxide emissions, 2003

still be responsible for the emission of twice as much carbon dioxide as China and India *combined*. Over the longer period 1940–2025, which would take some account of responsibility for the current situation, US emissions would still be twice those of China and eight times those of India. There can, therefore, be no doubt that the single greatest responsibility for the output of carbon dioxide and the impact this will have on the world's climate lies with the United States.

Global warming: action

The scientfic consensus about what needs to be done is clear. In order to stabilise carbon dioxide levels at about 400 ppm (which would imply a minimum average temperature rise of 2°C and about twice this in the higher latitudes) global emissions would have to fall below 1990 levels within less than a decade at the latest. Not only that, they would then have to continue to decline to somewhere between a third and and a half of 1990 levels by the middle of the twenty-first century. Further reductions after that might might still be needed. The reason why such drastic reductions are needed to stabilise carbon dioxide at only 5 per cent above current levels is that much of the carbon dioxide already emitted has not yet had an effect on the climate and because the thermal expansion of the oceans will continue for a considerable period even at current concentration levels. This assessment also assumes that there are no positive feedback mechanisms in the world's climate system and that warming at about twice the level so far experienced will not trigger melting of the tundra in Siberia, and the release of substantial quantities of methane, or a rapid collapse of the polar ice sheets. Also it has to be emphasised that these are global emission reductions. If, on grounds of equity, the developing countries were to be allowed to increase energy use and therefore carbon dioxide emissions in order to raise the living standards of some of the poorest people in the world then the cuts that the industrialised countries would have to make would be even bigger.

The longer action is delayed the greater the inevitable rise in temperatures. If it took until the late twenty-first century to reduce emissions to below 1990 levels, and then make a further reduction to perhaps a third or a half of 1990 levels, atmospheric concentrations of carbon dioxide would rise to at least 650 ppm or 140 per cent above pre-industrial levels. This would mean an average rise in global temperatures of about 5°C (at least twice that in the high latitudes). It is difficult to believe that increases on this level would be other than catastrophic for most societies around the globe. In addition such an increase would almost certainly trigger off major feedback mechanisms that could raise temperatures still more.

Following the earliest work of the IPCC an attempt was made to construct a global agreement on greenhouse gas emissions. At the much-hyped but ultimately ineffective UN Conference on the Environment and Development held in Rio de Janeiro in 1992 (the twentieth anniversary of the first UN conference on the environment in Stockholm) a Framework Convention on Climate Change (FCCC) was agreed. This did not set any targets for reductions but created a mechanism under which negotiations would take place and also established the principle that the industrialised countries should make the first reductions because they were largely responsible for the current situation. After a long series of difficult negotiations an agreement was finally reached at Kyoto in Japan in December 1997. The Kyoto treaty committed the signatories to make a reduction in their carbon dioxide emissions of 5.2 per cent (against 1990 levels) by 2012. Most environmentalists believed that these reduction targets were far too small and would not deal with the scale of the problem the world faced. But at least a process had been established which might lead to further reductions.

The Kyoto treaty came into force and became legally binding

following ratification by Russia in February 2005. However, the problems inherent in the treaty had become apparent long before then. President Clinton signed the treaty on behalf of the United States but this was an empty political gesture because he knew that there was no chance that the treaty would be ratified by the Senate (as required under the American constitution). In February 2002 the newly elected President Bush announced that the United States would take no further part in the Kyoto process. He rejected any binding targets which, he claimed, might damage the US economy, particularly if developing countries such as China did not have to make reductions. Instead he set a so-called 'target' for reductions (against a projected emission level in 2012) that allowed the United States to increase its emissions by 12 per cent. Overall US carbon dioxide emissions in 2004 were 16 per cent above 1990 levels and grew by 2 per cent in that year – the highest figure ever recorded. Australia also refused to ratify the treaty. Countries not involved in the Kyoto process continued to pump out more carbon dioxide. Chinese emissions rose by a third in the ten years to 2002 and India's rose by almost 60 per cent.

The ineffectiveness of the Kyoto process is demonstrated by the fact that it is only binding on countries that are responsible for just under a third of the world's carbon dioxide emissions. Even if all these countries achieved the 5 per cent reduction, the impact on the overall level of emissions would be minimal – at best it might offset about a year's worth of the expected growth in global emissions. In practice most of the countries that did ratify the Kyoto treaty are unlikely to meet their targets for reductions. For example, British emissions of carbon dioxide fell by 10 per cent between 1990 and 2002 but this was achieved almost entirely through the switch from coal to natural gas generation of electricity (emissions from transport grew by almost a half in that period). Once the switch to natural gas was over the increases from other sources had a major impact. Carbon dioxide emissions rose by over 5 per cent between 1997 and 2004 so that by that year the reduction against 1990 levels was only 4 per cent. Even the government now admits that it will fail to achieve its Kyoto target by a considerable margin.

In the early years of the Kyoto treaty many countries found it possible to take a number of measures to eliminate some of the worst energy inefficiencies. For example, in 1998 an EU directive required European car makers to produce new cars by 2008 with a 25 per cent reduction in carbon dioxide output. However, it very rapidly became clear that the driving force of economic growth (even at about

2 per cent a year), linked to rising energy use (especially in vehicles and aircraft), would make it very difficult to even stabilise at 1990 levels of output let alone begin to make reductions. Democratic politicians, depending on an electorate wedded to increasing car use, more international air travel and rising consumption, were unwilling to face up to the very difficult, fundamental decisions that would be required to reduce carbon dioxide emissions. It was judged to be electorally impossible to place limits on the 'freedom' of voters to drive their cars and fly wherever they want, whenever they want. Instead, politicians hoped that there might be a technological solution to the problem in the not too distant future even though there are few ideas about what this could possibly be that would be able to deal with all the different sources of carbon dioxide output. Rhetoric about 'green growth' (a suitably vacuous concept) replaced any hard plans about what to do. In this situation politicians were also subject to extensive lobbying by powerful interest groups and corporations which had a strong interest in the preservation of the status quo. These groups were also prepared to fund scientists to do 'research' that might undermine the growing weight of evidence about the impact of global warming. In other countries, many of which were still only semi-industrialised and where the people's living standards were far below those of the rich countries of western Europe and North America, the commitment to economic growth as a symbol of national status and a solution to poverty was even greater.

The next major conference of the Kyoto countries was held in Montreal in late 2005. After a long and difficult series of negotiations they agreed that further cuts after 2012 should be the objective and that they would try and agree them for each country by 2008 (a highly optimistic timescale). However, the Kyoto process was kept alive mainly because none of the democratic politicians wanted to be seen as being responsible, at a time of increasing public awareness of the dangers of global warming, for its demise. Given that few, if any, countries will meet their original and modest targets it seems that these new targets will be essentially cosmetic because they will be unattainable without implementing policies that governments are not prepared to adopt. The US attended the Montreal conference because, although it was not a member of the Kyoto group, it was still part of the original FCCC group set up at Rio de Janeiro. It attended on the understanding that it would never agree to any targets. It therefore tried to derail the whole process and refused to take part in the further round of talks on Kyoto targets. Only at the last minute did it agree to continue as part of the wider group discussing global warming but only on the understanding that these discussions would not lead to enforceable targets.

A further meeting held in Nairobi in November 2006 was even more depressing. It was agreed that there would be no major review of the Kyoto framework. This means that developing countries are unlikely to be asked to make reductions which makes it even less likely that the United States will participate. A tentative date was agreed for 'discussions' rather than 'negotiations' on future reductions to begin but there was no timeframe for these 'discussions' and not even an outline of any likely cuts. It will take a major political effort to keep the Kyoto process alive beyong 2012.

Already other non-Kyoto groupings have formed that are determined not to accept any targets for reductions. Immediately after the Montreal conference the grandly named 'Asia-Pacific Partnership for Clean Development and Climate' held its first meeting in Australia in January 2006. Those involved were the United States, Australia, India, China and South Korea. Mining, energy and metal manufacturing corporations attended the conference but environmental groups were excluded. The programme of this group of countries is clear – there are to be no targets for reductions in carbon dioxide emissions, faith will be placed in as yet unknown technologies to solve the problem at some time in the future and meanwhile there will be a major expansion of nuclear power. Even the official Australian estimates available at the meeting suggested, correctly, that new technologies will, at best, do no more than cut the rate of growth in emissions by about a quarter. This will still produce a minimum average rise in temperatures of 4°C which would almost certainly have catastrophic effects across the world. The theme of the conference was well summed by the Australian Prime Minister, John Howard, when he said: 'Our societies require of us that we find solutions to these issues that maintain the momentum of economic growth.'

It is now about twenty years since the dangers inherent in global warming first became widely apparent. What has happened in that time? The output of carbon dioxide, methane and nitrous oxide has continued to rise inexorably year after year. Carbon dioxide concentrations in the atmosphere have risen from 345 ppm in 1985 to 381 ppm in 2005 – a rise of over 10 per cent. The increase in 2005 – 2.6 ppm – was the highest ever recorded. A long and tortuous series of international negotiations has produced reduction targets that might, if they were achieved, offset about one year's worth in global emissions.

In practice these targets are very unlikely to be met. It seems that the underlying economic and social forces that have dominated the world in the last two hundred years are simply too strong to be altered in the decisive way that would be needed to make the sort of reductions in carbon dioxide output that climate scientists regard as essential if catastrophe is to be avoided.

17
THE SHADOW OF THE PAST

The foundations of human history lie in the way in which ecosystems work. All living things on earth, including humans, form part of these complex webs of interdependence between different plants and animals. When the first direct ancestors of modern humans emerged about two million years ago they operated mainly as herbivores, but they were also carnivores – scavenging dead animals and undertaking a small amount of hunting. Their numbers were therefore limited by the ability of ecosystems to support animals operating towards the top of the food chain. Human history is, at its most fundamental level, the story of how these limitations have been circumvented and of the consequences for the environment of doing so.

Population, technology and the environment

The most important departure from basic ecological constraints has been the increase in human numbers far beyond the level that could be supported by natural ecosystems. The first steps were the spread of humans around the globe from their place of origin in east Africa. This was made possible by their development of various technologies that enabled them to survive in almost every global terrestrial ecosystem. Even so the number of people in the world 10,000 years ago was probably no more than about four million. Ecological constraints were broken by the development of agriculture – the creation of an artificial environment in which selected plants could be grown and domesticated animals tended. This was a watershed in human history and the last 10,000 years of history have been shaped by an agriculture-based boom that has sustained a rise in numbers from four million to over six billion. However, for all but the last two hundred years the pace of population growth was low and often interrupted by food shortages, famine and disease. The rapid population growth of modern times only began around 1750 but it has been enough to increase the world's population more than six-fold. By the 1990s the earth had to support about ninety million extra people every year – an increment the same size as the total population only 2,500 years ago.

All of these extra people have had to be fed, housed, clothed and provided (to a varying extent) with goods and services. Even if the level of consumption per head had not increased, this huge rise in numbers was bound to place increasing demands on the earth's resources. Many of these demands were driven by basic human needs. As more land was needed to grow food, more natural ecosystems were destroyed. Timber was needed to build houses and wood was required for cooking and heating and so forests were cleared. Metal ores were needed to make tools and other items and so the earth's mineral resources were consumed. People had to be clothed and so land had to be cultivated for crops such as cotton, animals kept to provide wool and leather and wild animals hunted for their skins and fur. The increasing demand for resources resulting from the growth in numbers not only placed strains on the environment, it also forced the development of ever more complex techniques requiring more effort and the greater input of energy. This common pattern can be traced in three examples – agriculture, clothing and writing materials.

The transition from gathering and hunting to agriculture required a much higher level of effort in sowing, weeding, watering, harvesting and storing crops. Domesticated animals required tending, milking, shearing, fencing, shelter and the provision of pasture and winter feed. In return for this much greater effort a higher output of food could be obtained from a smaller area. However, the huge population growth of the last two centuries could only be sustained by much higher levels of inputs into agriculture in the form of machinery, irrigation, fertilisers and pesticides. Each of these areas required large inputs of energy and resources. Modern agriculture could not be sustained without these huge energy inputs. Similarly the first humans were able to clothe themselves using the skins of the animals they had scavenged or killed. As numbers rose this was not possible and textiles were made from natural fibres such as wool, cotton and flax. However, this required the use of land to grow crops and provide pasture for animals as well as the extra effort needed in the spinning and weaving of the raw materials. These sources, together with the hunting of animals for their fur, provided enough clothes for thousands of years. By the late nineteenth century this was no longer possible as human numbers rose dramatically. There would not have been enough land to feed the human population whilst leaving enough space for crops and animals to provide clothing. In addition, most fur-bearing animals had been driven to the point of extinction. Only the manufacture of artifical fibres from chemicals has enabled the world's population to be clothed. However, these complex manufacturing processes consume large amounts of energy and resources. The earliest forms of writing were recorded on baked clay tablets or plant materials such as papyrus. The only other source was parchment and vellum from animal skins but the amount available was very limited because so few animals could be maintained. As demand for writing materials increased it was necessary to substitute what was widely regarded as an inferior product, paper, which was invented in China about 2,000 years ago. However, paper was originally made mainly from natural fibres and old rags. That form of production was not sufficient to meet modern needs and so paper is now made mainly from wood pulp. The modern output of paper requires large-scale tree felling, the creation of monocrops of pine and eucalyptus and large amounts of energy.

From one perspective this invention of new techniques, the use of more complicated production processes and the use of more resources can be viewed as *progress* – the increasing ability of human societies to modify the environment and utilise its resources in order to meet their growing needs. From an ecological perspective this process has a very different interpretation. Human history can be seen as a succession of ever more complex and environmentally damaging ways of meeting the same basic human needs. There may not have been any alternative given the rise in human numbers and the impact of new technology but that does not alter the fact of the greater amount of environmental damage involved in all these processes.

The pace of change

It was the first great transition in human history – the development of agriculture and the rise of settled communities – that set human societies on this path. It took about 10,000 years before the second great transition began. This involved the use of fossil fuel energy sources, the creation of industrialised and mass consumption, urbanised and high-energy-dependent societies. It took place in little more than 200 years and has compounded all of the earlier environmental problems on an unimaginable scale. The impact of the second great transition on both societies and the environment has also been magnified by the growth of inequality in the world on a scale never experienced before.

The agricultural states and empires were all at roughly the same level of development and the differences in wealth between them were marginal. Western Europe was certainly one of the poor regions of the world until after 1500. However, it was the way in which the world economy was created by western Europe and its offshoots that produced development, wealth, industrialisation, high consumption and very high levels of resource and energy consumption in one part of the world and poverty and underdevelopment for a majority of the world's people. Over the last 250 years inequality in the world has increased enormously and the early stages of industrialisation in a few countries such as China and India have not altered this fact.

The course of human history over the last two centuries has produced change at a rate never before experienced and brought together a series of interlinked problems that almost defy solution because of their complexity. The pace of change in every area (population growth, industrialisation, consumption and energy use) in the nineteenth century, though unprecedented, was still very slow compared to what happened in the twentieth century. The last century has witnessed rapid growth in many areas and it is the conjunction of all these changes that has produced the current acute state of the world's environment. It has raised real doubts about whether these rates of growth can be sustained into the future.

The world in the twentieth century

	Increase 1900–2000
World population	x 3.8
World urban population	x 12.8
World industrial output	x 35
World energy use	x 12.5
World oil production	x 300
World water use	x 9
World irrigated area	x 6.8
World fertiliser use	x 342
World fish catch	x 65
World organic chemical production	x 1,000
World car ownership	x 7,750
Carbon dioxide in atmosphere	+ 30%

In the last few decades of the twentieth century there was a growing disillusion with some of the consequences of modern development and a greater interest in the idea of conservation and protection of the environment. Groups devoted to campaigning on a variety of issues emerged and environmental issues were injected into both domestic and international politics. In a number of important areas changes were achieved – the elimination of smoke pollution in cities, reductions in industrial pollution, international agreements on acid rain and CFC production. However, the impact of these ideas was limited when set against the dominant ideology of economic growth, technological innovation and mass consumption. Many of the measures adopted in the late twentieth century were little more than cosmetic when set against the driving forces of population growth, the need to increase food production, ever greater industrial output and increased energy consumption. Indeed it was quite possible to interpret many of the measures – wildlife conservation in limited areas, licensing pollution up to certain levels, let-out clauses for scientific whaling, limited help for organic farming while continuing massive subsidies to conventional agriculture and the drive for 'green consumerism' – as being no more than ways of shoring up the existing system rather than the first steps towards a more environmentally benign approach.

The environmental problems now facing the world are deeply rooted in human history. Many have developed over a long period of time but it is in the last two centuries that they have widened and intensified their impact dramatically. Over the same period the political history of the world has produced a large number of unequal states, all asserting their right to independence and national sovereignty and their need to meet their national interests as they define them. At the same time these states are obliged to operate within an international system characterised as much by conflict and competition as by co-operation. This means that, given the highly unequal distribution of the world's wealth and power that has evolved over the last five centuries, it is extremely difficult to deal with problems that cut across national boundaries and with those that entail significant economic and social costs for some countries. Environmental problems are experienced by different countries and different regions of the world in very different ways and perspectives therefore differ accordingly. These problems are further compounded by the emergence of world financial systems and transnational corporations that are more powerful than most governments. Of the hundred largest financial entities in the world less than half are now states and the majority are corporations. These organisations and systems are effectively beyond the control of individual governments and so even if the latter were willing to take difficult environmental decisions they would often lack the power to implement them.

What do past experiences suggest about the stability and sustainability of modern societies given the way they have evolved? For thousands of years humans have lived with the consequences of environmental degradation without incurring economic, social and political breakdown. In some cases societies did succumb to these pressures but the decline and eventual collapse were often prolonged. For example the decline of the first civilised societies in Sumer took place over a thousand years and generations living through this process would probably not have been aware that their society was locked into irreversible decline. Even when the collapse came relatively rapidly (as on Easter Island and with the Maya), the implications of the period of rising problems might not have been fully apparent to those alive at the time. The problem for modern societies is the scale and interlinked nature of the multitude of environmental challenges they face, the speed with which economic, technological and social change is taking place and the lack of effective mechanisms to deal with these issues.

Although this is a historical account of environmental problems it is worthwhile considering how the various environmental pressures that have evolved in the last few centuries might develop in the next few decades. Forecasting is a hazardous business but the main outlines, if not the exact details, of many of the trends that will affect the environment in the first half of the twenty-first century are already clear.

Prospects: population and food

The almost four-fold increase in the world's population in the twentieth century will not be repeated in the twenty-first century. However, population will continue to grow at a significant rate because of the 'momentum' built in by the increases in the second half of the twentieth century. The UN's 'medium' projection is based upon a 30 per cent reduction in fertility combined with a 20 per cent increase in life expectancy in the developing world. In the industrialised world it assumes very small increases in fertility and life expectancy. This would produce a world population of seven billion by 2013 (this is almost inevitable given current population patterns) rising to eight billion by

2028 with growth then slowing to produce a total of nine billion by about 2050. This is a relatively conservative figure but would still amount to an eleven-fold increase in world population in 300 years. (Projections for the second half of the twenty-first century assume very low rates of growth, producing a total of ten billion in 2100.)

All but about one per cent of the world's population growth in the next fifty years will take place in the developing world. In 1950 of the ten largest countries in the world four were in Europe and two of the others, the United States and Japan, were industrialised. In 2050 only the United States among the currently industrialised countries will be in the top ten and India will have overtaken China as the most populous country in the world. With half of the world's population already living in cities much of this population growth will result in a large increase in the size of urban areas with all the associated environmental problems that this will entail. The lack of sanitation and safe water will mean that without further action about two and a half to three million people a year will die by 2020 (a total of about fifty million) from easily preventable diseases. If the UN Millennium Development Goal of cutting the number of people who lack access to safe drinking water by a half is met (which currently seems unlikely) then the death toll will fall to about one million a year but this would still mean a total death toll of over thirty million by 2020. Projected population growth will produce acute overcrowding in some countries. The current population of Nigeria is expected to more than double from just over 120 million to about 260 million by 2050. This would be the equivalent of squeezing the current population of the United States into an area little bigger than Texas.

The expected increase in the world's population by about 50 per cent in the next fifty years has major implications for the world's food supply. In order to supply these extra people with the same amount of food as average food consumption today the minimum increase in agricultural output required will also be 50 per cent. This is on the assumption that there is no increase from current levels in either the industrialised or developing world's food consumption. This increase in production can only be obtained by a combination of two possibilities. First, more land can be brought into production. However, the best agricultural land is already in production and world grain production per person has been falling since 1984. This new land could only be obtained by further deforestation (mainly in the tropics) with major consequences for biodiversity, soil erosion and climate change. Second, the inputs into agriculture in the form of irrigation, artificial fertilisers, pesticides and machinery could be increased still further. This path is beset with problems. The world is now losing every year as much agricultural land to the effects of irrigation (water-logging and salinisation) as the amount of newly irrigated land brought into production. It is also unlikely that the world's water supply will be able to sustain the current level of irrigation, together with greater industrial and domestic use, let alone support an increased level of irrigation. The use of fertilisers and pesticides is now nearing the limits of the possible – the increase in agricultural productivity in the 1990s was half that achieved only thirty years earlier. In addition the increased use of fertilisers and pesticides will require a large increase in energy inputs.

Given these limitations there must be real doubts about whether the almost inevitable rise in population in the next fifty years can be sustained. Any increase in food consumption above current levels in either the rich or poor parts of the world would only mean that an even bigger increase in total output would be required. The attempt to obtain this increased output is also likely to have a major environ-mental impact in terms of forest destruction, soil erosion and increased energy consumption. Any significant changes in the world's climate will only compound all of these problems. Many of the world's trans-national chemical and agribusiness corporations advocate the widespread use of genetically modified crops and cloned animals as the only way of coping with these difficulties. The experience of the 'Green Revolution' suggests that this is not the appropriate solution. Apart from the potential (and largely unknown) risks entailed in releasing these seeds into the environment their use will take place in a specific economic and social context. Farmers will become even more dependent on the seed companies and the chemical companies that supply the pesticides that can be used with these seeds. Like the high-yield varieties used in the 'Green Revolution' these seeds require high inputs of water, fertilisers and pesticides and so their use will be restricted to rich farmers, thereby further widening inequalities in society. Given the power structures in the world it seems more likely that this course will be adopted in preference to any restructuring of the world's food system. This will mean that the net flow of food from the poorest countries to the richest with over-consumption in the rich countries (with its associated problems of obesity, diabetes and heart disease) matched by malnutrition and starvation in the poorest will continue.

Resources

The ability to utilise more resources is fundamental to the continuation and expansion of industrial society especially as countries such as China, India and Brazil industrialise on a substantial scale. In 1972 the Club of Rome published a controversial book, *The Limits to Growth*, which forecast a collapse in industrial output and an uncontrollable decline in population within a century largely as a result of the exhaustion of resources (and increasing pollution). The computer models on which the study was based were highly simplistic and the forecast was too heavily weighted by the very high levels of growth in the world economy in the twenty years after the Second World War.

Experience since the 1970s has shown that the immediate threats to the world do not come from a shortage of resources, although at some point these finite resources are bound to be exhausted. Apart from one or two specialised metals, known reserves of minerals are estimated to last for at least another century and in most cases even longer. New reserves are likely to be discovered (as they have been in the past) and recycling and substitution of materials can be extended far beyond current levels – as shortages become apparent these solutions will become economically more attractive. The main problem is that the trends of the twentieth century towards the exploitation of ever lower grades of ore will continue and this will require increased energy inputs to mine and refine these ores.

The fundamental problem facing the world's resources is the current highly inequitable way in which they are used. The United States contains about 5 per cent of the world's population yet it consumes every year about 40 per cent of the resources used in the world. Overall the industrialised countries use, per head of the population, twenty times the aluminium, seventeen times the copper and ten times the iron ore of the poorest. At the same time their average GDP per head of the population is about $25,500 compared with $1,250 for the low and middle income countries that contain 85 per cent of the world's population. Some of the differences are far greater than these averages would suggest. The number of people who have to live on less than $1 a day rose from 1.2 billion in 1987 to 1.5 billion in 2000 and within the next decade the total is likely to rise to 1.9 billion or more than a quarter of the world's population. If some attempt was made to combat the glaring inequalities in the world and the standard of living of the poorest countries were to rise to current European (not American) standards then the world's consumption of resources would have to rise more than 150-fold. It is unlikely that there are enough resources in the world to sustain this level of consumption. Even if there were the consequences, in terms of industrial pollution would be disastrous. For example, if China were to achieve Japanese levels of car ownership (still far below US levels) then there would be 640 million vehicles in the country compared with thirteen million today. This would increase the number of vehicles in the world by over 80 per cent compared with current levels. Apart from the resource consumption and pollution this would involve it would also have severe effects within China. New roads and parking areas would be required which would, given the experience of the industrialised countries, take up about thirteen million hectares of land. This is equivalent to the land area that grows half of China's current rice crop. Where would the extra land come from and how would the population be fed?

However, this 150-fold increase in resource consumption does not allow for any further rise in consumption in the rich world nor does it take account of the fact that the population of the poorest countries in the world is expected to rise by 50 per cent by 2050. Once these factors are taken into account an even greater consumption of resources would be required to alleviate world poverty. Since any reduction in consumption in the rich world would not be feasible politically, these constraints suggest that it will not be possible to sustain a reasonable standard of living for all the world's people. It is highly unlikely that the evolution of mass-consumption societies in the industrialised world can be replicated elsewhere.

Energy

These resource constraints are magnified by the linked problem of energy use. Modern industrialised, urbanised, high-technology, mass-consumption societies are totally dependent on very high energy use. Their evolution over the last 200 years has involved the exploitation of the world's vast, but finite, supplies of fossil fuels – at first coal, then oil and finally natural gas. The crucial, but difficult, question is how long these supplies will last.

Coal was the primary fuel of the nineteenth century and its use continued to increase throughout the twentieth century although its importance declined dramatically in the industrialised world. The best estimate is that there is enough coal in the world to last for at least the next two centuries, even at higher rates of consumption. The primary problem with coal is the high cost of extraction and transportation. Then there are the environmental problems that result from burning it – not just in terms of carbon dioxide output but also in the sulphur dioxide it produces and the acid rain that results. Coal is also a relatively inflexible fuel and is primarily used in iron and steel production and electricity generation. It plays little or no role in the increasingly important transportation sector.

Oil was the primary fuel of the twentieth century. Its use increased 300-fold but the level of known reserves rose despite this higher consumption. That pattern will not be repeated in the twenty-first century. Indeed there are increasing signs that the world is entering a new stage in its energy history characterised by the beginning of the end of the great oil-based energy boom. Since 1970 more oil has been extracted each year than has been discovered. Four-fifths of global oil production comes from fields discovered before 1973 and in most of these rates of extraction have fallen steadily. The problem lies in determining the level of recoverable reserves in the world. For many decades these have been estimated at about forty years of consumption (despite the increases in use) partly because oil companies do not spend money prospecting for new fields once known reserves are around this level. However, three-quarters of the world's known recoverable oil lies in 370 giant fields (all discovered before the early 1960s) whose reserves are accurately known. No major additions are likely in the reserves of these fields. These facts have led many analysts to argue that about 90 per cent of discoverable oil in the world has already been found. This would mean that the peak of world oil production will occur in the first decade of the twenty-first century or shortly thereafter. The decline in the importance of oil is likely to take place over a considerable period of time and as the price rises other poor-quality sources such as oil shales, heavy oil and tar sands could be utilised. Nevertheless the effects of declining oil production will be felt across the world in the next few decades.

Natural gas became increasingly important in the second half of the twentieth century. The amount of known reserves tripled between 1975 and 2000 although extraction only doubled in this period. The result was to increase known reserves from forty years' worth of consumption to sixty years. Nevertheless production will peak long before reserves become exhausted – the most likely date is around 2020 followed by a long decline in output. The decline in natural gas output will occur in conjunction with the fall in oil production and this will therefore greatly exacerbate the problems that will have to be faced.

In these circumstances energy will be a major constraint in the coming decades. If the twentieth century rise in world energy use were replicated in the twenty-first century then by 2100 each person in the world would, on average, be using 40 per cent more energy than each person in the United States currently uses. That will not happen. Indeed it is difficult to see how the energy consumption required to sustain the rise in living standards and resource consumption to bring the poorest parts of the world up to European levels set out in the previous section could be sustained. The world is clearly approaching a major crossroads. The profligate high-energy societies of a small minority of the world's population that developed in the twentieth century cannot be replicated for all of the world's people. Indeed it will be increasingly difficult to sustain it for the few far into the future.

Global warming

The problems caused by the huge increase in energy consumption in the last 200 years (nearly all of which has come from fossil fuels) will have a substantial impact on the world long before the problems of declining oil and natural gas production have to be faced. In the last two centuries the amount of carbon dioxide in the atmosphere has increased from about 270 ppm to 381 ppm – a rise of just over 40 per cent. Given the rate at which additional carbon dioxide is now being put into the atmosphere (between 2 and 2.5 ppm every year) this figure will rise to over 400 ppm by about 2012. The scientific consensus is that this will produce an average rise in global temperatures of at least 2°C and above this level 'the risks to human societies and ecosystems grow significantly'. To stabilise concentrations at around 400 ppm would require emissions to drop below 1990 levels within the next few years and decrease steadily thereafter to only a small fraction of current output.

What are the prospects of achieving this goal? Energy efficiency can be increased in every area from generation and distribution to domestic and industrial use. Cars could be far more fuel-efficient, particularly in the United States. However, past experience suggests that increasing efficiency does not reduce overall consumption; in fact it tends to increase it. What other solutions might be possible? A major increase in nuclear power, even ignoring the environmental and security risks involved, cannot make more than a minor impact on emissions and it would take a couple of decades before these stations would be fully operational. Fusion power is still in its pre-protoytpe stage and it will probably be several decades before even the first plant can produce more electricity than it consumes (even discounting the energy used in building it) and be available in a commercial form. In 2006 the European Union, the United States, Japan, Russia and China agreed an initial ⇔10bn project to build a prototype fusion plant in the south of France. It is hoped (perhaps optimistically) that this might be completed by 2015. If this project goes well it is hoped that a demonstration commercial power station might be available by 2040. Even on this timescale it would take several more decades (well into the late twenty-first century) before fusion made any impact on world energy consumption. It is technically possible to remove carbon dioxide from the waste gases of power stations and industry, although the technology would need some time to develop on a major scale. The problem is what to do with the carbon dioxide when it has been captured. The world's annual production of carbon dioxide is currently about 27,000 million tonnes. If this were frozen at -80°C (which would require a huge amount of energy) it would form a mountain over a kilometre high and almost twenty kilometres in diameter. If carbon dioxide were stored as a gas it would have to be sequestered somewhere it could not escape into either the atmosphere or the oceans. It is difficult to see what technical solutions to this problem are possible in the near future. Even if they were found they would not deal with the problem of carbon dioxide emissions from vehicles and planes which currently account for about 30 per cent of the total. Other more exotic technical solutions have been proposed such as erecting huge 'umbrellas' around the earth to reflect the sun's energy back into space but these are unlikely to be feasible or carried out. All of these technological solutions are probably a minimum of twenty years away and even once they were available it would be about another twenty years before their use was widespread across the world. This means that technology is unlikely to make any major impact on global carbon dioxide emissions in the first half of the twenty-first century.

These technological problems mean that there are no easy solutions to the problem of global warming. This is one of the reasons why governments have found even the extremely modest targets contained in the Kyoto treaty difficult to achieve and why many refuse to participate. The prospect of carrying out deeper cuts after 2012 seems remote. Even if these reductions were made they would do no more than slightly slow up the rate of growth in emissions as long as key states, in particular the United States but also China and India, continue to increase their carbon dioxide emissions with little or no restraint. Given the deeply ingrained nature of high energy consumption in modern societies, the difficulty, especially in democratic societies, of carrying through any major reductions in consumption and the lack of any easy technical solutions in the next few decades, the prospects for the world's climate look bleak. The most likely scenario over the next three or four decades is a pessimistic one of 'business as usual' with few effective controls over greenhouse gas emissions and a continued rise in energy consumption worldwide. In its 'Energy Outlook' for 2006 the International Energy Agency forecast that the most likely scenario was a 53 per cent increase in world energy use by 2030 and that fossil fuels would make up over 80% of that increase. The course on which societies have been embarked for the last two hundred years will not be altered easily or quickly.

In these circumstances it seems inevitable that the world will see a continuing rise in carbon dioxide concentrations. Currently about 2.5 ppm are added to the atmosphere every year – more than twice the level of only thirty years ago. As energy consumption and fossil fuel use continue to increase this figure will also continue to rise. Most scientists now agree that there is little prospect of avoiding a rise to 450 ppm in the atmosphere which would be about 75 per cent above pre-industrial levels. That will probably occur in the next twenty to twenty-five years. Beyond this point concentrations would rise very quickly indeed. Unless some checks are instituted on carbon dioxide emissions atmospheric concentrations would rise to between 650 and 970 ppm by 2100, which would be two or three times above pre-industrial levels. The lower of these figures assumes that annual output remains at roughly current levels so continued growth in the world economy and energy consumption and the lack of easy technological solutions make the higher figure a much more likely outcome. This would imply an average temperature rise of at least 5°C but perhaps twice that in the high latitudes. Long before these levels are reached the impact of global warming will be significant with climatic change on a major scale, the collapse of ice sheets, rising sea levels and the possibility of feedback effects (especially the release of methane from previously frozen tundra in Siberia), which would lead to even greater temperature rises and much more catastrophic disruption.

The complex of problems

The problems that continued global warming will bring about will have their impact in a world already facing a multitude of other environmental problems stemming from the past – deforestation, soil erosion, desertification, salinisation, water shortages on a large scale, loss of wildlife and plants and urbanisation. Global warming will exacerbate all of these problems. In addition continued industrial expansion and further increases in consumption will lead to higher levels of air and water pollution around the world. Continued increases in population (even at a lower rate than the recent past) will put more pressure on resources and agriculture. All of these problems have to be addressed in a world characterised by gross inequalities in the distribution of wealth, income and power and where political mechanisms are losing control of key economic and financial decision-taking. The scale of the problems facing the world are interrelated and unprecedented. Any significant climatic changes brought about by global warming might well push economic, social and political systems into acute instability and possibly breakdown.

The fact that a breakdown has not so far occurred does not guarantee that it will not happen. Many societies in the past believed that they had a sustainable way of life only to find some time later that this was not the case. By the time they had to face the crisis they were unable to make the social, economic and political changes necessary for survival. The problem for all human societies has been to find a way of extracting from the environment their food, clothing, shelter and other goods in a way that does not render it incapable of supporting them. Some damage is clearly inevitable and the environment can withstand some degradation. The challenge for human societies (which few, if any, have clearly recognised) has been to anticipate the point at which the environment is being badly damaged by the demands placed upon it and find the political, economic and social means to respond accordingly. Some societies have succeeded in finding the right balance, some have failed.

The problem facing modern societies stems from the way they have evolved and, in particular, the momentous changes that have occurred in the last two hundred years. The achievements of modern industrial, urban, high-consumption, high-energy-consumption societies have been remarkable. However, the other side of the coin is that the scale of the environmental problems they have created as a consequence of these achievements is unprecedented and of a complexity that almost defies solution. From a wider, historical perspective it is clearly far too soon to judge whether modern industrialised societies are environmentally sustainable.

GUIDE TO FURTHER READING

General

Clapp, B., *An Environmental History of Britain*. London: Longman, 1994

Ehrlich, P., Ehrlich, A. & Holdren, J., *Ecoscience: Population, Resources, Environment*. San Francisco: Freeman, 1977.

Goudie, A., *The Human Impact: Man's Role in Environmental Change*. Oxford: Blackwell, 1981

Grigg, D., *The Agricultural Systems of the World: An Evolutionary Approach*. Cambridge: Cambridge University Press, 1974.

McNeill, J., *Something New Under the Sun: An Environmental History of the Twentieth Century*. New York: Penguin, 2000.

Meyer, W., *Human Impact on the Earth*. Cambridge: Cambridge University Press, 1996.

Simmons, I., *Changing the Face of the Earth: Culture, Environment, History*. Oxford: Blackwell, 1989.

Thomas, W. (ed.), *Man's Role in Changing the Face of the Earth*. Chicago: Chicago University Press, 1956.

Wilkinson, R., *Poverty and Progress: An Ecological Model of Economic Development*. New York: Praeger, 1973.

Worster, D. (ed.), *The Ends of the Earth: Perspectives on Modern Environmental History*. Cambridge: Cambridge University Press, 1988.

Chapter 1: The Lessons of Easter Island

Heyerdahl, T., *Aku-Aku: The Secret of Easter Island*. London: Allen and Unwin, 1958.

Irwin, G., *The Prehistoric Exploration and Colonization of the Pacific*. Cambridge: Cambridge University Press, 1992.

Jennings, J., *The Prehistory of Polynesia*. Cambridge: Harvard University Press, 1979.

Metraux, A., *Easter Island*. London: André Deutsch, 1957.

Chapter 2: The Foundations of History

Colinvaux, P., *Why Big Fierce Animals Are Rare*. Princeton: Princeton University Press, 1978.

Kump, L., Kasting, J. & Crane, R., *The Earth System*. Englewood Cliffs: Prentice Hall, 2004.

Lovelock, J., *Gaia: A New Look at Life on Earth*. Oxford: Oxford University Press, 1979.

Lovelock, J., *The Ages of Gaia: A Biography of Our Living Earth*. Oxford: Oxford University Press, 1988.

Odum, E., *Fundamentals of Ecology*. Philadelphia: Saunders, 1971.

Chapter 3: Ninety-nine per cent of Human History

Butzer, K., *Environment and Archaeology: An Ecological Approach to Prehistory* (2nd ed.). London: Methuen, 1972.

Clark, J., *World Prehistory: In New Perspective* (3rd ed.). Cambridge: Cambridge University Press, 1977.

Dennell, R., *European Economic Prehistory: A New Approach*. London: Academic Press, 1983.

Durrant, J. (ed.), *Human Origins*. Oxford: Oxford University Press, 1989.

Fagan, B., *The Journey From Eden: The Peopling of Our World*. London: Thames and Hudson, 1990.

Fagan, B., *People of the Earth: An Introduction to World Prehistory* (8th ed.). New York: Longman, 1995.

Gamble, C., *The Paleolithic Settlement of Europe*. Cambridge: Cambridge University Press, 1992.

Lee, R., & DeVore, I., *Man the Hunter*. Chicago: Aldine, 1968.

Lewin, R., *Human Evolution*. Oxford: Blackwell Science, 1989.

Martin, P. & Wright, H., *Pleistocene Extinctions: The Search for a Cause*. New Haven: Yale University Press, 1967.

Mellars, P. & Stringer, C., *The Human Revolution: Behavioural and Biological Perspectives on the Origin of Modern Humans*. Edinburgh: Edinburgh University Press, 1989.

Sahlins, M., *Stone Age Economics*. Chicago: Aldine, 1972.

Shick, K. & Toth, N., *Making the Silent Stones Speak: Human Evolution and the Dawn of Technology.* London: Weidenfeld & Nicolson, 1993.

Wenke, R., *Patterns in Prehistory: Humankind's First Three Million Years.* New York: Oxford University Press, 1990.

Chapter 4: The First Great Transition

Adams, R., *The Evolution of Urban Society: Early Mesopotamia and Prehispanic Mexico.* Chicago: Aldine, 1966.

Clutton-Brook, J., *The Walking Larder: Patterns of Domestication, Pastoralism and Predation.* London: Routledge, 1988

Cohen, M., *The Food Crisis in Prehistory: Overpopulation and the Origins of Agriculture.* New Haven: Yale University Press, 1977.

Crawford, H., *Sumer and the Sumerians.* Cambridge: Cambridge University Press, 1991.

Gebauer, A. & Price, T., *Transitions to Agriculture in Prehistory.* Madison: Prehistory Press, 1992.

Gledhill, J., Bender, B. & Larsen, M., *State and Society: The Emergence and Development of Social Hierarchy and Political Centralization.* London: Routledge, 1988.

Harris, D. & Hillman, G., *Foraging and Farming: The Evolution of Plant Exploitation.* London: Routledge, 1988.

Henry, D., *From Foraging to Agriculture: The Levant at the End of the Ice Age.* Philadelphia: Philadelphia University Press, 1989.

Higgs, E., *Papers in Economic Prehistory.* Cambridge: Cambridge University Press, 1972.

Higgs, E., *Palaeoeconomy.* Cambridge: Cambridge University Press, 1975.

Maisels, C., *The Emergence of Civilization: From Hunting and Gathering to Agriculture, Cities and the State in the Near East.* London: Routledge, 1990.

Megaw, J., *Hunters, Gatherers and the First Farmers Beyond Europe.* Leicester: Leicester University Press, 1977.

Postgate, J., *Early Mesopotamia: Society and Economy at the Dawn of History.* London: Routledge, 1992.

Rindos, D., *The Origins of Agriculture: An Evolutionary Perspective.* Orlando: Academic Press, 1984.

Smith, B., *The Emergence of Agriculture.* New York: Scientific American Library, 1995.

Ucko, P. & Dimbleby, G., *The Domestication and Exploitation of Plants and Animals.* Chicago: Aldine, 1969.

Ucko, P., Tringham, R. & Dimbleby, G., *Man, Settlement and Urbanism.* London: Duckworth, 1972.

Zohary, D. & Hopf, M., *Domestication of Plants in the Old World: The Origin and Spread of Cultivated Plants in West Asia, Europe and the Nile Valley.* Oxford: Oxford University Press, 1988.

Chapter 5: Destruction and Survival

Butzer, K., *Early Hydraulic Civilization in Egypt: A Study in Cultural Ecology.* Chicago: Chicago University Press, 1976.

Carter, V. & Dale, T., *Topsoil and Civilization.* Norman: University of Oklahoma Press, 1974.

Culbert, T., *The Classic Maya Collapse.* Albuquerque: University of New Mexico Press, 1973.

Hughes, D., *Ecology in Ancient Civilizations.* Albuquerque: University of New Mexico Press, 1975.

Hughes, D., *Pan's Travail: Environmental Problems of the Ancient Greeks and Romans.* Baltimore: Johns Hopkins University Press, 1994.

Rzoska, J., *Euphrates and Tigris: Mesopotamian Ecology and Destiny.* The Hague: Kluwer, 1980.

Tainter, J., *The Collapse of Civilizations.* Cambridge: Cambridge University Press, 1988.

Thirgood, J., *Man and the Mediterranean Forest: A History of Resource Depletion.* London: Academic Press, 1981.

Yoffe, N. & Cowgill, G., *The Collapse of Ancient States and Civilizations.* Tucson: University of Arizona Press, 1988.

Chapter 6: The Long Struggle

Braudel, F., *Capitalism and Material Life 1400–1800.* London: Fontana, 1975.

Chao, K., *Man and Land in Chinese History: An Economic Analysis.* Stanford: Stanford University Press, 1986.

Crosby, A., *The Columbian Exchange: Biological and Cultural Consequences of 1492.* Westport: Greenwood Press, 1972.

Elvin, M., *The Pattern of the Chinese Past.* Stanford: Stanford University Press, 1973.

Garnsey, P., *Famine and Food Supply in the Graeco-Roman World: Responses to Risk and Crisis.* Cambridge: Cambridge University Press, 1988.

Gribbin, J., *Climatic Change*. New York: Scribner, 1979.

Grigg, D., *Population Growth and Agrarian Change: An Historical Perspective*. Cambridge: Cambridge University Press, 1980.

Ladurie, Le Roy, *Times of Feast, Times of Famine: A History of Climate Since the Year 1000*. London: 1972.

Lamb, H., *Climate, History and the Modern World*. Andover: Methuen, 1982.

Livi-Bacci, M., *A Concise History of World Population*. Oxford: Oxford University Press, 2001.

McEvedy, C. & Jones, R., *Atlas of World Population History*. New York: Facts on File, 1979.

Newman, L. (ed.), *Hunger in History: Food Shortage, Poverty and Deprivation*. Oxford: Blackwell, 1990.

Post, J., *The Last Great Subsistence Crisis in the Western World*. Baltimore: Johns Hopkins University Press, 1977.

Rotberg, R. & Rabb, T., *Hunger and History: The Impact of Changing Food Production and Consumption Patterns on Society*. Cambridge: Cambridge University Press, 1985.

Smith, T., *Agrarian Origins of Modern Japan*. Stanford: Stanford University Press, 1959.

Walter, J. & Schofield, R., *Famine, Disease and the Social Order in Early Modern Society*. Cambridge: Cambridge University Press, 1989.

Watson, A., *Agricultural Innovation in the Early Islamic World: The Diffusion of Crops and Farming Techniques 700–1100*. Cambridge: Cambridge University Press, 1983.

Wigley, T., Ingram, M. & Farmer, G., *Climate and History: Studies in Past Climates and their Impact on Man*. Cambridge: Cambridge University Press, 1981.

Woodham-Smith, C., *The Great Hunger*. New York: Harpers & Row, 1963.

Chapter 7: Ways of Thought

Capra, F., *The Turning Point: Science, Society and the Rising Culture*. New York: Simon & Schuster, 1982.

Glacken, C., *Traces on the Rhodian Shore: Nature and Culture in Western Thought from Ancient Times to the End of the Eighteenth Century*. Berkeley: University of California Press, 1967.

Passmore, J., *Man's Responsibility for Nature: Ecological Problems and Western Traditions*. New York: Scribner, 1974.

Pollard, S., *The Idea of Progress: History and Society*. London: CA Watts, 1968.

Porritt, J., *Seeing Green*. Oxford: Blackwell, 1984.

Porritt, J., *Playing Safe: Science and the Environment*. London: Thames & Hudson, 2000.

Thomas, K., *Man and the Natural World: Changing Attitudes in England 1500–1800*. New York: Pantheon, 1983.

Worster, D., *Nature's Economy: A History of Ecological Ideas*. San Francisco: Sierra Club Books, 1977.

Chapter 8: The Rape of the World

Busch, B., *The War Against the Seals: A History of the North American Seal Fishery*. Montreal: McGill–Queen's University Press 1985.

Cherfas, J., *The Hunting of the Whale*. London: Bodley Head, 1988.

Crosby, A., *Ecological Imperialism: The Biological Expansion of Europe 900–1900*. Cambridge: Cambridge University Press 1986.

Drake, J., *Biological Invasions: A Global Perspective*. Chichester: Wiley, 1989.

Ellis, R., *Men and Whales*. New York: Hale, 1991.

Elton, C., *The Ecology of Invasions by Animals and Plants*. Chicago: University of Chicago Press, 2000.

Hall, S., *The Fourth World: The Heritage of the Arctic and its Destruction*. New York: Knopf, 1987.

Innis, H., *The Fur Trade in Canada*. Toronto: University of Toronto Press, 1956.

Jackson, G., *The British Whaling Trade*. Hamden: Shoe String Press, 1978

Martin, J., *Treasure of the Land of Darkness: The Fur Trade and its Significance in Medieval Russia*. Cambridge: Cambridge University Press, 1986.

Ritchie, J., *The Influence of Man on Animal Life in Scotland*. New York: Macmillan, 1920.

Schorger, A., *The Passenger Pigeon: Its Natural History and Extinction*. Madison: University of Wisconsin Press, 1955.

Stoett, P., *The International Politics of Whaling*. Vancouver: University of British Columbia Press, 1997.

Thompson, H. & King, C., *The European Rabbit: The History and Biology of a Successful Colonizer*. Oxford: Oxford University Press, 1994.

Veale, E., *The English Fur Trade in the Later Middle Ages*. Oxford: Oxford University Press, 1966.

Wishart, D., *The Fur Trade of the American West 1807–1840*. Lincoln: University of Nebraska Press, 1979.

Chapter 9: The Foundations of Inequality

Binder, P., *Treasure Islands: The Trials of the Ocean Islanders*. London: Bland and Briggs, 1977.

Blackburn, R., *The Making of New World Slavery: From the Baroque to the Modern 1492–1800*. London: Verso Books 1997.

Curtin, P., *The Rise and Fall of the Plantation Complex: Essays in Atlantic History*. Cambridge: Cambridge University Press, 1986.

Dean, W., *Brazil and the Struggle for Rubber*. Cambridge: Cambridge University Press, 1987.

Fernandez-Arnesto, F., *Before Columbus: Exploration and Colonization from the Mediterranean to the Atlantic 1229–1492*. London: 1987.

Frank, A., *World Accumulation 1492–1789*. London: Palgrave MacMillan, 1978.

Headrick, D., *The Tools of Empire: Technology and European Imperialism in the Nineteenth Century*. Oxford: Oxford University Press, 1981.

Klee, G., *World Systems of Traditional Resource Management*. London: E. Arnold, 1980.

Lanning, G. & Mueller, M., *Africa Undermined: Mining Companies and the Underdevelopment of Africa*. Harmondsworth: Penguin, 1979.

Mackenzie, J., *Imperialism and the Natural World*. Manchester: Manchester University Press, 1990.

Thornton, J., *Africa and Africans in the Making of the Atlantic World 1400–1680*. Cambridge: Cambridge University Press, 1992.

Tucker, R. & Richards, J., *Global Deforestation and the Nineteenth Century World Economy*. Durham, NC: Duke University Press, 1983.

Wallerstein, I., *The Modern World System* (3 vols.) New York: Academic Press, 1974–1989.

Ward, R., *Man in the Pacific Islands: Essays on Geographical Change in the Pacific Islands*. Oxford: Oxford University Press, 1972.

Williams, M. & Macdonald, B., *The Phosphaters: A History of the British Phosphate Commissioners and the Christmas Island Phosphate Commission*. Carlton: Melbourne University Press, 1985.

Wolf, E., *Europe and the People Without History*. Berkeley: University of California Press, 1982.

Wolf, R., *The Economics of Colonialism: Britain and Kenya 1870–1930*. New Haven: Yale University Press, 1974.

Chapter 10: Disease and Death

Cohen, M., *Health and the Rise of Civilization*. New Haven: Yale University Press, 1989.

Kiple, K. (ed.), *The Cambridge World History of Human Disease*. Cambridge: Cambridge University Press, 1993.

Kiple, K. (ed.), *Plague, Pox and Pestilence: Disease in History*. London: Weidenfeld & Nicolson, 1997.

McKeown, T., *The Origins of Human Disease*. Oxford: Blackwell, 1988.

McNeill, W., *Plagues and Peoples*. New York: Doubleday, 1976.

Oldstone, M., *Viruses, Plagues and History*. New York: Oxford University Press, 1998.

Thornton, R., *The American Indian Holocaust and Survival: A Population History*. Norman: University of Oklahoma Press, 1987.

Chapter 11: The Weight of Numbers

Blaikie, P., Brookfield, H., *Land Degradation and Society*. London: Routledge, 1987.

Blaxter, K. & Robertson, N., *From Dearth to Plenty: The Modern Revolution in Food Production*. Cambridge: Cambridge University Press, 1995.

Dauvergne, P., *Shadows in the Forest: Japan and the Politics of Timber in Southeast Asia*. Cambridge: MIT Press, 1997.

Eckholm, E., *Losing Ground: Environmental Stress and World Food Prospects*. Oxford: Elsevier, 1978.

George, S., *How the Other Half Dies: The Real Reasons for World Hunger*. Harmondsworth: Penguin, 1977.

Gleick, P., *Water in Crisis*. Oxford: Oxford University Press, 1993.

Gleick, P., *The World's Water 2004–2005*. Washington DC: Island Press, 2004.

Grigg, D., *The World Food Problem 1950–1980*. Oxford: Blackwell, 1985.

Grigg, D., *The Transformation of Agriculture in the West*. Oxford: Blackwell, 1992.

Meyer, W. & Turner, B., *Changes in Land Use and Land Cover: A Global Perspective*. Cambridge: Cambridge University Press, 1994.

Pearse, A., *Seeds of Plenty, Seeds of Want: Social and Economic Implications of the Green Revolution*. Oxford: Oxford University Press, 1980.

Rich, B., *Mortgaging the Earth: The World Bank, Environmental Impoverishment and the Crisis of Development*. Boston: Earthscan, 1994.

Richards, J. & Tucker, R., *World Deforestation in the Twentieth Century.* Durham, NC: Duke University Press, 1988.

Sen, A., *Poverty and Famines: An Essay on Entitlement and Deprivation.* Oxford: Oxford University Press, 1981.

Warnock, J., *The Politics of Hunger.* London: Routledge, 1987.

Worster, D., *Dust Bowl: The Southern Plains in the 1930s.* New York: Oxford University Press, 1979.

Chapter 12: The Second Great Transition

Albion, R., *Forests and Sea Power: The Timber Problem of the Royal Navy 1652–1862.* Cambridge: Harvard University Press, 1926.

Clark, J., *The Political Economy of World Energy: A Twentieth Century Perspective.* London: Prentice Hall, 1990.

Daniels, G. & Rose, M., *Energy and Transport: Historical Perspectives on Policy Issues.* London: Sage, 1982.

Etemad, B. & Luciani, J., *World Energy Production 1800–1985.* Geneva: Droz, 1991.

Foley, G., *The Energy Question.* Harmondsworth: Penguin, 1976.

Gimpel, G., *The Medieval Machine: The Industrial Revolution of the Middle Ages.* New York: Holt, Rinehart and Winston, 1977.

Maczak, A. & Parker, W., *Natural Resources in European History.* Washington, DC: Resources for the Future, 1978.

Schurr, S. & Netschert, B., *Energy in the American Economy 1850–1975).* Baltimore: Johns Hopkins University Press, 1960.

Smil, V., *Energy in World History.* Boulder: Westview Press, 1994.

Smil, V., *Energy at the Crossroads: Global Perspectives and Uncertainties.* Cambridge: MIT Press, 2003.

Thompson, F., *Victorian England: The Horse Drawn Society.* London: Bedford Coll.,1970.

Thompson, F., *The Horse in European Economic History: A Preliminary Canter.* Reading: University of Reading Press, 1983.

Chapter 13: The Rise of the City

Berry, B., *Comparative Urbanisation: Divergent Paths in the Twentieth Century.* London: Macmillan, 1981.

Chandler, T. & Fox, G., *3,000 Years of Urban Growth.* New York: Academic Press, 1974.

Chudacoff, H., *The Evolution of American Urban Society.* Englewood Cliffs: Prentice Hall, 1975.

Dyos, H. & Wolff, M., *The Victorian City: Images and Reality*. London: Routledge, 1973.

Gugler, J., *The Urban Transformation of the Developing World*. Oxford: Oxford University Press, 1996.

Hall, P., *The World Cities*. London: Wiedenfeld & Nicolson, 1977.

Hohenberg, P. & Lees, L., *The Making of Urban Europe 1000–1950*. Cambridge: MIT Press, 1985.

Kornhauser, D., *Urban Japan: Its Foundations and Growth*: London: Longman, 1976.

Lowder, S., *Inside Third World Cities*, London: Routledge, 1986.

Mumford, L., *The City in History: Its Origins, Its Transformations, and Its Prospects*. London: HBJ, 1961.

Sjoberg, G., *The Preindustrial City*. New York: Free Press, 1960.

Sutcliffe, A., *Metropolis 1890–1940*. London: Mansell, 1984.

Whitehand, J. *The Making of the Urban Landscape*. Oxford: Blackwell, 1992.

Chapter 14: Creating the Affluent Society

Davis, D., *A History of Shopping*. London: 1966.

Flink, J., *The Car Culture*. Cambridge: MIT Press, 1975.

Freund, P. & George, M., *The Ecology of the Automobile*. Montreal: Black Rose Books, 1993

Galbraith, J., *The Affluent Society*. New York: Houghton, 1958.

Hirsch, F., *Social Limits to Growth*. Cambridge: Harvard University Press, 1976.

Chapter 15: Polluting the World

Ashby, E. & Anderson, M., *The Politics of Clean Air*. Oxford: Oxford University Press, 1981.

Ashworth, W., *The Late, Great Lakes: An Environmental History*. New York: Knopf, 1986.

Bridgman, H., *Global Air Pollution: Problems for the 1990s*. London: Wiley, 1990.

Carter, F. & Turnock, D., *Environmental Problems in Eastern Europe*. London: Routledge, 1993.

Commoner, B., *The Closing Circle*. London: Cape, 1972.

Goldman, M., *The Spoils of Progress: Environmental Pollution in the Soviet Union*. Cambridge: MIT Press, 1972.

Huddle, N., Reich, M. & Stiskin, N., *Island of Dreams: Environmental*

434 · A New Green History of the World

Crisis in Japan. New York: Autumn Press, 1975.

McCormick, J., *Acid Earth: The Global Threat of Acid Pollution*. London: Earthscan, 1989.

McKibben, B., *The End of Nature*. London: Penguin, 1990.

Wylie, J., *The Wastes of Civilization*. London: Faber and Faber, 1959.

Chapter 16: The Threat to Global Systems

Boyle, S. & Ardill, J., *The Greenhouse Effect*. London: New English Library, 1989.

Gribbin, J., *The Hole in the Sky: Man's Threat to the Ozone Layer*. London: Corgi, 1981.

Houghton, J., *Global Warming*. Cambridge: Cambridge University Press, 2004.

McGuffie, K. & Henderson-Sellars, A., *A Climate Modelling Primer*. Chichester: Wiley, 2005.

INDEX